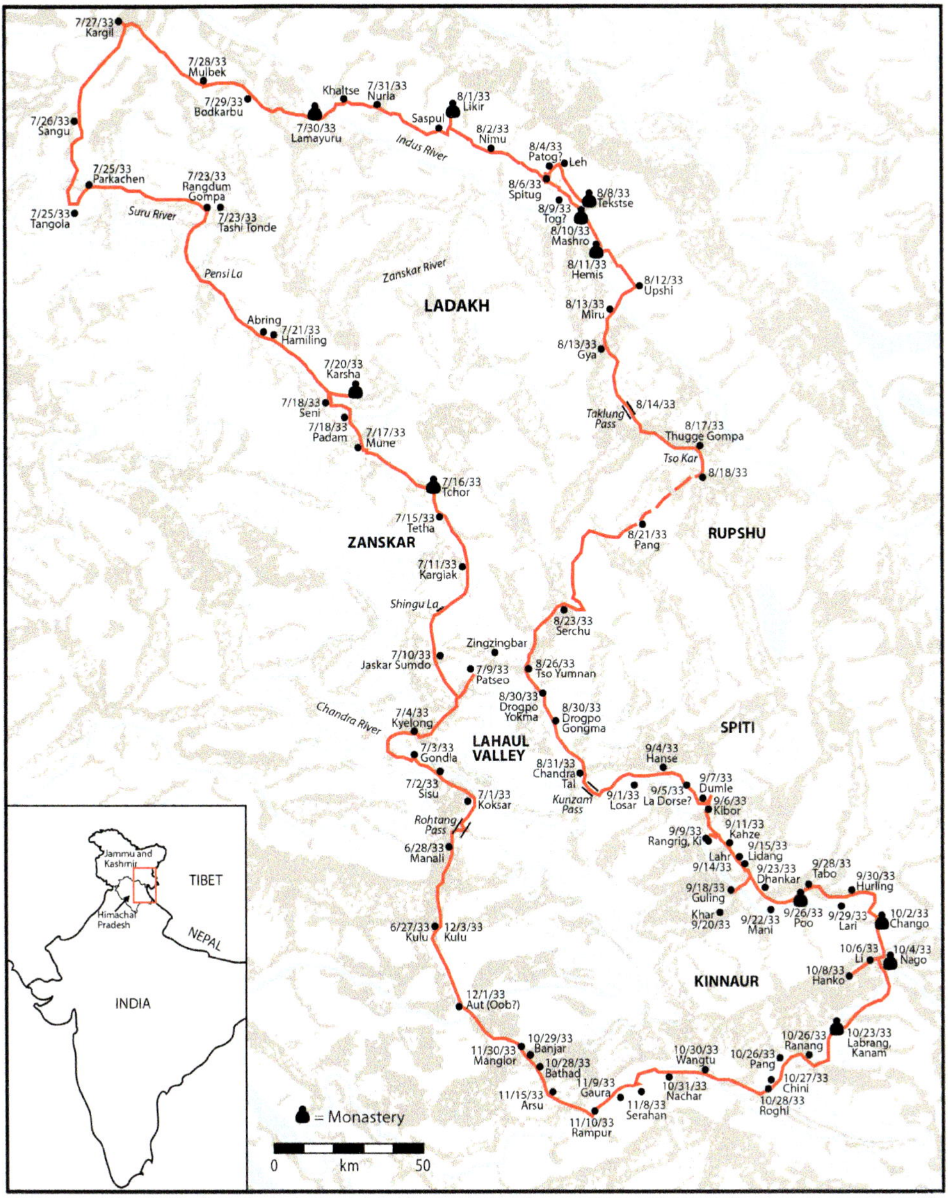

Frontispiece. Map depicting Koelz's route from June 27 through December 3, 1933. The dates reference stopping points on his route; monasteries where Koelz acquired thangkas are marked.

Anthropological Papers
Museum of Anthropology, University of Michigan
Number 98

The Himalayan Journey of Walter N. Koelz

The University of Michigan Himalayan Expedition, 1932–1934

by

Carla M. Sinopoli

Ann Arbor, Michigan
2013

©2013 by the Regents of the University of Michigan
The Museum of Anthropology
All rights reserved

Printed in the United States of America
ISBN 978-0-915703-80-7

Cover design by Katherine Clahassey

The University of Michigan Museum of Anthropology currently publishes two monograph series, Anthropological Papers and Memoirs, as well as an electronic series in CD-ROM form. For a complete catalog, write to Museum of Anthropology Publications, University of Michigan, 4013 Museums Building, 1109 Geddes Avenue, Ann Arbor, MI 48109-1079, or see www.lsa.umich.edu/umma/publications.

Library of Congress Cataloging-in-Publication Data

Sinopoli, Carla M.
  The Himalayan journey of Walter N. Koelz : the University of Michigan Himalayan expedition, 1932-1934 / by Carla M. Sinopoli.
     pages cm. -- (Anthropological Papers. Museum of Anthropology, University of Michigan ; number 98)
  Includes bibliographical references.
  ISBN 978-0-915703-80-7 (alk. paper)
  1. Koelz, Walter, 1895-1989--Diaries. 2. Ethnological expeditions--Himalaya Mountains Region--History. 3. Ethnological expeditions--China--Tibet Autonomous Region--History. 4. Koelz, Walter, 1895-1989--Ethnological collections--History. 5. University of Michigan. Museum of Anthropology--Ethnological collections--History. I. Title.
  GN635.H55S56 2013
  305.80095496--dc23
                            2012051417

The paper used in this publication meets the requirements of the ANSI Standard Z39.48-1984 (Permanence of Paper)

Front Cover: Near Phutkal Monastery, Zanskar (Koelz Collection, Bentley Historical Library, University of Michigan).

Contents

List of Tables	*vi*
List of Plates	*vi*
Acknowledgments	*ix*

Walter Koelz and His Himalayan Expedition   1

References Cited   59

Himalayan Expedition Diary Transcription   63

Koelz Diary Entries   71

June–July 1933	71
August 1933	105
September 1933	127
October 1933	145
November 1933	163
December 1933	179
January 1934	191
February 1934	211
March 1934	223
April 1934	245
May 1934	271

Appendix   281
   Descriptions of Thangkas in the George Lauff,
   the Harley H. Bartlett, and the Mr. and Mrs.
   Alexander Ruthven Collections
    by T. Joseph Leach and Rebecca Bloom

Tables

1. Sources of paintings in the Koelz Collection, *28*
1. Koelz's diary 1933 summary, *56*

Plates

Frontispiece Map depicting Koelz's route
xii   Expedition on landscape

1. Walter Koelz on yak back, *7*
2. Dorje, *8*
3. Kashmiri shawl from the Koelz Collection, *17*
4. Bronze sculptures in the Koelz Collection, *21*
5. Carved book cover collected in Lahaul, Himachal Pradesh, India, *22*
6. Vajra (thunderbolt scepter) and bell, *22*
7. Ga'u, or amulet boxes, *23*
8. Jewelry from the Koelz Collection, *24*
9. Wooden stamps, used to seal doors or storage containers, *25*
10. Pack animals and team members crossing a bridge, *26*
11. Vairochana Buddha thangka, from Poo monastery, Spiti, *29*
12. Ngari Tshang Rinpoche with Walter Koelz, *32*
13. Vajrabhairava, the wrathful form of Manjushri, *33*
14. Palden Lhamo ("Glorious Goddess"), *35*
15. Amitayus Buddha (the Buddha of Infinite Life), *36*
16. Tara, the female embodiment of compassion, *37*
17. Bhaisajyaguru Buddha (Medicine Buddha), *38*
18. Amitabha Buddha, the Buddha of Infinite Light, *39*
19. Palden Lhamo ("Glorious Goddess"), *40*
20. Vajradhara Buddha, a Primordial Buddha, *41*
21. Torma offering, *42*
22. Tenpai Nyima, the fourth incarnate Panchen Lama, *43*
23. Guhyasamajo Aksobyavajra mandala, *44*
24. Samvara mandala, *45*

25. Shadbhuja Mahakala, *46*
26. Tsong kha pa, founder of the Gelug sect of Tibetan Buddhism, *47*
27. Amitabha Buddha, the Buddha of Infinite Light, *48*
28. Vajrabhairava, the wrathful form of Manjushri, *49*
29. Shakyamuni Buddha, *50*
30. Changsem Sherab Zangpo, a student of the lama Tsong kha pa, *51*
31. Shakyamuni Buddha, *54*
32. Final page of Koelz diary from U-M Himalayan Expedition, *56*
33. Koelz diary entry, July 14–16, 1933, *64*
34. Man on road near Phuktal monastery, Zanskar, *88*
35. Women's waist ornament, *89*
36. Thangka from Karsha monastery, *95*
37. Women's waist ornament, dolcha, *100*
38. Wooden door seal, *102*
39. Copper container for monk's alms bowl, *103*
40. Phonograph performance, *107*
41. Round perag ornament from Nirmu, Ladkah, *109*
42. Mashro lama at Talognath, *113*
43. Man with yak, *120*
44. Purba or ritual peg or blade, *124*
45. Member of Koelz's team on mountain top by stone cairn, *128*
46. Woman gathering fuel, *130*
47. Sunkyil, the expedition's dog, near tent, *134*
48. Team on mountain top, *140*
49. Walter Koelz crossing a river, *149*
50. Wooden plate with silver work added, *153*
51. Snuff bottle with silver work added in Colombo, Sri Lanka, *155*
52. Small carving from Labrang, Kinnauar, *157*
53. Silver ornamented copper tea kettle, *167*
54. Women's dress ornament, *167*
55. Rup Chand with owl, *174*
56. Inlaid box of Bidri ware, purchased in Amritsar, *193*
57. Phulkari textile, *193*
58. Jewelry purchased near Khaufer Lake, Pakistan, *204*
59. Walter Koelz holding vulture, *212*
60. Virupaksha temple at Vijayanagara along Tungabhadra River, *233*
61. Knives from Colombo, Sri Lanka, *238*

Appendix Plates

Thangka 2007-22-1 Avalokiteshvara, *283*
Thangka 2007-22-1, detail of monk holding peacock feathers, *284*
Thangka 2007-22-1, detail of Hayagriva, *284*
Thangka 2007-22-2 Akshobhya Buddha, with preserved covering cloth, *285*
Thangka 2007-22-2 Akshobhya Buddha, *286*
Thangka 2007-22-3 Arhat Subhuti, detail, *287*
Thangka 2007-22-3 Arhat Subhuti, *288*
Thangka 47157 Shakyamuni Buddha with jataka scenes, *290*
Thangka 47157, detail showing joins of painting segments, *291*
Thangka 47158 Shakyamuni Buddha in Bhumisparsha mudra, detail, *294*
Thangka 47158 Shakyamuni Buddha in Bhumisparsha mudra, *295*
Thangka 47159 Go Lotsawa, detail, *296*
Thangka 47159 Go Lotsawa, *297*
Thangka 47160, detail of Bhavaviveka, *298*
Thangka 47160 Bhavaviveka, *299*
"Repairs" in lower right corner of Bhavaviveka (47160) painting, *300*
Thangka 47308, detail of lower figures, *301*
Thangka 47308 Vajrabhairava (Yamantaka) with consort Vajra Vetali, *302*
Thangka 47309 Tsong kha pa, *304*
Thangka 47309, detail of Shakyamuni Buddha, *305*
Thangka 47309, detail of Yama Dharmaraja, *305*
Thangka 47310 Tsong kha pa, *307*
Thangka 47310, detail of Tsong kha pa, *308*
Thangka 47310, detail of Yama Dharmaraja and Gelug teacher, *308*

Acknowledgments

I have spent much of the last year and a half immersed in the words, stories, and objects of Walter Koelz: both in preparing this book and in laying the groundwork for a larger work on Himalayan collecting and American university museums that I hope will build upon it. During that time I have accumulated many debts and it is with great pleasure I take this opportunity to acknowledge the many institutions, colleagues, and friends who have supported this project.

The Bentley Historical Library on the University of Michigan campus is a remarkable resource, with extraordinary archives that document the history of the university and State of Michigan. The Bentley is the caretaker of the Walter Koelz papers, including the diary volumes transcribed here, as well as Koelz's correspondence, newspaper articles, legal documents, and many of the photographs he and his team took during their travels. Karen Jania and the reference staff at the Bentley Library have generously made these and other related collections available and have provided a lovely haven for archival research.

In the Museum of Anthropology, this book would not have existed (literally) without the forceful advocacy and superb editorial labors of Jill Rheinheimer. Kay Clahassey worked her usual wonders with the photos and graphics, and showed her usual patience with my last-minute requests. Collection Manager Karen O'Brien helped me locate objects and track down information and museum records on the Koelz Collection. My gratitude and admiration to these three fine people.

The University of Michigan Center for South Asian Studies provided financial support for printing the book through its U.S. Department of Education Title VI grant, allowing us to include many more color images than would otherwise have been possible. I am grateful to former Center Director Juan Cole and current Director Farina Mir for their support of this project.

I owe an enormous debt of gratitude to Donald S. Lopez, Jr., the A.E. Link Distinguished University Professor of Buddhist Studies. Don has been extraordinarily patient and generous with the ignorant questions of an archaeologist and museum curator trying to venture into the complexities of Tibetan Buddhism; he has saved me from more than a few egregious errors. All the ones that remain are questions I did not think to ask him. U-M doctoral student Joe Leach, who shares my interests in Walter Koelz, has helped in innumerable ways with this project—translating the Tibetan entries in Koelz's diary, preparing descriptions for the appendix, and in his valiant attempts to follow Walter Koelz's route through Ladakh and Zanskar in the summer of 2012. Becky Bloom's work is also visible throughout the appendix. Thanks so much to you all.

Janet Hinshaw, in the university's Museum of Zoology Bird Division, provided me with detailed information on the birds that Koelz collected during the 1932–1934 expedition, helping me immeasurably as I tried to plot Koelz's route and decipher his handwriting. And Robert Linrothe of the Art History Department at Northwestern University has also been a generous and willing answerer of my ignorant questions.

This book is appearing concurrently with a new exhibition of thangkas and sacred objects at the University of Michigan Museum of Art—the first time these paintings have been exhibited in more than 30 years. I thank the Museum's Director Joe Rosa, Deputy Director Ruth Slavin, and, especially, Natsu Oyobe, Associate Curator of Asian Art, and the wonderful staff (especially Kate Holoka, Lisa Bessette, and Orian Neumann) of the "other UMMA" for their support of the exhibition. Thanks also to conservator Ann Shaftel for mounting the paintings for exhibition—someday, we will be able to give them the conservation attention they deserve! A simultaneous exhibit on Walter Koelz the collector is being held at the Ella Sharp Museum in nearby Jackson, Michigan; it has been a pleasure working with Charlie Aymond, who knew Walter, and with Ella Sharp Collections Manager Judy Horn. They have shared my enthusiasm for these materials, and it was delightful to view the amazing collections I am privileged to curate through their eyes. Charlie also had some wonderful stories to tell of how meeting Walter Koelz as a young man from rural Michigan opened a new world to him, and of the friendship that grew out of these early encounters.

I spent much of a blissful sabbatical year in Santa Fe, New Mexico, transcribing the Koelz diaries and delving into the story of this unusual man and the world he inhabited. My friends Norman and Barbara Weber-Yoffee were wonderful neighbors and patient listeners. Abby Stewart and David Winter were especially enthusiastic supporters of this project and I thank them both for their encouragement and friendship.

I never had the opportunity to meet Walter Koelz, and there were times as I worked on his diary that I felt that this was a good thing. He was irascible and difficult, and, as I discuss in the introductory chapter, his collecting practices would certainly not pass ethical muster today. Nonetheless, I owe him a debt of gratitude; for Koelz, like other early collectors, and the collections they made, are "good to think." And for that I thank him.

Expedition on landscape (location and date unknown) (Walter Koelz Collection, Bentley Historical Library, University of Michigan).

Walter Koelz and His Himalayan Expedition

In November 1932, Walter Norman Koelz left his home in Waterloo, Michigan, and traveled to British India to undertake a collecting expedition for the University of Michigan. Koelz, who had earned a PhD in zoology at the university in 1920, was at heart and in fact an explorer and adventurer, a self-defined naturalist who harkened back to his legendary eighteenth- and nineteenth-century colonial predecessors. For nearly thirty years, Koelz traveled the world, acquiring zoological and botanical specimens for museums and purchasing art and ethnographic objects for himself. In the University of Michigan Himalayan Expedition, however, Koelz was specifically charged to collected Tibetan material culture for the university's Museum of Anthropology.

Throughout his life, Walter Koelz was also a prodigious diarist. At the end of each day, whether in Michigan, India, or Iran, he recorded his daily activities—where he went, what he ate, who he met, and what he acquired. This volume presents Walter Koelz's diary entries from the University of Michigan Himalayan Expedition, recounting his travels from late June 1933 through his return to Michigan in May of 1934. Unfortunately, the volume preceding this, describing the first seven months of the expedition, has been lost; however, the sections published here encompass the major portion of his journey and virtually all the collecting activities that generated the University of Michigan Himalayan collection of sacred art and material culture.

By the time Koelz began his journey back to Michigan in spring 1934, he had fulfilled the charge he was given, and had acquired a remarkable collection of Western Himalayan art and artifacts. Today, the University of Michigan's Koelz Collection counts among one of the most important assemblages of Western Himalayan material culture in North America. In this chapter, I briefly summarize the history of the collection and the biography of the man who made it. I then draw on Koelz's own words to tell the story of some of the objects in the collection, tracing their journeys from Buddhist monasteries or private homes in the Himalayas to the collections of a university research museum in the United States Midwest. While I will use these stories to say something about the objects themselves—their forms, meanings, and traditional uses—I am in this essay more concerned about understanding them as parts of a larger collection, which itself has a complex biography: related to the man who made it, the historical late colonial context in which it was made, and the institution in which it is housed.

The Walter N. Koelz Collection and the University of Michigan Museum of Anthropology

The Walter N. Koelz Collection in the Asian Division of the University of Michigan Museum of Anthropology (UMMA) consists of 583 objects that Koelz collected for the museum during travels in British India from 1932 through 1934. The museum was quite young when Koelz journeyed to India to collect on its behalf, having been formally established in 1922 and moved into its first physical home in a new specially-constructed museum building in 1928. It was created to be a "research museum," focusing on the broad sweep of early twentieth-century anthropology, from human evolution to the study of contemporary "native" (that is, colonized) cultures. Its primary mission was to collect field specimens and foster scholarly research, allowing faculty and students to directly engage with the material evidence of diverse cultures (Ruthven 1929:7).

Over succeeding decades, the museum's collections grew enormously; today, the museum cares for more than 4500 discrete collections curated in twelve divisions and totaling more than three million objects. The majority derives from curatorial fieldwork and is mostly archaeological; a smaller number of collections come from donations or bequests. A few collections were purchased by the university from private individuals in the museum's early decades. The Koelz Collection is unusual in that it was the result of an explicit commission from the Regents of the University of Michigan, who channeled (donor) funds to Koelz to conduct an expedition in the Himalayas specifically to collect material culture for the museum.

While available records provide little information on the motivations behind the university's sponsorship of the Himalayan Expedition, they lie almost certainly in the personal energies and ingenuity of Walter Koelz (1895–1989) himself and his strong desire to return to a region that he had first visited in 1930 and 1931. Koelz's journey to the Himalayas began via the Great Lakes and Greenland. A native of rural Waterloo, Michigan, Koelz earned a doctorate in zoology from the University of Michigan in 1920, focused on Lake Huron whitefish (Koelz 1920). His first major scientific expedition was to the Arctic, where he served as naturalist for the famous U.S. Navy–National Geographic MacMillan-Byrd Arctic Expedition.[1] This experience seems to have solidified his lifelong passion for travel to distant and exotic places. Subsequent trips, which occurred almost continuously from 1930 through 1953, would take him to South Asia and Iran (Copeland 1983: ix). In all these places, Koelz made enormous biological and cultural collections for numerous institutions (in addition to the University of Michigan Museums of Zoology, Anthropology, and Herbarium, he collected for the Smithsonian Institution, Field Museum of Natural History, American Museum of Natural History, New York Botanical Gardens, among others), as well as for himself and his friends. When he returned to Michigan in 1954, Koelz lived in his natal house in Waterloo, surrounded by elaborate gardens planted with specimens he had collected in his travels and rooms full of the exotic treasures he had acquired.

Walter Koelz was a fascinating individual. I know him primarily from his diaries and correspondence, including his extant writings from the Michigan Himalayan Expedition that are published in this book, and, to a lesser extent, from his publications (which were fairly limited), and from the reminiscences of his friends. He was, as an editor of a portion of his 1939–1941 Persian diary observed, in at least some respects "quintessentially an American of his time" (Wright 1983: x), with the prejudices (and also impatience and condescension) one might expect to accompany this description. Another friend described him, however, as among the last of the Victorian adventurers (*Ann Arbor News*, November 29, 1989). And he was that too—a traveler and adventurer with a passion for solitary exploration (with his "native" collaborators) in exotic lands, and with wide-ranging knowledge of zoology, botany, and geology and keen interests in art and material culture. He could write quite sentimentally (though seldom beautifully) about landscapes, flowers, and Shakespeare, and occasionally (though far less often) about people. Overall though, he does not come across in his writings as a particularly sentimental person. While he was entranced by the Himalayan landscape, which he described "as really powerful, over-powering, stupendous, everything,"[2] he certainly did not belong to the legions of Himalayan travelers entranced by romantic visions of monastic Buddhism (or indeed of any religion) or the spiritual wonders of the mystic East. Indeed, when offered a book in July 1933 by a seller of "contraband things" he did not buy it, noting that "it was on religion and as incomprehensible in Tibetan as it would have been in English" (July 12, 1933, diary entry). His interests in Buddhist religious art seem to have been primarily aesthetic, though he clearly also took pleasure in the challenge of the chase and, especially, in outcompeting other foreign collectors (for example, the famous Italian Orientalist Giuseppe

---

1. The Museum of Anthropology curates a number of ethnographic objects from Greenland that Koelz acquired during this expedition.
2. Letter to H.H. Bartlett, November 2, 1930, Koelz Collection, Bentley Historical Library, University of Michigan.

Tucci, who was in the region at the same time) in acquiring quality goods, and at the lowest possible price!

As both collector and naturalist, Koelz was heir to a lengthy tradition of eighteenth- and nineteenth-century predecessors. The great scientific expeditions of individuals such as James Cook, Charles Darwin, Charles Wilkes (Joyce 2001), and—in the Himalayas—Brian Hodgson (Waterhouse 2004), among many others, were intimately connected with European and American imperialism. They were also closely tied to other developments, including: the growth of universal museums in imperial metropoles; the post-enlightenment commitment to scientific knowledge and the cataloging of the known world; and the creation and formalization of academic disciplines, including anthropology. The naturalists on such expeditions were typically charged with studying and collecting a wide range of materials, including zoological, botanical, and geological specimens (often with a particular eye to potential economic resources). They, or other expedition members, were also expected to record information on the peoples they encountered: documenting languages, lifestyles, beliefs, and objects, and speculating on their ranking within racial frameworks and along the then widely accepted scale from savagery to civilization (Conn 1998; Joyce 2001). The implements, art works, and sacred objects (and human remains) collected by such expeditions came to be housed side by side with birds, reptiles, and insects in national metropolitan museums, and increasingly over time, in smaller regional and university museums.

Walter Koelz differed in some important ways from his predecessors, both in the much later date of his travels as compared to the early naturalists (though he was certainly not alone in his time) and in having been, for much of his career, a solitary and highly entrepreneurial operator rather than a member of a large, well-funded, national or imperial, expedition. By the time Koelz was collecting, most of his contemporaries in the natural and social sciences had become narrowly focused within increasingly formalized academic disciplines and university departments (see Conn 1998). Although he was trained as a zoologist, and specifically an ichthyologist, Koelz resisted being pigeon-holed into a single discipline and his biography reveals that he was temperamentally unsuited to working long within the constraints of any institution. His periods of formal employment were brief and inevitably ended in contention. Nonetheless, while Walter Koelz's Museum of Anthropology Himalayan collection can be viewed as the product of a fiercely independent renaissance man, it can also be seen as the product of larger history—of the geopolitics of India in the decades before Independence; of the growing competition to acquire Buddhist art among European and American institutions; and of the often disturbing story of the removal of valued heirlooms and sacred objects from institutions and individuals.

Background to the Expedition: 1930–1931

Koelz first traveled to the Himalayas in May 1930, when he undertook a position with the Urusvati Institute of Himalayan Studies. The institute, located in the western Himalayan Kulu Valley (Himachal Pradesh, India), was founded in 1928 by Russian expatriates Nicholas and Helena Roerich. Nicholas Roerich, an internationally renowned painter, explorer, archaeologist, peace activist, and theosophist, had himself led an expedition in the region with his wife Helena, a self-proclaimed mystic and clairvoyant, from 1925 through 1928 under sponsorship from the United States government (Drayer 2005). During the expedition, the Roerichs became entranced with the region's natural beauty and spirituality. Their expedition collected enormous quantities of biological, archaeological and ethnographic objects and at its end in 1928 Nicholas and Helena Roerich returned to Kulu to found the Urusvati Institute as a base for future Himalayan studies.

The institute, a branch of the New York-based Roerich Museum, had a particular interest in collecting botanical specimens, especially those with medicinal value. Its director, George Roerich (eldest son of Nicholas and Helena), contacted Dr. Elmer Drew Merrill, then director of the New York Botanical Garden, seeking a recommendation for a botanist they could hire to make their collections. In a decision he would come to regret, Merrill recommended Walter Koelz, a young zoologist from the University of Michigan.[3] Koelz was neither a botanist nor had he any experience in Asia when he was tapped to head the Botanical and Biological Section of the Urusvati Institute's Department of Natural Sciences and Applied Research.[4] He did have, as noted earlier, experience as a naturalist in Greenland and an intense desire for travel.

To take up the Roerich position, Koelz resigned from an appointment as Assistant Curator of Fishes in the University of Michigan Museum of Zoology—a post he briefly held from July 1 through December 1, 1929.[5] With this resignation, Koelz permanently left behind a conventional academic trajectory for the life of an independent field worker. He would maintain various titular appointments and ongoing relationships with the university throughout much of his life, and before heading off to work with the Roerichs (and after a two-and-a-half-month interlude as the ichthyologist in the university's Institute for Fisheries Research), the regents appointed Koelz as an (unsalaried) "Honorary Collaborator in Asiatic Research."

Koelz arrived at the Urusvati Institute's headquarters in the Kulu Valley in late May 1930 and immediately began collecting. His initial appointment was for one year; this was quickly extended for a second year scheduled to end in April 1932. As noted, the primary charge of his work for the institute was to conduct ethnobo-

---

3. Merrill most likely learned of Koelz from Harley H. Bartlett, director of the University of Michigan Herbarium.
4. February 28, 1930, letter on file, Koelz Collection, Bentley Historical Library, University of Michigan; also Roerich Institute Annual Report 1929–1930.
5. University of Michigan Regents Proceedings, 1929–1932:44, 141.

tanical research, specifically focusing on collecting medicinal plants and gathering specialized knowledge of their names and uses, which was largely restricted to "a few initiated lamas" (Annual Report 1929–1930:82). The institute had ambitious goals for their botanical work, noting in their 1929–1930 Annual Report:

> The great humanitarian possibilities and momentous interest of this line of research of the Institute are clearly evident to anyone who had the chance of surveying the vast and virgin field presented by the Himalayan highlands. The Institute plans also to undertake research in the field of cancer, for we have reason to believe that new, potent cures can be found in this vast and unexplored domain. [1929–1930:82]

Koelz's work with the institute began idyllically. In a November 5, 1930, letter to Michigan botanist Professor Harley H. Bartlett, Koelz wrote: "The life has been heavenly. There is no question about anything I need: help or equipment, and if I don't do good work, I ought to be shot."[6] In a subsequent letter to Bartlett, he noted his positive impressions of the Roerichs, describing them as "striving, hard-working souls, who *never* refuse help to anyone" (emphasis in original). He added, "You were right. They aren't bolechiviks [sic], but are interested in pure Buddhism."[7] Koelz would soon come to doubt all these assessments.

Descriptions of Koelz's first season of work were published in the institute's first Annual Report in an article summarizing his prodigious collecting efforts from May through December 1930. He began in the Kulu Valley; in July, he crossed the Rohtang Pass to collect in Lahaul, and then in the fall returned again to the Kulu Valley and from there traveled to Rampur and the Upper Sutlej Valley. These were all places he would revisit again on the University of Michigan expedition. Over these eight months, Koelz collected more than 10,000 plant specimens and 300 birds, as well as an unspecified number of mammals and insects. While many of these specimens were destined for the institute's collections, others were shipped to a number of international institutions. The institute's Annual Report recorded that the University of Michigan received 3000 plants and an entomological collection; the National Museum of Natural History in Paris received 2000 plants; the New York Botanical Garden received 3000 plants and a collection of seeds; and a collection of seeds was shipped to the U.S. Department of Agriculture (Annual Report, 1930:80).

We do not have the diary from Koelz's first season of work for the institute. However, the rich descriptions in the Annual Report indicate that he kept one, and, as noted earlier, he certainly kept detailed diaries throughout all his subsequent travels. Many of these later diaries are now curated with the Koelz papers in the University of Michigan Bentley Historical Library.[8] Koelz also

---

6. Letter on file, Koelz Collection, Bentley Historical Library, University of Michigan.
7. Undated document, Koelz Collection.
8. Portions of the diaries from his 1939–1941 Persian travels were published in 1983 (Walter Koelz, *Persian Diary 1939–1941*, Anthropological Papers, no. 71 [Ann Arbor: Museum of Anthropology, University of Michigan, 1983]).

took numerous photographs of landscapes and people he encountered, several of which appear in the published report. He continued this practice too throughout subsequent travels.

In his second season of work for the Urusvati Institute, Koelz undertook three expeditions: to Kangra, the Western Himalayas (Rupshu, Ladakh, Zanskar), and Rampur (Annual Report, 1931). A portion of his diary describing these trips was published in the institute's *Urasvati Journal* (Koelz 1931). The excerpt tracks his travel through Lahaul from June 7 through July 14, 1931 (Koelz 1931) and reveals that Koelz's Himalayan routines, routes of travel, and the wide-ranging scope of his collecting activities were already well established; they include stories of the acquisition of birds, plants, and objects, interspersed with descriptions of the landscape, people, meals, and the many challenges of traveling on horse- and yak-back through the region (Plate 1).

Also in place by 1931 were several key members of Koelz's team: men who would subsequently accompany him on the University of Michigan expedition. Most important was Thakur Rup Chand, a respected member of a local Lahauli elite family. Rup Chand became Koelz's partner, traveling companion, and collaborator for the next three decades, eventually joining him in Michigan and

Plate 1. Walter Koelz on yak back (Walter Koelz Collection, Bentley Historical Library, University of Michigan).

Plate 2. Dorje (Walter Koelz Collection, Bentley Historical Library, University of Michigan).

working for the University of Michigan Herbarium.[9] Chand's presence on the Himalayan expeditions greatly facilitated Koelz's access to local rulers, religious leaders, and monasteries, and was of key importance to generating the Museum of Anthropology collection.

A number of other men also participated in Koelz's work for the Roerichs, working as servants, guides, collectors of plant and animal specimens, and specimen preparators. Several of these individuals—Dorje (Plate 2), Wang Gyel, Rinchen Gyaltsen, and Tashi—also subsequently joined the University of Michigan expedition and contributed importantly to its success.

Koelz's interests in acquiring material culture are amply evident in the 1931 journal excerpt as well as in his correspondence. He appears to have been particularly interested in acquiring paintings and other sacred objects, as well as tea-tables and jewelry. On June 20, 1931, he wrote with some frustration that

---

9. The two men had a bitter falling-out in the 1970s, and Chand successfully sued Koelz after their split, for failure to pay him for his work on their joint expeditions.

he had "Rs. 350 to spend on native jewelry and I have not seen a piece" (Koelz 1931:96), and later, while camped along Lake (Tso) Kyagar on July 5, he wrote:

> We tried desperately to buy something as a souvenir, but there was not a thing but rubbish, and very little of that, except one nice turquoise and silver ring which the owner would not part with on reasonable terms. [Koelz 1931:110]

His efforts were not always unsuccessful. On July 1, while camped near the large lake Tso Marari in Ladakh, he wrote:

> A Tibetan who was taking his flock to Nirma Mad stopped for a visit and I bought an old rosary made from human skull bones, a nicely covered old sandalwood image, a few tiny old tankas, a holy medal and a mold for human ashes. [Koelz 1931:105]

A few days earlier he had acquired from a "mendicant Tibetan" lama "a human thighbone trumpet and an old string of lama beads," observing that "such eagerness to possess coin I have seldom seen" (July 28; Koelz 1931:103).

While the above references reveal that some sacred objects were available for purchase—if perhaps, in the last case, due to the poverty of their owner—occasional references also reveal that Koelz was already well aware of the anxiety and reluctance that many individuals felt about parting with sacred items: "Our lama[10] went to the monastery to-day and discovered several good tankas and images I had not seen. The monks told him they had not shown them to me for fear I would carry them off" (July 6, 1931; Koelz 1931:112).

Even in monasteries, though, reluctance was not a universal response, nor was it equally applied to all objects. Paintings of deities were quite different than pieces of furniture or ornaments. Thus, on July 13 and 14, 1931, Rup Chand and Koelz visited the large monastery at Hanle where Koelz reports being warmly welcomed by the abbot. There they

> bought an ancient curiously carved tea table from the custodian of the monastery. The carving is bold and graceful and of a totally different character from that of the tables nowadays manufactured. The top was soaked in generations of butter imbibed from the tea that had been spilled by the guests of the ages and the gay paint that the people in this country apply to all carvings has been toned to grey-black by similar agencies. [Koelz 1931:121]

The above commentaries begin to point to the varied and complicated stories behind Koelz's acquisition of various objects from individuals and institutions—stories of exchange, coercion, and gifting; and stories of shame and resistance as well as of generosity and pleasure. I explore some of the stories behind the acquisition of objects during Koelz's 1932–1934 expedition below.

---

10. An individual previously described as having helped Koelz to acquire thangkas in 1930 (and as a drunkard) (Koelz 1931:88).

In addition to insights on his acquisitions, Koelz's diaries from 1931 also contain, as noted earlier, observations on local practices and peoples that the expedition encountered: both local residents and fellow travelers. These included villagers, pastoralists, merchants, monks, pilgrims, British and Russian officers, missionaries, and fellow collectors who shared the treacherous passes, foot paths, and campsites with Koelz and his team. Koelz had mastered some Tibetan and Urdu (and perhaps small bits of other local languages and dialects occasionally mentioned in his diaries), and subsequently wrote that "I am able to converse with the people" (Koelz 1979:4). However, it is unclear just how fluent he was versus how much he relied on Rup Chand and others for cultural and linguistic translation. Certainly, many of his ethnographic observations indicate a less than nuanced understanding of the people and events he observed. Nonetheless, beyond informing on what things Koelz thought important to record, his writings contain much of interest concerning a tremendously dynamic landscape—which, at least during the five or so months per year when the mountain passes were accessible and travel across the area possible, was a landscape of movement and encounters among many different peoples (see Rizvi 1999).

Finally, this early diary excerpt, like his subsequent writings, reveals a number of characteristics of Koelz's personality. These include his strong confidence in his aesthetic judgments—which pieces were "rubbish" and which were fine "*objets d'art*" and worthy of his collecting efforts. Also evident are his stubbornness and determination in bargaining over the prices of objects and his willingness to bully reluctant owners to part with their possessions. Many other aspects of his temperament can also be glimpsed: his love of the landscape; his fascination with gardens, crops, and the region's natural vegetation, particularly blooming flowers and fruit trees (indeed he seems particularly obsessed with the latter and brought many seeds home to plant in his garden in Michigan); and his often judgmental, acerbic, and patronizing attitude toward many of the people he encountered (including members of his own team).

Also becoming evident by late 1931 were Koelz's difficulties in working for others. In March 1932, he left the Roerichs' employ. This was far from a happy parting. The breakdown of Koelz's relations with the Roerichs can be tracked in a series of letters from Walter Koelz to University of Michigan botanist Harley H. Bartlett and from Helena Roerich in Kulu to her New York coworkers, and in increasingly terse notes exchanged between Koelz and Helena and Nicholas's son George Roerich between December 1931 and March 1932.

From the Roerichs' perspective, Koelz was an insolent and ungrateful employee: who devoted more time to collecting zoological specimens than to the botanical ones he was hired to focus on; who unethically used his institute salary to collect for the University of Michigan and other institutions;[11] and

---

11. A practice, however, previously acknowledged in the institute's own 1929–1930 report, discussed above.

who jeopardized the institute's precarious relation with the British colonial government through a variety of improper actions.[12] To Koelz, the Roerichs were liars and charlatans "who are collecting money and cloaking activities under the guise of scientific investigation";[13] who did not understand the nature of scientific research; and who were trying to constrain his autonomy and force him to leave the Himalayas before the end of his contract. Other factors were also at play: the Roerichs' fear that Koelz would establish a rival institute in the region;[14] the colonial government's anxiety about the presence and political goals of the Russians and their institute, and their (initially) much friendlier relations with Koelz, likely in the hope that he would provide colonial officers with inside information on the activities of the institute; Koelz's close relations with Rup Chand and his local assistants; and the intense personal dislike that developed between the irascible scientist and the mystic Helena Roerich, to name a few.

Whatever the truths of the many claims and counterclaims contained in the correspondence, after several months of conflict the institute formally terminated its relations with Koelz. Koelz returned to the United States in April 1932. The acrimony did not stop there, however, and the Roerich Museum soon brought a suit against Koelz in the New York State Supreme Court. The suit sought $100,000 from Koelz in recompense for the harm done by his libelous speech and actions, which, the suit alleged, had exposed the institute "to public ridicule, contempt, and shame and disgrace and intended to degrade and did degrade the plaintiff in its good name and reputation."[15] In July 1932, the legal case was settled out of court and "without costs to either party."[16] As part of the settlement, Koelz was required to send a letter to the Roerich Museum formally renouncing any prior criticisms of the institute and attesting to its "bona fide" and worthy

---

12. In a March 20, 1932, letter to Nicholas and Helena Roerich, Sina Lichtman, vice president and trustee of the Roerich Museum, wrote: "I cannot tell You how speechless with indignation we are, reading about Koeltz's treacherous and vile conduct. He must be suddenly gone mad, but since his form of insanity is becoming dangerous in as much as he slanders our Institution and its officers, spreading lies where he can he can be sued for criminal libel, as we have plenty of proofs to this effect. We are at present conferring with our lawer [sic] and as soon as Dr. Koeltz arrives here, he will face at once the charges against him. He is fit either for a lunatic asylum (his degradation makes him a dangerous person to be left at liberty) or for a prison. To all his horrible conduct the steeling of Institute's property—the guns, adds a touch of criminal tendencies. Dr. Merril [sic] already wrote strong letters to Dr. Bartlett and we are coming in touch with other authorities in scientific institutions so as to check Koeltz's [sic] slander. If he only knew how effectively and absolutely he killed himself in the scientific world, he would have given a deep thought before launching a career of immorality and criminal slander" (Roerich Museum archives, Ref. No. 205874; http://www.roerich.org/archive/correspondence/correspondence.html).

13. Letter to H.H. Bartlett, January 11, 1932, Koelz Collection, Bentley Historical Library, University of Michigan.

14. Helena Roerich wrote that Koelz was attempting to purchase property in the Kulu Valley in order to establish a rival institute and supply specimens to American universities (letter to New York office, January 6–7, 1932, Roerich Museum archives, Ref. no. 202398; http://www.roerich.org/archive/correspondence/correspondence.html).

15. Roerich Museum against Walter Koelz, document on file, Koelz Collection.

16. Document on file, June 28, 1932, Koelz Collection.

activities.[17] The 1932 Annual Report of the Urusvati Institute contains only a terse note: "Dr. W. Koelz is no longer connected with the Himalayan Research Institute" (1933:500).

The University of Michigan Expedition

In late 1931, as his relations with the Urusvati Institute were disintegrating, Koelz began to explore ways to remain in the Himalayas under the auspices of the University of Michigan. In a letter to Bartlett dated December 22, 1931, he wrote:

> I have signed a contract to stay until April 1932. Why can't the University keep me on here after that? This is absolutely confidential. If a word of it came here my life would be hell. I can speak Urdu and am studying Tibetan and have help that I couldn't duplicate anywhere. I know the country and how to manage the people and could make your collections outstanding in North America. At the end of a year and a half we have 3800 nice bird skins, 3200 plant numbers, and a hundred mammals, mostly big game. I could do better than that for you. The plants you could have as many duplicates as you needed. I should want a salary plus expenses amounting in all to $10,000 a year, that sum guaranteed for 10 years. From various institutions money could be raised in return for plant sets if friends or departments or Regents couldn't find the money. Please try hard. It's a shame that this lovely material shouldn't go into good hands. These people, whatever else, don't understand biological research.[18]

Once back in Michigan and with his legal troubles with the Roerichs resolved, Koelz continued to seek out resources that would allow him to return to India for the university. Funds were soon forthcoming, though not only for making biological collections. In their September 1932 meeting, the university regents acknowledged a gift of $500 from Walter R. and Margaret Watson Parker, important Detroit-based collectors of Asian and American art, for a "Himalayan Expedition" to be led by Koelz (University of Michigan Regents Proceedings, 1929–1932:16). Additional funds came from the university's Charles L. Freer Endowment Fund, created in honor of the deceased Michigan industrialist and art collector.[19] With these donors lined up, a major charge of the expedition was to collect art and artifacts, and not just birds and plants. The regents appointed Koelz a:

---

17. Letter from Walter Koelz to Roerich Museum, July 1, 1932, document on file, Koelz Collection, Bentley Historical Library, University of Michigan.
18. Document on file, Koelz Collection.
19. Freer was a wealthy Detroit industrialist and a close friend of Margaret Watson Parker. He was a prodigious collector of Asian and American art, particularly the work of James Whistler, and eventually donated his private collection of more than 30,000 objects to the Smithsonian Institution, where it constitutes the core of the Freer Gallery. He also had close relations with the University of Michigan Museum of Art (then, Alumni Memorial Hall), where he had exhibited some of his personal collections in 1910. Freer died in 1919, and the Freer Fund was established in his name to support research in Art History.

Research Fellow on the Charles L. Freer Fund with a stipend of $5,500 annually effective with the beginning of the present University year. Dr. Koelz is to work in Tibet and upon the Freer Collections in Washington. The proposal has the endorsement and the financial cooperation of Dr. Lodge, Curator of the Freer Gallery in Washington.

The gift of Dr. and Mrs. Walter R. Parker toward the expenses involved in conducting the study has already been recorded in the *Proceedings* of this meeting (page 16), and the president stated to the Board that Mr. George G. Booth for the Cranbrook Foundation, proposed to contribute an additional sum of $1000. The project in general received the approval of the Regents, and the President was requested to express the thanks of the Board to Mr. Booth. [University of Michigan Regents Proceedings, 1932–1936:16]

Although the university regents approved the expedition, there is no reason to believe that the university had any particular interest in the Himalayan region. Indeed, it is unlikely that the expedition would have ever occurred without the provision of funds by private donors. Nonetheless, the regents and the university administration were clearly interested in acquiring scientific collections for their museums, and viewed the museums and their collections as providing essential resources through which students could engage distant worlds in their Ann Arbor classrooms and laboratories. That museums and collections were part of the vision of the University of Michigan from its beginnings is evident in the formal creation of a "cabinet of natural history" simultaneously with the university's foundation in Ann Arbor in 1837 (University of Michigan Regents Proceedings 1837–1864:2). For this expedition, though, it is almost certainly the case that Koelz himself was the force propelling it and that he had personally encouraged friends and donors to provide the resources to allow him to return to the Himalayas (and to Rup Chand?), this time formally on behalf of the university. It is also likely (though I have not been able to confirm this) that Koelz also collected for private individuals (including the Parkers?) while on this trip, and he certainly collected for himself.

With funds in hand, Koelz left for India on November 2, 1932, with the charge to collect Tibetan artifacts for the Museum of Anthropology and biological specimens for the Museum of Zoology, the University Herbarium, and the Cranbrook Institute of Science.[20] The expedition lasted until March 1934. That it was originally intended to last even longer is evident from a diary entry of February 23, 1934, in which Koelz wrote of a telegram he had received from Professor John G. Winter[21] in Ann Arbor: "Doctor Ruthven and I believe your return in May necessary. Cannot now secure funds for further stay in India be-

---

20. Though in December 1934, even before Koelz returned with the collections, the Cranbrook Institute of Science formally designated the University of Michigan as the repository for his collections, to be housed "in the appropriate museum." They specified only that the objects, which would become the property of the university, be identified with a special tag acknowledging the joint sponsorship of the expedition (Regents Proceedings, 1932–1936:505).

21. John Garrett Winter: Director of Fine Arts 1928–1936, and of the Museum of Classical Archaeology 1929–1950.

cause depression and collections demand museum attention. Winter." As a result, Koelz, accompanied by Rup Chand, departed India on March 17, beginning a lengthy homeward boat journey that would take them to Sri Lanka, Japan, and Hawaii, before finally landing in California on April 25 and from there traveling by train back to Michigan, arriving home on May 5, 1934.

In his eighteen months in British India, Koelz spent two stints in low-altitude cities and rural regions of northern India[22] when the snow-covered mountains were inaccessible for travel. These lasted from December 1932 to early April 1933 and from December 1933 through January 1934. While in the lowlands, Koelz continued to acquire bird and plant specimens. He also spent considerable time scouring antique stores in the cities of Lahore, Amritsar, and Delhi, seeking out textiles, paintings, and other objects. Between these two periods, from mid-April through early December 1933, Koelz and his team conducted a lengthy tour through the mountainous regions of Ladakh, Zanskar, Spiti, and Rupshu, recapitulating travels he had taken in previous years. In a letter to expedition patron Margaret Watson Parker dated August 5, 1933,[23] Koelz described his route (see frontispiece):

> If you should be interested in our peregrinations, they can be traced on a map. From Kulu, Manali, across Rotang Pass, Kyelang, across Shingo La in Lahul into the province of Zankskar, Kargiak, Padam, Seni, Phe, Abring, across Pensi La to Rangdum, then through the province of Purig down the Rangdum River to Karghil, from Karghil, Bodkarbu, Lamayuru, to the Indus River at Kalatse, thence along the big river to Leh. The population in all that area is sparse and the places you will find marked conspicuously on the map are not cities.

It was during his travels in the mountains that most of the objects in the Museum of Anthropology collection were acquired (with the exception of the Kashmiri shawls and phulkari embroideries, which were primarily acquired from urban antique dealers; see Beardsley and Sinopoli 2005). Koelz described his collecting work in an unpublished manuscript prepared after his return:

> In the winters of 1932–33 and 1933–34, collections of textiles in the various old culture centers of India were examined, a number of pieces of the best work were acquired. In the summer of 1933, a survey was continued of the art treasures of the great monasteries of West Tibet, including the Gelukpa district of Spiti and Upper Kunawar. Some 500 objects, chiefly old paintings, old brass, copper, and wood work, and modern silver work and jewelery were acquired in this area.[24]

More detailed descriptions of individual acquisitions are found in his diaries, and are discussed below.

---

22. Including regions in modern India *and* Pakistan.
23. Manuscript on file, Koelz Collection, Bentley Historical Library, University of Michigan.
24. Manuscript on file, Koelz Collection.

In January 1934, Koelz was out of the hills and preparing to ship the year's collection home. He had, as noted, amassed an impressive assemblage of both objects and biological specimens. Most were packed into twelve large wooden crates that were shipped from Karachi on the *President Polk* on January 26, 1935 (Koelz carried some objects, including several of the finest paintings, home by hand). The crates arrived in New York a month later, on February 27, and reached Ann Arbor in late March. Inside the twelve boxes were more than 600 other ethnographic and historic artifacts, including iron arrow points, silver rings and ornaments, brass and copper teapots, wood seals, bronze and stone images, icon cases, thangka paintings (Buddhist religious paintings on cloth), shawls, and other textiles. With these were more than 5000 zoological specimens (mostly birds) and 4000 dried plants. Koelz's packing list distinguished objects intended for the museum from those that were part of his personal collections, a distinction that did not sit well with his university sponsors (see below).

Describing the shipment in a letter to Professor Winter at the University of Michigan, Koelz wrote:

> The contents of the boxes are for the most part *objets d'art* and natural history specimens collected this year. There are, however, some ethnological collections made by me in previous years and some other of my own property from previous collection, including a few shawls, two rugs, a few pieces of silver, and a few paintings. Some of the better paintings and rarer textiles I retain for the present. They are of both this year's and former collections. You will find a large plant collection, one 4000 numbers, and a very nice lot of bird skins. There are some mammal skins of the particular species desired by the Museum. Emphasis has been laid on the acquisition of both study and exhibit material in the field of Indian and Tibetan art. There are about 40 very good Tibetan paintings bot [bought] this year and about 30 of mine. This I consider a most choice assortment. . . . The textile collection consists principally of shawls, Old Persian and Kashmir. Roughly 40 were bot [bought] this year. There are a few pieces of old Persian brocade, old silk print, velvet print, silk and cotton saris, etc. The acquisition of good textiles is a slow process, more laborious even than the body-exhausting trek in the mountains after the Tibetan paintings. I am accumulating experience and should be able to do better work in the future. All the objects sent are not considered beautiful by me. Some have been bot [bought] as illustrative of type of design, execution, or otherwise. I shall be glad, however, if you agree with me that some of the things are really beautiful.[25]

And this was just a portion of Koelz's Himalayan collecting. In addition to the plants, zoological specimens, and other objects previously collected in the same region for the Urusvati Institute and other institutions (Koelz 1931), after the Michigan expedition he returned to the region from 1936 to 1938 to collect for the United States Department of Agriculture's Bureau of Plant Industry.

---

25. Letter dated January 13, 1934, on file, Museum of Anthropology, Asian Division, University of Michigan.

In his travels in South Asia as elsewhere, Koelz collected widely, both the specific materials required by the expedition's sponsors and a range of other objects and specimens.[26] He retained many of the latter as personal property, though he sold others to institutions (Payne 2001).[27] His Himalayan biological specimens are now held by numerous institutions, including University of Michigan and Michigan State University museums, the Smithsonian Institution, New York Botanical Garden, American Museum of Natural History, and Field Museum of Natural History, among others. His Himalayan materials alone totaled more than 10,000 plant species (Stewart 1945:408) and 24,000 birds, and his lifetime collecting activities yielded several times these numbers.

As a scholar, Koelz's contributions to biology may be his most important legacy. While he was far from the only North American naturalist collecting specimens in the Himalayas in the early twentieth century, he was among the most prodigious. The University of Michigan Museum of Zoology has more than 32,000 bird specimens that he collected in Asia, including 176 type specimens (Trow 1957; Payne 2001).[28] His plant collections are equally important and include several previously undocumented species. These include a high-altitude grass subsequently named *Stipa Koelzii*, samples of which were collected by Koelz, or more likely, members of his team, in Gya, Ladakh, on August 13 and 14, 1933 (though these are not mentioned in his diary, where plant collecting activities seldom appear) and designated as a new species by botanist R.R. Stewart (1945:441).

Koelz did not publish a great deal, though he did produce a number of scholarly publications on his zoological (Koelz 1937, 1939a, 1939b, 1940, 1950, 1951a, 1951b, 1952, 1954) and botanical (Koelz 1979) work, in addition to a few more descriptive pieces describing his travels (Koelz 1931, 1935, 1957, 1983). He never published on his material culture collections, nor did he make scholarly contributions to the history of South Asian art and Buddhist studies. However, that he was not uninterested in making such contributions is apparent from a detailed study that he conducted on Kashmiri shawls (Plate 3), based on the textiles he had collected for the University of Michigan.

Koelz began working on his 200-plus page manuscript shortly after his return to the United States in 1934. However, it was not completed before he headed off on his most extensive Asian travels, which lasted from 1936 to 1953. He took up the manuscript again when he returned to Michigan in 1955, though he never saw it through to publication. The unpublished manuscript was eventually deposited in the Bentley Historical Library archives and was not published

---

26. By today's standards, the Roerichs were certainly correct in questioning Koelz's ethical practices, particularly that he collected "choice objects" for himself at the same time that he was paid by institutions to be acquiring on their behalf.
27. Payne (2001) noted the Museum of Zoology purchased 25,000 bird specimens from Koelz in 1955, and received an additional 1000 (which had been kept in his home) from Koelz's estate after his death in 1989.
28. As well as 3350 bird specimens from his earlier Arctic expedition, which the museum purchased from Koelz in 1929 (Payne 2001:229).

until 2005, more than fifteen years after Koelz's death (albeit without essential illustrations, which have not been located; in Beardsley and Sinopoli 2005). Today, this work—*Kashmir Shawls: A Study of a Collection of Textiles Made in the Indian Empire*—is most interesting as an example of an early attempt to describe and classify these elaborate textiles. Koelz drafted his manuscript decades before the first major published scholarly work on Kashmiri shawls (i.e., John Irwin's 1973 *The Kashmiri Shawl*, which is today the canonical work in the field). Like his successors, he successfully identified the key historical sources that would underlay all subsequent work on the topic. But unpublished and hidden in an archive, Koelz's writings on the topic have had no impact on the scholarly study of shawls. The shawls he collected have, and scholars continue to examine and publish them as important exemplars of artistic styles and periods (e.g., Beardsley and Sinopoli 2005; Cohen et al. 2012; Rizvi and Ahmed 2009).

Plate 3. Kashmiri shawl from the Koelz Collection (late-nineteenth-century piecework shawl; acquired in Delhi, UMMA 17318).

What is perhaps most interesting about Koelz's study is the insight it provides on his approach to the objects he collected. He certainly appreciated the beauty of many of the shawls he acquired; yet, he makes clear that beauty was not his only interest. As noted earlier, for shawls and other objects, he claimed to seek a "representative" range of specimens that were important for their scientific value as well as for being works of art, noting in a letter to Bartlett, "then there are some examples, not beautiful, but needed for study" (December 31, 1933).[29] In his writings on the shawls, Koelz's approach was very evidently that of a biologist, influenced by Linnaean classifications and the Darwinian concept of "descent with modification." He noted that since the shawls were made in many places and over a long period of time, changes in their design "cannot have escaped the general law of local variation" and that the identification of regional variants should be possible through "the inductive path of scientific procedure" (Koelz 2005 [1934]:20).

For the Buddhist thangka paintings, amulet boxes, and other objects in the Koelz Collection, it is the objects themselves that are important, as Koelz produced no scholarship on them. He was, as noted earlier, not particularly interested in Buddhist studies, iconography, or the specifics of Himalayan Buddhist history. In this he was unlike many of his contemporaries. However, the objects he collected remain of considerable importance today for three main reasons. First, and perhaps in keeping with his scientific training, they are (reasonably) well documented in that we can trace many of the paintings to the specific monasteries or locations where they were acquired. Second, as with his shawl collecting, Koelz collected a range of objects of varying artistic qualities and preservation. That said, this claim toward representativeness should not be overstated, as despite his post hoc acknowledgment of its importance, his diary reveals that acquiring representative images was *not* a primary goal for his thangka collecting practices. Instead, the paintings' aesthetic qualities were far more important. Thus, Koelz noted in the above-referenced December 31, 1933, letter to Bartlett that he "bought with great care and after long and wide inspection, believing that one first class specimen is desirable above any number below that standard."[30] And in many instances recounted in the journal, Koelz failed to acquire objects he was interested in, either because of disputes over their price or the unwillingness of their owners to part with them. And sometimes he settled for lesser quality objects in order to get the few higher quality ones he wanted.

Intentional or not, the paintings Koelz acquired *are* diverse and in one fascinating case at least, he was able to acquire a sizeable group of seventeen paintings (some quite damaged) from a single monastery (Likir monastery, purchased on August 3, 1933; discussed below), which, if not necessarily "representative," do

---

29. Letter on file, Koelz Collection, Bentley Historical Library, University of Michigan.
30. December 31, 1933, letter on file, Koelz Collection.

nonetheless constitute a fascinating assemblage. In the end, both the documentation and the breadth of the collection make the Koelz Collection of interest as an entire assemblage as well as for the individual *"objets d'art"* it contains. Finally, a third factor contributing to the importance of the Koelz Collection lies in its source in the western fringes of Tibet, a region often not well represented in other collections of Tibetan Buddhist art.

As noted above, Koelz collected many more objects during his India travels than are currently in the Museum of Anthropology collections. Some he collected for, or sold to, friends and sponsors. Others were part of the extensive personal collection that filled his Waterloo home. These included a number of spectacular thangka paintings, as well as shawls, carpets, paintings, pottery and the like. Over the last decades, a number of these objects have been donated to the Museum of Anthropology by his friend George Lauff. In addition, a few other thangkas in the Asian Division's collections were almost certainly collected by Koelz. These were donated by Mr. and Mrs. Alexander Ruthven[31] and Harley H. Bartlett, director of the University Herbarium (see below). Despite efforts by the Museum of Anthropology to acquire Koelz's personal collection after his death, the majority was auctioned off by Christie's, Sotheby's and Boos auction houses, with the nearly two million dollars in proceeds bequeathed to the Nature Conservancy. As a result, many of the most spectacular objects that Koelz collected are now dispersed around the world.

Even so, the University of Michigan collection remains remarkable and important, not least as the product of a single individual. Below, I introduce some of the objects that Koelz collected that now are in the Museum of Anthropology collections. I focus particularly on the Buddhist religious art, especially the thangka paintings, and more on the stories behind their acquisition than on their iconography. My goal here is not to present a catalog of the objects in the collection[32] but to trace its origins as "a collection." As I have argued above, it is certainly the case that the Museum of Anthropology's Walter N. Koelz Himalayan Collection can be viewed as the peculiar product of the efforts of a fiercely independent renaissance man. However, it was also the product of a larger history—of the geopolitics of India in the decades before Independence and of the colonial officers, spies, and collectors that moved through the rugged Himalayan landscape. And it is a story of the intense competition among European and American institutions to acquire Buddhist art, and of the often disturbing efforts to remove valued heirlooms and inalienable sacred objects from institutions and individuals. I will not touch on all these dimensions here, focusing instead primarily on the last topic, drawing on Koelz's own words concerning how he acquired the objects in the collection.

---

31. Alexander Ruthven was president of the University of Michigan from 1929 to 1951, and had formerly been director of the Museum of Zoology.
32. Which has already been done by Carolyn Copeland (1980); see also http://webapps.lsa.umich.edu/umma/exhibits/Koelz_Collection_2010/index.html.

The Walter Koelz Collection

As noted above, throughout his life Walter Koelz was an eclectic and passionate collector—of biological specimens, art objects, and other artifacts that caught his interest.[33] From his youth, he collected objects wherever he traveled. After finishing his PhD in the early 1920s, he spent time in New Mexico caring for a dying friend and, while there, acquired Native American pottery, basketry, textiles, and jewelry. And, as this diary reveals, on his homeward journey from the Himalayas he scoured antique shops at every stop: Madras, Colombo, Japan, San Francisco, Chicago. It is not clear what drove Koelz's passion for collecting; his modest upbringing in a German immigrant family in rural Michigan could not predict either his passion for things or for travel. What is abundantly clear is his strong aesthetic sense and fierce appetite to possess "quality" objects, however he came to define them.

During his Himalayan travels, the objects that Koelz most desired were Buddhist religious paintings or *thangkas*.[34] These images, often referred to as scroll paintings since they could be rolled up and easily transported, are made of mineral pigments painted on cotton cloth and mounted in a silk brocade frame. They depict buddhas, bodhisattvas, historical figures, and wrathful deities of the Tibetan Buddhist tradition, typically following strict iconographic conventions (Jackson and Jackson 1984). Thangkas play important roles in ritual and meditation practices of Tibetan Buddhism. Hung in great monasteries and modest homes and carried by traveling monks, thangkas are the focus of offerings to the deities depicted; they assist devotees in devotional practices as they attempt to visualize themselves as a particular buddha or bodhisattva; and they are used as icons in requests for blessings or for removal of hardships (Jackson and Jackson 1984:9–11). More than mere images, thangkas are consecrated before use; once consecrated, they are understood to be a living presence, inhabited by the principle of enlightenment called the Truth Body of the Buddha (Donald S. Lopez, personal communications). As such, they are sacred and inalienable, removed from the world of commodities; while consecrated thangkas can be gifted, they should never be sold.

Nonetheless, sold they were, and by the time Koelz was acquiring thangkas, many hundreds of such objects had been collected by earlier European travelers and colonial officers (Harris 2012; Lopez 1998; Martin 2012). Koelz's diary provides rich details of the ways he acquired his paintings—purchasing already alienated paintings from Tibetan traders; acquiring them from householders or secretive monks; and at two important monasteries, enlisting the aid of a

---

33. He was also, it seems, a collector of people. As I began working on this project and telling colleagues and friends about it, I was struck by how many of them—academics, artists, collectors, gardeners—reminisced about having visited Walter in his Waterloo home, eating his homemade apple pie, admiring his gardens and collections, and listening to stories of his adventures.
34. Also spelled tankas, tangkas.

powerful lama to force their reluctant resident monks to part with holy objects. I discuss some of these transactions below.

In addition to thangkas, Koelz collected a range of other objects. Many were religious: small bronze sculptures of the historical Buddha (Shakyamuni) and other figures (Plate 4), wooden printing blocks for making prayer flags, a beautiful wooden book cover for a sacred text (Plate 5), a bronze *vajra* (ritual object in the form of a thunderbolt) and bell (Plate 6), and two human bone trumpets used in exorcism rituals, among others. He was particularly interested in acquiring *ga'u*, bronze or silver containers made to hold small images of the Buddha or slips of paper inscribed with prayers or blessings. Worn as jewelry by women and carried by men, ga'u were believed to protect their owners from dangerous mountain spirits and other harms as they traveled through the treacherous Himalayan landscape (Clarke 2004). Koelz acquired seventy ga'u of a wide range of forms and styles (Plate 7). Interestingly, only two of the ga'u in the collection still contain their sacred contents, suggesting that their sellers had removed the objects that give them their power before selling them to Koelz.

Plate 4. Bronze sculptures in the Koelz Collection: *left*, Shakyamuni Buddha (UMMA 17004); *right*, standing figure (UMMA 17008).

Plate 5. Carved book cover collected in Lahaul, Himachal Pradesh, India. From left to right are a female yogini, Amitayus, the Buddha of Infinite Life, and Rechung, one of the two disciples of the famous eleventh–twelfth century teacher Milarepa, date 1600–1899 (UMMA 17014).

Plate 6. Vajra (thunderbolt scepter; UMMA 17233, *below*) and bell (UMMA 17234, *right*). The most important implements of Tibet Buddhist ritual, the vajra represents compassion (the male principle), while the bell represents wisdom (the female principle).

Not all the objects Walter Koelz collected had religious significance. He acquired brass vessels and metal spear points. He collected jewelry—rings, earrings, necklaces, garment fasteners, brooches, breastplates, and so on (Plate 8)—sometimes commissioning local silversmiths to make new objects for him or to add new ornamentation to old ones. And he collected twenty wooden

Plate 7. Ga'u, or amulet boxes, held sacred images or texts and protected their wearers from danger (*clockwise from top left:* UMMA 17102, 17062, 17093, 17058).

stamps—small wooden plaques with carvings on front and back, which were pressed into soft clay to seal doors or storage containers, and were rapidly disappearing from use when Koelz was in the region (Plate 9).

With the exception of the thangkas, which could be rolled, all the objects that Koelz collected were quite small. They had to fit into the trunks and bundles carried by the horses and yaks on which Koelz and his team traveled (Plate 10)—and which occasionally fell into treacherous streams (July 23, 1933, diary entry) or down mountain slopes, endangering their precious contents. In the lowland cities, Koelz was able to collect larger objects, and it was there that he acquired the many Kashmiri shawls and embroidered textiles that are also in the collection.

Plate 8. Jewelry from the Koelz Collection: turquoise earrings (17132); coral, turquoise, and silver necklace (UMMA 17134); rings (UMMA 17184, 17186); broaches (UMMA 17147, 17151). Turquoise and coral are the most important stones in Tibetan jewelry and both are believed to have protective and medicinal powers.

Plate 9. Wooden stamps, used to seal doors or storage containers (*clockwise from top left:* UMMA 17026, 17028, 17042, 17032).

Plate 10. Pack animals and team members crossing a bridge (Walter Koelz Collection, Bentley Historical Library, University of Michigan).

The hunt for objects was at least as exciting for Koelz as the hunt for rare birds, if not more so. His diary contains rich details on his triumphs in acquiring objects he particularly desired, as well as his frustrations when his attempts failed (though it is often not possible to link specific objects in the collection with specific events described in the diary). In the remainder of this chapter, I elaborate on some of Koelz's acquisitions, focusing particularly on the objects he considered most important: the thangkas.

The Koelz Thangkas

The twelve wooden shipping crates that left India in January 1934 arrived at the University of Michigan on March 19, a month and a half before Koelz himself. Museum staff began unpacking Koelz's shipment shortly after it arrived, and accessioned the collection as # 865. Lacking a detailed invoice from Koelz, they prepared an "unpacking list"[35] using the numbers Koelz had attached to objects, K1 through K664. Of these 664 objects, 152 were designated as belonging to Koelz, including 12 books, among which were Goethe's *Faustus* and a Pashtu dictionary, as well as art objects. The remainder were intended for the university's collections.

Within the boxes were 82 thangkas. Of these, 48 were accessioned into the Museum of Anthropology Koelz Collection; the remaining 34 were designated as belonging to Koelz. In addition, Koelz carried 20 or so thangkas in his hand baggage. Altogether then, he had purchased some 100 thangkas. It is not possible to ascertain how many of these were collected during the period of the University of Michigan expedition versus how many were acquired during Koelz's previous employment in the region; however, a memo survives suggesting that university officials were not at all pleased that Koelz intended to keep some of the finest paintings for himself.[36]

The acquisition locations that Koelz reported for the 48 thangkas in the Museum of Anthropology Koelz Collection are summarized in Table 1. Detailed catalog information on each of the paintings was published in Copeland (1980) and is not repeated here.[37] As discussed in more detail below, Likir and Karsha monasteries in Ladakh contributed the largest groupings of thangkas (17 and 10, respectively), though in each case, these represented only a small and not necessarily representative portion of the monastery's thangkas. Other monasteries where Koelz purchased multiple paintings were Thikse ($n = 4$), Lamayuru ($n = 2$), and Nago ($n = 2$). Four paintings in the collection are reported as coming from Labrang; at least some of these appear to have come from households rather than a monastery, as Koelz wrote of viewing several paintings and other interesting things in two Labrang houses (October 23, 1933, diary entry).

The paintings Koelz donated to the University of Michigan broadly represent Western Himalayan painting traditions from the seventeenth through nineteenth centuries (plus one likely much earlier, perhaps thirteenth century, painting, of Vairochana, UMMA 17461; Plate 11). Beyond this, it is hard to evaluate the criteria Koelz used to value paintings or to determine those he wished to

---

35. List on file, Asian Division, Museum of Anthropology, University of Michigan.
36. An unsigned memo on file in the Asian Division, Museum of Anthropology, University of Michigan, addresses possible responses the university might pursue to secure the entire collection. In a diary entry of May 11, 1934, Koelz records a conversation with President Alexander Ruthven about this issue; ultimately, Koelz retained all the paintings he had originally designated as his personal property.
37. See also http://webapps.lsa.umich.edu/umma/exhibits/Koelz_Collection_2010/index.html.

Table 1. Sources of paintings in the Koelz Collection.

Location	#	UMMA Catalog Number and Image
Ladakh	26	
Likir monastery	17	17454 Palden Lhamo
		17457 Amitayus
		17458 Tara
		17459 Bhaisajyaguru
		17460 Amitabha
		17463 Palden Lhamo
		17464 Vajradahtu Buddha
		17466 Torma
		17470 Tenap Nyima
		17471 Guhyasamaja Aksobyavajra Mandala
		17476 Mandala of Samvara
		17477 Shabhuja Mandala
		17478 Tsong kha pa
		17479 Amitabha
		17493 Vajrabhairava
		17495 Shakyamuni
		17498 Changse Sherab Zangpo
Thikse monastery	4	17480 Arhat
		17481 Arhat Badhra
		17494 Arhat Rahula
		17475 Lamas
Lamayuru	2	17452 Shakyamuni
		17488 Lama
Hemis monastery	1	17482 Begtse Chen
Mashro	1	17490 Amitayus
Unknown	1	17462 Bhaisajyaguru, Manjushri, Tara, and Aryavalokisteshvara
Zanskar	11	
Karsha monastery	10	17465 Vajrabhairava
		17467 Shakyamuni, Tsongkhapa, Tara, Sadaksari Avalokiteshvara
		17469 Padmasambhava
		17472 Vajrabhairava
		17473 Vajrabhairavaa
		17474 Shakyamuni
		17487 Lama
		17482 Manjushri
		17496 Shakyamuni
		17497 Tara
Tetha monastery	1	17456 Begtse Chen
Kinnaur	7	
Labrang	4	17453 Avalokiteshvara
		17455 Begtse Chen
		17483 Chakrasamvara Mandala
		17484 Padmsambhava
Nago monastery	2	17468 Sadaksari Avalokiteshvara
		17485 Ngawang Namgyal
Chango monastery	1	17486 Padmasambhava
Spiti	3	
Poo	1	17461 Vairochana
Pin Valley	1	17489 Vajrasattva
Unknown	1	17451 Shakyamuni
Unknown	1	17491 Amitayus

Plate 11. Vairochana Buddha thangka, from Poo monastery, Spiti; late twelfth to thirteenth century (?) (UMMA 17461).

acquire. He did value age, and often mentioned his lack of interest in acquiring recent paintings. He did not seem to have particular sectarian or iconographic preferences. Instead, he was driven by a strong personal aesthetic, which, though idiosyncratic, has also stood the test of time. Below I present the story behind Koelz's acquisition of the two largest groupings of thangkas in the Museum of Anthropology collection, from the monasteries of Likir and Karsha, before briefly describing some additional paintings from Koelz's personal collections (acquired by Koelz) that have been subsequently added to the museum's collections (see also the Appendix).

The Quest for the Likir Thankgas

On the morning of August 1, 1933, Walter Koelz and his team set out from their campsite near the village of Nurla on their way to Likir monastery. From Nurla, they traveled east, following the Saspol road along the bank of the turbulent Indus River before turning north along the small tributary that led up toward the ancient monastery. Describing the day's travels, Koelz wrote "Came the pass road to Likir. At the crest of the pass a magnificent view toward Alchi of the range that separates Ladakh and Zankskar. It is very green in the high valleys of the range and snow patches rest on the peaks. Toward Likir, a splendid view of the fields in the valley with the monastery on a knoll at the top. Toward all other directions peak after peak."

Koelz was very eager to reach Likir. Rup Chand had told him that there were magnificent thangkas to be had there: "17 of the loveliest old things, the likes of which there are not in this country" and Koelz wanted them badly. To assure that he would be successful in acquiring the paintings, Koelz sought the support of Lama Ngari[38] Tshang Rinpoche, the titular head of Likir and several other Gelugpa monasteries in Ladakh and Zanskar.[39] His efforts to engage the powerful lama's support had begun three weeks earlier. On the morning of July 10, as Koelz's team was preparing to cross the Shingu La Pass into Zanskar, they encountered a group of fifteen Zanskari travelers crossing in the opposite direction. The Zanskaris informed Koelz that Ngari Tshang Rinpoche was at Seni, some 140 kilometers (about 90 miles) to the northwest, and that he would be returning to Tibet in the fall. Koelz's expedition rapidly proceeded toward Seni, hunting birds and gathering plants along the way. On July 18, they set up camp above Seni village. That evening, Koelz wrote in his diary, "we will rest tomorrow and meet the big lama."

Koelz and his team spent the morning of July 19 drying some of the many plant specimens they had collected on their hurried journey. After lunch, Koelz sent his assistant Rinchen Gyaltsen to the lama, writing later that day that "the servant was well received, offered tea, and a red rag was tied around his neck." Koelz followed soon thereafter with Dorje, another member of his team. He

---

38. The western area of the Himalayas, including Ladakh, Spiti, and Zanzkar, has been known as Nga-ri since the tenth century (Snellgrove and Skorupski 1977:81).
39. Koelz always wrote this name in abbreviated Tibetan script, transliterated as "C ri tsha C's," an abbreviation for nga RI TSHang; according to Rob Linrothe (personal communications), this is almost certainly Ngari Rinpoche, who is the nominal head of many Gelugpa monasteries in Ladakh and Zanskar; the present Ngari Rinpoche is the younger brother of the Dalai Lama.

began the visit by entertaining the lama and his entourage with American music played on the phonograph that he carried with him in his travels.[40]

Over the course of the next few days, Ngari Tshang Rinpoche and Koelz (Plate 12) had a number of conversations in which Koelz repeatedly expressed his desires to acquire the Likir thangkas. These began after the phonograph performance during their initial July 19 encounter when, Koelz wrote, he had "expressed hope [to the lama] he would give me the Likir tankas." Instead, Ngari Rinpoche encouraged him to go to the nearby monastery at Karsha and "take out what you want" (July 19, 1933, diary entry). Koelz went to Karsha on July 20 accompanied by a servant of the "big lama." At the monastery, "an impregnable fortress with houses for 100 or more people," Koelz encountered some thirty lamas who were working to repair a portion of the ancient monastery that had recently collapsed. The lamas welcomed him warmly and Koelz noted that among them were two men who knew Rup Chand's family.[41] The thangkas the lamas first showed him were, however, "ordinary Tibetan-made of some age, but nothing I cared to spend money for." Koelz insisted that the monks look for additional paintings, mentioning in his diary that he had noticed a jumbled pile of discarded paintings in a gate near the monastery's entrance, so knew that more were to be had.

> Finally a dirty pile of rubbishy clothes was bought and among them, a beautiful old thing in repairable condition. I spurred on the search and three more of the very oldest type came up. They were willing to let me have all at my price, so I said rupees 100 for the five. Then they said the best and largest couldn't be given or ill would befall them. I suggested that the dirt of the ages on his face couldn't have pleased the demon it represented and anyhow how much more could happen—a wall had fallen already. The things were wrapped up and then the lamas with much excitement got their picture taken. The big lama's servant said he had never put in such a hard day in his life.[42] He had to walk seven to eight hours with us. He objected now to my having the big demon tanka, but his master laughed and said I should take them. Such things he could give me in Lhasa by the donkey load. It was decided all were to be photographed again in the morning. [July 20, 1933, diary entry]

The Museum of Anthropology collection contains ten paintings from the Karsha monastery. While we cannot identify which five were acquired on this day, the "big demon tanka" is undoubtedly UMMA 17465 (Plate 13). This unusu-

---

40. The use of phonographs was a common practice of many early twentieth-century expeditions, and images of entranced natives encountering western technology were a common trope of early anthropological photography (Mueggler 2012:175–76); Koelz makes frequent references in his diary to playing the phonograph in monasteries and villages and of their inhabitants' responses to various genres of American popular music.

41. Rup Chand was not traveling with Koelz at this time, and was instead recovering from sciatica, and had arranged to meet up with Koelz later in the summer in Spiti (Koelz letter to Mrs. Margaret Watson Parker, August 5, 1933, Koelz Collection, Bentley Historical Library, University of Michigan). He met up with the team on August 27 at Kolung.

42. Presumably, this conversation occurred in front of Ngari Tshang Rinpoche after Koelz returned to Seni.

Plate 12. Ngari Tshang Rinpoche with Walter Koelz (Walter Koelz Collection, Bentley Historical Library, University of Michigan).

ally large and striking painting depicts Vajrabhairava, the wrathful tantric form of Manjushri (bodhisattva of wisdom), who is shown in sexual union with his consort Vajravetali and is depicted with 9 heads (the main face being that of a buffalo), 34 arms, and 16 legs, standing above Yama, the Lord of Death, who he has conquered. It is small wonder that the lamas were worried about its removal.

Despite, or perhaps encouraged by, his success at Karsha, on the morning of July 21 (and after preparing 16 bird specimens) when Koelz returned to Ngari Tshang Rinpoche to take the photographs, he again asked the lama to give him permission to buy the Likir thangkas. He offered "600 rupees for 15, 100 rupees bonus, and him a goat." Ngari Tshang Rinpoche agreed and wrote a letter to "the head lama at Likir ordering him to let me take any 15 tankas I wanted" (July 21, 1933, diary entry). In addition, he ordered Lama Ishi Gering from Rangdum monastery to accompany Koelz to Likir to see that his commands were fulfilled.

Plate 13. Vajrabhairava, the wrathful form of Manjushri, bodhisattva of wisdom; acquired at Karsha monastery (UMMA 17465).

The payments committed and letter received, Koelz's party quickly left Seni and began the 10-day, 300-plus-mile journey to Likir, stopping at Abring, Tashi Tonde, Rangdum, Shakkar, Gyama Tongdze, Parkachen, Tangola, Sangu, Kargil, Mulbek, Bodkarbu, Lamayuru, and Nurla along the way. Throughout their trip, team members continued to hunt animal specimens, collect plants, and acquire objects from monasteries and individuals. On July 24, for example, the expedition stopped at a monastery near Tashi Tonde, where Koelz purchased an "old tea table." He also looked for one thangka he had seen there on an earlier visit, though he did not find it. He further noted that many monks (including several from Likir) were gathered "reading in full dress in the main hall so I couldn't go in looking for tankas" (July 24, 1933, diary entry).

Koelz and his team arrived at Likir on the evening of August 1. At the monastery, Koelz discovered that the monks had been informed of his impending arrival and had a room "fitted up for me with a sort of a cradle for a bed and everything covered with rugs"; the expedition's horses and team members camped in a grove beneath the monastery. Koelz handed over Ngari Tshang Rinpoche's letter before making up a bed for himself on the balcony of his room where he could take in the spectacular views "toward the Alchi Mountain wall with the new snow on the peaks and the melting mist clouds, and the scattered green fields nearby."

On August 2, Koelz was woken at sunrise by music emanating from the hall where monks had gathered for the morning worship. He met the monastery's head lama who

> read his letter this morning and said his abbot ordered him to give me 15 tankas of my choosing. I now again went through the monastery and came out with some 15 very old and very beautiful ones—so old that they are not registered in the monastery list. Of these 15, five were large and in fair condition. The others were damaged badly, two large and eight small. These I was willing to consider as *three* and to take the seven neden chudruk[43] for the lot and make 600 rupees and 100 rupees tip and shell in payment. It was clear now my choice wasn't pleasant and I was urged and made to make a third inspection of the collections but not much more could be gleaned. There were many new tankas, many with beautiful brocade, but I remained firm in my choice. They said now they couldn't give the seven. Very well then, give me a letter saying they wouldn't obey the superior's command. That they wouldn't do. Finally I said I would take the 15 old things at 250 rupees and 30 rupees baksheesh more tips and a letter saying they wouldn't give me the seven. This they finally did and after 20 rupees more tips I left at 4 pm with my tankas. All day of negotiation. The monastery presented me with a huge basin of each rice, flowers, kumanis, and ten eggs. I took the eggs and a meal of rice. They treated us well: gave wood and pasture for the horses. . . . The monk from Rangdum whom I brought along, foreseeing the monastery's refusal to obey orders signed the letter as a witness. The monks didn't want to sign it, but two finally did. [August 2, 1933, diary entry]

Their mission achieved (Plates 14–30), on the next morning, Koelz and his team continued their journey, heading east toward Nimu. Before breaking camp, he wrote: "The monks came again this morning and asked for more tip: I had given to the monastery in general, to the monks in general, to the caretaker, to the building of new houses, and this was now for Rinpoche. We made it clear that by their refusal to obey orders that had cost the monastery 420 rupees plus a valuable present."

---

43. *neten chudrug*: the 16 Arhats, disciples of the Buddha.

Plate 14. Palden Lhamo ("Glorious Goddess"), considered the protector of all Tibet; acquired at Likir monastery on August 2, 1933; nineteenth century (UMMA 17454).

Plate 15. Amitayus Buddha (the Buddha of Infinite Life); acquired at Likir monastery on August 2, 1933; seventeenth to eighteenth century (UMMA 17457).

(*facing page*) Plate 16. Tara, the female embodiment of compassion depicted in the form of White Tara; acquired at Likir monastery on August 2, 1933; seventeenth to eighteenth century (UMMA 17458).

(*facing page*) Plate 17. Bhaisajyaguru Buddha (Medicine Buddha); acquired at Likir monastery on August 2, 1933; seventeenth to eighteenth century (UMMA 17459).

Plate 18. Amitabha Buddha, the Buddha of Infinite Light; acquired at Likir monastery on August 2, 1933; nineteenth century (UMMA 17460).

Plate 19. Palden Lhamo ("Glorious Goddess"), considered the protector of all Tibet; acquired at Likir monastery on August 2, 1933; date unknown (UMMA 17463).

Plate 20. Vajradhara Buddha, a Primordial Buddha from whom all Buddhist teachings are believed to arise; acquired at Likir monastery on August 2, 1933; nineteenth century (UMMA 17464).

Plate 21. Torma offering. Torma is a cone-shaped ritual food offering made of barley dough decorated with butter; red torma offerings are made to wrathful protector deities; acquired at Likir monastery on August 2, 1933; nineteenth century (UMMA 17466).

(facing page) Plate 22. Tenpai Nyima (1782–1853), the fourth incarnate Panchen Lama; acquired at Likir monastery on August 2, 1933; nineteenth century (UMMA 17470).

Plate 23. Guhyasamajo Aksobyavajra mandala; acquired at Likir monastery on August 2, 1933; eighteenth century (UMMA 17471).

Plate 24. Samvara mandala. The central figure is Samvara, a fierce emanation of Akshobya Buddha; acquired at Likir monastery on August 2, 1933; date unknown (UMMA 17476).

Plate 25. Shadbhuja Mahakala, a wrathful emanation of the bodhisattva Avalokiteshvara; protector of monks of the Shagpa Kagyi and Gelugpa sects of Tibetan Buddhism; acquired at Likir monastery on August 2, 1933; eighteenth century (UMMA 17477).

Plate 26. Tsong kha pa (1357–1419), founder of the Gelug sect of Tibetan Buddhism; acquired at Likir monastery on August 2, 1933; date unknown (UMMA 17478).

Plate 27. Amitabha Buddha, the Buddha of Infinite Light; acquired at Likir monastery on August 2, 1933; date unknown (UMMA 17479).

Plate 28. Vajrabhairava, the wrathful form of Manjushri, bodhisattva of wisdom; acquired at Likir monastery, August 2, 1933; nineteenth century (UMMA 17493).

Plate 29. Shakyamuni Buddha; acquired at Likir monastery on August 2, 1933; nineteenth century? (UMMA 17495).

Plate 30. Changsem Sherab Zangpo, a student of the lama Tsong kha pa (1357–1419), founder of the Gelug sect of Tibetan Buddhism; acquired at Likir monastery, August 2, 1933; date unknown (UMMA 17498).

The Story Behind the Story

I have recounted the story of Koelz's acquisition of the Likir paintings in some detail because it is simultaneously distinctive *and* emblematic of many of the transactions that he engaged in during his search for thangkas. Below I expand upon some of the issues that this story evokes.

At both Likir and Karsha, resident monks were coerced into selling their paintings to Koelz through the mediation of a single powerful lama: Ngari Tshang Rinpoche. He primarily resided in Lhasa, and does not appear to have had close connections or relations with the local monks at either monastery. However, as head of Gelugpa monasteries in Ladakh and Zanskar, he had authority over them. In the spring and summer of 1933, Ngari Tshang Rinpoche was traveling through the region to raise funds ("50,000 rupees in Zankskar alone," July 15, 1933, diary entry), receive followers, and host religious gatherings and ceremonies. Indeed, in the days before meeting up with Ngari Tshang Rinpoche, Koelz and his team had encountered multiple groups of porters from his entourage who were carrying the wealth the lama had collected back to Tibet. For example, on July 16, 1933, the expedition "met 14 people and two yaks laded with Ngari Tshang's things" on their way to Lhasa. On the following day, Koelz's team traveled with the lama's group and that evening Koelz hosted them to dinner at his camp. Before parting ways the next morning, one of Ngari Tshang Rinpoche's men told Koelz that the lama was soliciting permission from the Dalai Lama to allow Koelz to accompany him to Tibet where "I could get as many paintings and objects as I ever dreamed of" (July 17, 1933, diary entry).

At both Likir and Karsha, Koelz's descriptions reveal the anxiety and reluctance of the resident monks to part with their paintings. At Karsha, the greatest unease focused on a single painting: that of the wrathful deity Vajrabhairava. Despite the monks' anxiety, Koelz was able to acquire this painting. Even Ngari Tshang Rinpoche's emissary who had accompanied him to Karsha later expressed dismay that the painting had been removed from its rightful home.

At Likir, the monks defied the orders of their superior and successfully refused to part with seven of the paintings that Koelz had selected.[44] Koelz used their resistance as justification to pay far less than he had originally committed for the paintings that he did acquire: instead of the pledged 700 rupees (600 rupee purchase plus 100 tip) for fifteen paintings, he paid 300 (250 rupee purchase plus 50 rupee tip) for seventeen paintings. Further, by forcing two of the Likir monks into signing a letter that confirmed that they had disobeyed their superior, he no doubt increased their fear by making them personally vulnerable within the monastic hierarchy.

---

44. The exact content of these seven *neden chudruk* or "16 Arhat" paintings is not evident from Koelz's descriptions; Rob Linrothe (personal communications) has suggested that the set most likely contained four paintings with four arhats each, with "Buddha and bodhisattvas in the center, and the 'outermost' paintings including the four lokapala (two on each side) and one of the two Arhat attendants (Dharmatala and Hvashang) on each side."

Ngari Tshang Rinpoche was the most influential lama who collaborated in Koelz's efforts to acquire paintings. But he was not the only one. On July 3, 1933, while camped at Sisu, Koelz wrote: "the Sisu lama came this morning and gave us a letter to the chagzot[45] of Ki Gompa who will be able to help us much in Spiti if he is so disposed. The lama will also instruct his son to stay and help us."

In other cases, individual monks interacted with Koelz without the knowledge of their superiors. For example, on July 31, 1933, while camped near Lamayuru, Koelz wrote of a monk who

> went back and forth half the night to my tent bringing tankas. He never brought over three at a time. Kept me awake during the process. He always wanted to be sure I wouldn't say a word or the head lama would skin him. I found among his bringings two very old specimens, one beautiful, one interesting. If I stayed a day many things would come he said, but they don't know the difference between a good and bad tanka and it would have been a big chance. They are going to an old monastery in the nulla and bring things to Nurla tonight.

The two specimens referenced here are likely the painting of Shakyamuni Buddha, seated on a throne, flanked by his disciples Shariputra and Maudglayayana (Plate 31; UMMA 17452; the "beautiful"?), and a very dark painting depicting a seated figure of a lama (perhaps Padmasambhava, the early teacher who brought Buddhism to Tibet in the eighth century; UMMA 17488). This painting (the "interesting"?) likely originally had a silver background, which is now tarnished to a near black; today, only the dramatic red hat of the Ngyima tradition worn by the dimly visible seated figure stands out.

In all these transactions, subterfuge and unease are evident—appropriately so, as the commodification of sanctified thangkas is unacceptable in Buddhist beliefs. Thangkas can be gifted, but once they have been consecrated it is a great sin to traffic in religious images (Donald Lopez, personal communications). Once they have been consecrated, the only acceptable way to dispose of even damaged images is through ritualized burning (McGowan 2008:60). The spiritual and, in the case of the Lamayuru lama, personal risks to those who sold images were considerable.

We cannot be certain of the motivations of the lamas who sold sacred objects to Koelz. Pressure from superiors, such as described above, was a factor for some. For others, desire for personal gain appears to have been a motivating factor. While this may at times have been a result of simple greed, for many, the selling of heirlooms and sacred objects appears to have been forced on the sellers by poverty and need. In addition, as noted earlier, a further factor facilitating Koelz's access to local people and their goods was his partnership with Rup Chand, whose family was well known and widely respected throughout the region. This relationship may have contributed to some thangkas, whether or

---

45. Lama who manages the administrative affairs of a monastery; one of two head lamas in a monastery.

Plate 31. Shakyamuni Buddha; acquired from a monk from Lamayuru Monastery, July 31, 1933 (UMMA 17452).

not they were in fact paid for, being reconceptualized as "gifts" freely given to Walter Koelz and his companion, removing them from the spiritually dangerous realm of commodification.

Not all lamas encountered by Koelz were willing to sell him their monastery's treasures. Singly and in groups, Koelz's diary records many individuals who refused to part with their monastery's property. Koelz was initially optimistic that he would be able to acquire thangkas from Tektse Monastery in Zanskar, writing after an initial visit there on August 7, 1933, that "they have at least 50 of the very old ones, uncared for as usual. . . . I am delighted beyond words. There is good reason to believe the acquisitions can be engineered." He returned to the monastery the following day to begin to negotiate the purchase with the five monks he had been informed would have to be consulted about any decision to sell him paintings. His hopes were dashed the next day as he was told that "no tankas of any sort could be given a sahib."

Finally, not all the thangkas Koelz acquired came from monasteries. Some had already been commoditized and were sold to Koelz by traveling merchants who had acquired them through unknown means. Others belonged to households and individuals willing to part with family property, each for their own reason.

Valuing the Invaluable

Koelz was proud of his bargaining acumen and in his diary he bragged about the low prices he paid and about his willingness to walk away even from art works he greatly desired over minor disagreements on price. I have alluded above to some of the prices that Koelz paid for paintings: 100 rupees for the five (or was it ten?) Karsha thangkas and 300 rupees for the seventeen Likir thangkas. While the coercive contexts of those particular transactions might suggest these were unusually low, they appear instead to lie within the norm of his expenditures for thangkas. On the end page of the diary volume that begins on June 25, 1933, and ends on January 18, 1934, Koelz maintained a running list of his expenses, in which he recorded payments for food, tips, wages, and other expenditures. Beneath this list he compiled a summary budget for 1933, grouping his expenses under several broad categories (Plate 32). This summary is presented in Table 2.

In the notebook, Koelz summed these numbers to 7514 rupees; the total is in fact 7739. While most of the categories are self-explanatory, it is not clear to me what Koelz meant by the heading "Museum" (perhaps his salary?). The total amount that Koelz paid for thangkas is 1152 rupees, presented as the sum of 759 and 393 rupees. I do not know why he subdivided this and a few other entries. It may be that separate numbers distinguish purchases Koelz made on behalf of the University of Michigan from his personal acquisitions or acquisitions for others, but this is at best a guess, and the list of individual expenditures is not complete enough to evaluate this.[46]

---

46. In addition, several entries contain multiple numbers separated by dashes, for example

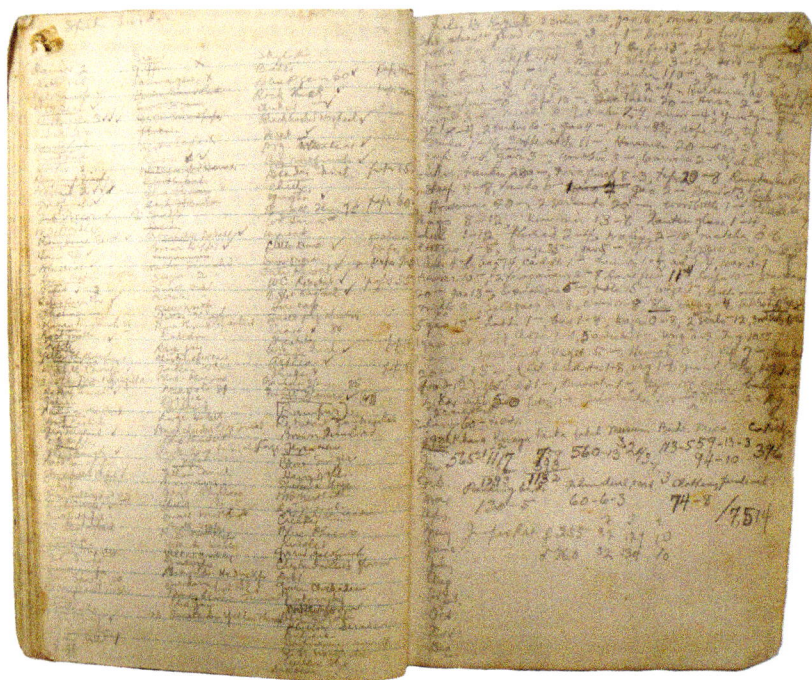

Plate 32. Final page of Koelz diary from the University of Michigan Himalayan Expedition (June 25, 1933–January 18, 1934). The left-hand page is a list of bird species, while the right-hand page contains Koelz's expense records from the Himalayan Expedition (Walter Koelz Collection, Bentley Historical Library, University of Michigan).

Table 2. Koelz's diary 1933 summary.

Khana[1]	Kiraya[2]	Tanka	Baksh. (tips)	Museum	Birds	Misc.	Cartridges	Packing Birds	Film Develop. and Cost	Clothes
565	1117	759	560	3243	113	59	396	130	60	74
	176	393				94				
	1293	1152								

[1]board (Urdu)
[2]wages (Urdu)

under Misc. 59-13 and 94-10. It is also not apparent what these mean, though I suspect that the first number refers to expenses in rupees, and the second to expenses in a different currency (whether British pound or U.S. dollars). I have presented only the first—rupee—number in Table 2.

The running record above this summation includes the prices Koelz paid for many (though not all) of the thangkas. These include: 10 rupees for the painting from Tetha (UMMA 17456: Begtse Chen); 16 rupees for the two tankas from (Lama) Yuru (UMMA 17452, 17488); 6 rupees for a thangka purchased at Rangdum (unknown); 20 rupees for two thangkas from Spitug; 20 and 15 rupees for two thangkas at Hemis (UMMA 17482?), and 100 rupees for thangkas purchased at Tiktse (probably UMMA 17475, 17480, 17481, 17494). These numbers reveal that Koelz rarely spent more than 20 rupees for a painting. The text of the diary also contains an occasional reference to prices that Koelz was unwilling to pay for paintings: 100 rupees per painting at Kiomo (September 4, 1933) and 70 rupees per painting at Poo (September 28, 1933). In both these cases, he recorded that these prices were unacceptable and he did not pursue any purchases.

As noted by historian Janet Rizvi (1999:279–82), it is difficult to meaningfully chart the value of the rupee in the Himalayas in the early twentieth century. On September 30, 1933, the *New York Times* reported the Indian rupee as valued at U.S. $0.365. The 300 rupees that Koelz paid for the seventeen Likir thangkas thus cost $109.50,[47] and the 100 rupees for the Karsha thangkas equaled $36.50.[48] Koelz's total reported 1933 expenditures for paintings of 1152 rupees totaled $420.48 in 1933 dollars.[49]

These amounts are not in and of themselves terribly meaningful in the fluid monetary landscape of the Western Himalayas. An alternate way to evaluate them is to compare them to the other expenses that Koelz reported for his expedition. The totals in Table 2 indicate that the purchase of thangkas accounted for some of Koelz's greatest expenditures, coming in only slightly less than wages (which included the hire of animals and their porters as well as salaries for his assistants) and below the ambiguous "museum" category. While the prices Koelz paid for thangkas seem miniscule by twenty-first-century standards, in the early twentieth-century Himalayas, this comparison reveals that they were valuable and valued objects.

*Other Thangkas Acquired by Koelz in the
Museum of Anthropology Collection*

Over the years, several other thangkas collected by Koelz have come to the Museum of Anthropology. Three were donated in 2007 by George Lauff, a close friend of Walter Koelz and executor of his estate. Lauff purchased two of the paintings at auction following Koelz's death (2007-22-1, 2007-22-2),[50] and

---
47. $1929.76 in 2012 dollars (CPI inflation calculator, http://146.142.4.24/cgi-bin/cpicalc.pl).
48. $643.25 in 2012 dollars (CPI inflation calculator, http://146.142.4.24/cgi-bin/cpicalc.pl).
49. $7410.28 in 2012 dollars (CPI inflation calculator, http://146.142.4.24/cgi-bin/cpicalc.pl).
50. Koelz willed his estate to the Nature Conservancy, and after his death his collection of art objects was auctioned off by Christie's, Sotheby's and the Frank Boos Gallery.

received the third from Carolyn Copeland (2007-22-3), a specialist in Himalayan art, close friend of Koelz, and author of the only comprehensive catalog of the forty-eight Koelz thangkas.

Seven additional thangkas had come to the museum in earlier decades: four were gifted by Mr. and Mrs. Alexander Ruthven in 1958 (UMMA 47157–47160) and three by Harley H. Bartlett in 1960 (UMMA 47308–47310).[51] We do not have much detail on the history of these objects. Koelz had a close and warm relation with Harley Bartlett, and he may well have given these paintings to him. And although Koelz had a falling out with Ruthven, he does appear to have been close to his son Peter. After Peter's death in the late 1950s, the Ruthvens donated a number of antiquities that Peter had owned to the university's Kelsey Museum (L. Talalay, personal communications) and it is likely that the thangkas, donated at roughly the same time, had been given or sold to Peter Ruthven by Koelz. None of these paintings have been previously published, and are described here in the Appendix.

Conclusions

Walter Koelz's diary contains the unedited words of a unique individual, written in the early 1930s while he traveled in the western fringes of the Tibetan cultural region on an expedition sponsored by the University of Michigan. Much more could (and should) be written about both the man and the biological and material culture collections he made. In this chapter, I have briefly discussed the history of the Koelz Collection of Himalayan material culture that I have the privilege of curating, touching upon both how it was made and the objects contained within it. In the remainder of this volume, I let the man who made the collection speak for himself.

---

51. An additional thangka was donated by Conrad Roger in 1959, but I know of no direct relationship to Koelz. It is listed as having come from the Freer Museum at the Smithsonian.

References Cited

Annual Report of the Urusvati Himalayan Research Institute
1930   Annual Report of the Urusvati Himalayan Research Institute, 1929–1930. *Journal of Urusvati Himalayan Research Center* 1:71–90. Roerich Museum Press, New York. (http://lebendige-ethik.net/engl/3-Annual-Report.html)
1932   Annual Report of Urusvati Himalayan Institute, 1931. *Journal of Urusvati Himalayan Research Center* 2:149–63. (http://lebendige-ethik.net/engl/3-Annual-Rep.1931.html)
1933   Annual Report of Urusvati Himalayan Institute, 1932. *Journal of Urusvati Himalayan Research Center* 3:497–510.

Beardsley, Grace, in collaboration with Carla M. Sinopoli
2005   *Wrapped in Beauty: The Koelz Collection of Kashmiri Shawls*. Anthropological Papers, no. 93, Museum of Anthropology, University of Michigan, Ann Arbor.

Berger, Partricia
2003   *Empire of Emptiness: Buddhist Art and Political Authority in Qing China*. University of Hawai'i Press, Honolulu.

Clarke, John
2004   *Jewellery of Tibet and the Himalayas*. V&A Publications, London.

Cohen, Steven, Rosemary Crill, Monique Levi-Strauss, and Jeffrey Spur
2012   *Kashmir Shawls: The Tapi Collection*. The Shoestring Publisher, Mumbai.

Conn, Steven
1998 *Museums and American Intellectual Life, 1876–1926.* University of Chicago Press, Chicago.

Copeland, Carolyn
1980 *Tankas from the Koelz Collection, Museum of Anthropology, the University of Michigan.* Center for South and Southeast Asian Studies, University of Michigan, Ann Arbor.
1983 Biographical introduction. In *Persian Diary, 1939–1941*, by Walter N. Koelz, p. ix. Anthropological Papers, no. 71, Museum of Anthropology, University of Michigan, Ann Arbor.

Drayer, Ruth A.
2005 *Nicholas and Helena Roerich: The Spiritual Journey of Two Great Artists and Peacemakers.* Quest Books, Wheaton, IL.

Harris, Clare
2012 *The Museum on the Roof of the World: Art, Politics, and the Representation of Tibet.* University of Chicago Press, Chicago.

Irwin, John
1973 *The Kashmir Shawl.* Victoria and Albert Museum, London.

Jackson, David, and Janice Jackson
1984 *Tibetan Thangka Painting: Methods and Materials.* Snow Lion Publications, Ithaca, NY.

Joyce, Barry Alan
2001 *The Shaping of American Ethnography: The Wilkes Exploring Expedition, 1838–1842.* University of Nebraska Press, Lincoln.

Koelz, Walter N.
1920 *The Coregonine Fishes of Lake Huron.* Doctoral dissertation, Department of Zoology, University of Michigan, Ann Arbor.
1931 Diary of the 1931 expedition to western Tibet. *Journal of Urusvati Himalayan Research Center* 2:85–131.
1935 The roof of the world: From the diary of a traveler in Tibet. *Michigan Alumnus Quarterly Review* 41:283–92.
1937 Notes on the birds of Spiti, a Himalayan province of the Punjab. *Ibis* 14(1):86–104.
1939a New birds from Asia, chiefly from India. *Proceedings of the Biological Society of Washington* 52:61–82.
1939b Three new subspecies of birds. *Proceedings of the Biological Society of Washington* 52:121–22.
1940 Notes on the winter birds of the lower Punjab. *Papers of the Michigan Academy of Sciences* 25:323–56.
1950 New subspecies of birds from southwestern Asia. *Amer. Mus. Novit.* 1452:1–10.
1951a Four new subspecies of birds from southwestern Asia. *Amer. Mus. Novit.* 1510:1–3.
1951b New birds from India. *Journal of the Zoological Society of India* 3:27–30.

1952 New races of Assam birds. *Journal of the Zoological Society of India* 4(2):153–55.
1954 Ornithological studies. I. New birds from Iran, Afghanistan, and India. *Contr. Inst. Reg. Expl.*, Ann Arbor 1:1–32.
1957 Persian Diary. I. *Asa Gray Bulletin*, N.S. 3:133–76.
1979 Notes on the ethnobotany of Lahul, a province of the Punjab. *Quarterly Journal of Crude Drug Research* 17:1–56.
1983 *Persian Diary, 1939–1941*. Anthropological Papers, no. 71, Museum of Anthropology, University of Michigan, Ann Arbor.
2005 [1934] Kashmir Shawls: A Study of a Collection of Textiles Made in the Indian Empire. In *Wrapped in Beauty: The Koelz Collection of Kashmiri Shawls* (on CD-Rom), by Grace Beardsley in collaboration with Carla M. Sinopoli. Anthropological Papers, no. 93, Museum of Anthropology, University of Michigan, Ann Arbor.

Koelz Collection
   Bentley Historical Library, University of Michigan, Ann Arbor.

Lopez, Donald S., Jr.
1998 *Prisoners of Shangri-La: Tibetan Buddhism and the West*. University of Chicago Press, Chicago.

Martin, Emma
2012 Charles Bell's collection of 'curios': Acquisitions and encounters during a Himalayan journey. In *Narrating Objects, Collecting Stories*, ed. by Sandra H. Dudley, Amy Jane Barnes, Jennifer Binnie, Julia Petrov, and Jennifer Walklate, pp. 167–83. Routledge, London.

McGowan, Dianne
2008 Materialising the sacred. In *Negotiating the Sacred II: Blasphemy and Sacrilege in the Arts*, ed. by Elizabeth Burns Coleman and Maria Suzette Fernandes-Dias, pp. 55–66. Australian National University, Canberra.

Mueggler, Erik
2012 *The Paper Road: Archive and Experience in the Botanical Exploration of West China and Tibet*. University of California Press, Berkeley.

Payne, R. N.
2001 Systematic notes on Asian birds. 20. Recent additions to the list of type specimens of birds collected by Walter Koelz in the University of Michigan Museum of Zoology. *Zoologische Verhandlingen Leiden* 335:229–34.

Rizvi, Janet
1999 *Trans-Himalayan Caravans: Merchant Princes and Peasant Traders in Ladakh*. Oxford University Press, Delhi.

Rizvi, Janet, with Monisha Ahmed
2009 *Pashmina: The Kashmir Shawl and Beyond*. Marg Foundation, Mumbai.

Ruthven, Alexander
1929 Description of the building. *The University Museums Building of the University of Michigan*. University of Michigan, Ann Arbor.

Snellgrove, David L., and Tadeusz Skorupski
1977 *The Cultural Heritage of Ladakh*. Vol. 1, *Central Ladakh*. Prajna Press, Boulder.

Stewart, Ralph R.
1945 The grasses of northwest India. *Brittania* 5(4):404–68.

Trow, Elizabeth Sunderland
1957 Dr. Walter N. Koelz and Thakur Rup Chand, collaborators in Asiatic research, Botanical Gardens and Museum of Zoology, University of Michigan. *Asa Gray Bulletin*, N.S. 3(2):281–84.

University of Michigan Board of Regents
1837–64 *Proceedings of the Board of Regents (1837–1864)*, University of Michigan, Ann Arbor. Digital Library Production Service. (http://quod.lib.umich.edu/u/umregproc/)
1929–32 *Proceedings of the Board of Regents (1929–1932)*, University of Michigan, Ann Arbor. Digital Library Production Service. (http://quod.lib.umich.edu/u/umregproc/)
1932–36 *Proceedings of the Board of Regents (1932–1936)*, University of Michigan, Ann Arbor. Digital Library Production Service. (http://quod.lib.umich.edu/u/umregproc/)

Van Tyne, J., and W. Koelz
1936 Seven new birds from the Punjab. *Occasional Papers of the Museum of Zoology*, University of Michigan, Ann Arbor, Univ. Michigan 34:1–6.

Waterhouse, David M. (ed.)
2004 *The Origins of Himalayan Studies: Brian Houghton Hodgson in Nepal and Darjeeling, 1820–1858*. Routledge Curzon, London.

Whistler, H.
1925 The birds of the Kangra district. *Ibis* 68(3):521–81.

Wright, Henry T.
1983 Introduction to Iran, 1939–1941. In *Persian Diary, 1939–1941*, by Walter N. Koelz, pp. xi–xii. Anthropological Papers, no. 71, Museum of Anthropology, University of Michigan, Ann Arbor.

Himalayan Expedition Diary Transcription

The bulk of this book consists of the transcription of Walter Koelz's diary from the University of Michigan Himalayan Expedition, beginning on June 25, 1933, and ending on May 31, 1934, when Koelz and Rup Chand were back in Michigan. The beginning date was forced on me, as the diary containing the first seven months of the expedition has been lost. The end date is a few months after Koelz left the Himalayas, but includes continuing references to the collections and the thangkas (including the disconcerting information that he bought a set of watercolor paints, with twelve colors, and used these to "repair" the thangkas to his satisfaction!).

In transcribing this text, I have been true to Koelz's words and did not correct grammar or take other editorial liberties. I did spell out certain regularly used abbreviations (for example, "Rup Chand" for "R.C." or "rupees" for "Rs."), but have retained the various spellings that Koelz used (particularly for place names, such as "Zinzingbar," "Zingzingbar," and "Zingzinbar"). There are also occasional words I could not decipher, and other portions of the text that were illegible. Koelz wrote his entries in pencil on lined pages in small leather-bound notebooks (5 × 8"), which he carried with him throughout his journeys (Plate 33). Careful to preserve paper, he eschewed margins and filled all available bit of blank space. I have transcribed this text as carefully as I could, but do not doubt that there are occasional errors.

It is also certainly the case that had Walter Koelz himself lived to publish this text, he would have edited it significantly. Indeed, he did publish small portions

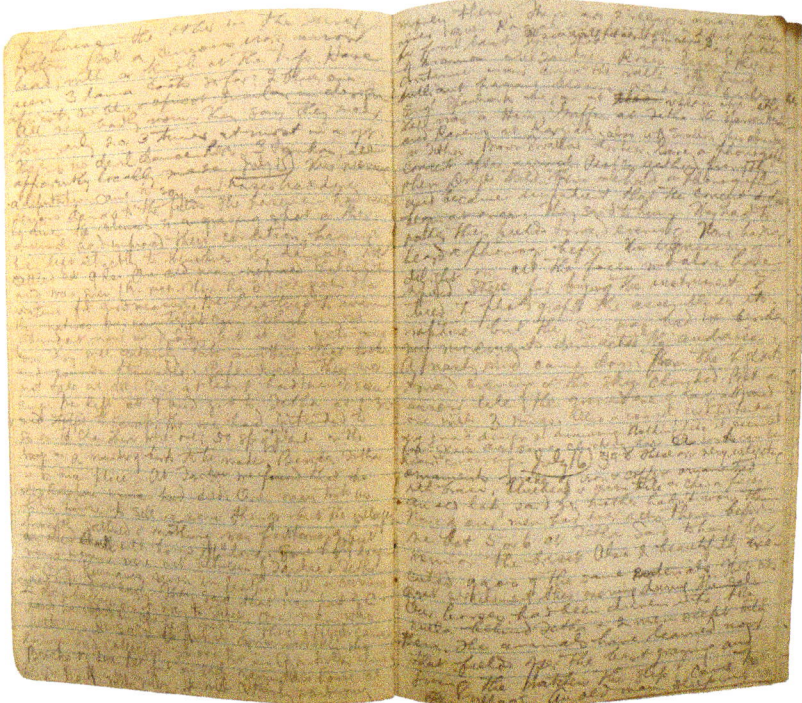

Plate 33. Koelz diary entry, July 14–16, 1933 (Walter Koelz Collection, Bentley Historical Library, University of Michigan).

of it in "The Roof of the World: From the Diary of a Traveler in Tibet" (1935), a ten-page article that includes entries from December 7 and 25, 1933, and January 19, April 14, April 28, May 15, June 30, July 1, July 23, August 11, August 19, and September 11, 1934. The first four entries reveal that Koelz did have the now-missing diary when he returned; it is not clear when or how this one was lost. Equally interesting, by comparing the published entries with the original text, we can see how Koelz himself edited his text for publication—elaborating on his experiences, reorganizing the often stream-of-conscious flow of his original text, and, on occasion, reordering specific events. Below, I present a few of these entries, side by side, with the unedited text on the left (in gray), and the edited version on the right (in black).

June 30, 1933. A native cigarette is made by rolling some green grass in a rhododendron leaf leaving both ends of the roll open. On top is put a live coal, and on this tobacco. The smoke is then inhaled (drunk) through the palm of the second hand. Failing a pipe, you can also drink smoke by planting the tobacco and fire in a hole in the ground. Rain came in the night. The morning and until three in the afternoon was clear. We tried in vain to induce some travelers to give us two horses to Khohsar so our horses could travel better, but no inducement prevailed. Rup Chand gave up his riding horse and we thus got to Sum (Kulu for mud; Snipe = Sum; Kukri = mud hen). All the horses are weak this year because of the late spring. Besides the Lahuli horses had short rations all winter because of the drought in the summer. Our horses could hardly get us to Khohsar. All but the botany party and a few Lahuli horses stopped at Sum too. From our tent we see the camps of sheep caravans taking grain (six camps), a Tibetan lama who has been on pilgrimage below from Chang Tang, a camp of gudjers[1] with their buffalos, and two or three horse camps like our own. The alpine flowers have just begun. In the bottom of the Beas Valley among the birch trees the lavender rhododendrons are laden with bloom. The air is fragrant near them. Beside them are patches of a rose pink Androsace, or of a yellow buttercup or of a white Anemone, or of the glorious Primula rosea or its lavender relative. The whole landscape is magnificent with rare tones of purple, pink, rose, magenta, royal blue, yellow, that is, if you look from near. The purple dwarf iris Kumaonensis is nearly out of bloom. Karma says his baby is free of diarrhea and is very hungry. Rain followed a thick wave of fog that came from below and continued from 3:30–5:30. Rup Chand

---

1. Koelz (1935:288) elaborates on this entry, describing "gudjers" as "the curious Indian people who subsist solely on their herds of buffaloes."

June 30. Tomorrow we expect to cross the Himalayas on the Rotang Pass (13,400 feet). All the horses are weak this year on account of the late spring; besides, the Lahuli horses had short rations all winter. The hay failed from the drought in the summer, otherwise we should have made the crossing today. But the poor beasts couldn't go farther than Sum (11,000 feet). Sum is the Kuluese word for mud. It is well named. The snow has just retreated from the shoulder that provides this happy camping spot where the strenuous journey across the mountain range can be broken. Countless rivulets trickle down from heavy snow banks that will cling to the mountain tops for at least another month, and magic flowers have sprung up in the wake of the snow's retreat. The yellow cowslips (our familiar Caltha palustris), and a salmon pink primrose (Primula rosea), are pressing the snow hard and are in bloom at its border. A huge patch of dwarf iris on a sunny slope where the snow melted first leaves a splash of purple in the variegated field of yellows, blues, and pinks.

A thousand feet below us, just out of its cradling spring below the crest of the pass, runs the infant Beas, one of the five mighty rivers that give the Punjab province its Persian name of "Five Waters." Among the silver birches that clothe its opposite bank huge clumps of lavender rhododendrons are in full bloom. Primroses, Dutchman's breeches, trilliums, jack-in-the-pulpits, and butter cups are blossoming in the fragrant shade beneath. One might be tempted to linger and enjoy to the fullest the bit of paradise before us, but we have learned that each day brings an experience equally rare.

Six of the sheep caravans carrying grain from India have stopped to camp with us. The little lad of twelve who started with us this morning, with his herd of a dozen sheep carrying kodra seeds (a kind of millet), has gone on to the other side. A lama, like Kim's lama, from the Great Plains of Tibet has pitched his streamer-laden tent a little

met Wang Gyel's father in Kulu and told him his son probably had mengi [?]. "I saw him out on the road and blew on him and sent him home. No matter." For mad dog bites, etc. a holy man is often summoned to blow on the patient, much as you blow dust away. Kuluese call it mantar kama. It's good for snakebites, boils, and for certain diseases for which there is no other cure. The same sort of practice is in vogue in Lahaul where it is called རྒ [ngag; mantras, speech]. The carved bones of which I have a sample are useful in the performance of such rituals here.

above and is staying on a few days. He has been visiting the holy places of India, which every Tibetan, whatever his station in life, plans sometime to see. It isn't alone the soul that is improved by these wanderings. Down in the valley is a camp of "gudjers," the curious Indian people who subsist solely on their herds of buffaloes. The poor naked beasts must be uncomfortable here in this icy air. The Gyawa Rimpochhe at Lhasa had a buffalo once for milk, I heard, and made silk clothes to keep it warm.

Tibetans have been coming today for treatment for malaria and one of our horsemen is also smitten with it. The poor devils know that the disease is somehow associated with heat, but they don't know that mosquitoes carry the germ. The horseman has a red string tied around each of his big toes, a cure for malaria that he got from a mountain priest. For mad dog bites, snake bites, boils, and certain other diseases for which there is no cure in the native pharmacopreia, the holy man may blow on the patient, much as you blow away dust. It is called "manterkarna" in Kulu, and in Lahul "ngak." Rare carved powder horns are used in the ceremony in Lahul. Karma's baby has responded perfectly to yesterday's salol treatment.

A native cigarette is made by rolling some green grass in a rhododendron leaf, leaving open both ends of the roll. On top of the grass is put a live coal and on top of this the tobacco. The smoke is then inhaled (smoke is "drunk" in the Hindustani and Tibetan languages) through the palm of the second hand. Failing a pipe, you can also "drink" tobacco by planting the tobacco and fire in a hole in the ground and arrange for an air current. Connoisseurs of delicate tobacco aromas will best appreciate this technique. Anyway, a well brought up Tibetan knows that smoking is a sin.

July 1, 1933. We left Sum at 8:45 and reached the top of the pass in three hours. There is bad going now. The snow begins a little above Sum and continues to near Koksar. It has melted and the horses sink in. In a few days it will be dangerous. The old horse tired below the pass and the trunk carrier gave out too later. We tried to get two horses at two and a half rupees each (less than six miles to go) but they refused, then changed their minds, but two horsemen undertook the work meanwhile. The white horse deprived of the troublesome boxes came gaily along. One animal stepped in the mud and besmirched his load but no damage. Got a horned lark from among three or four on the pass crest. Many flowers out, several I have never seen in all my clambering around the peaks. The distribution of the plants is strange. The soil and ecological conditions otherwise appear the same and yet certain species are restricted to small areas. You simply have to keep looking. We got to Koksar at three sans other difficulty. The sheep counter says so far 45,000 sheep and goats have passed here. They will continue to come til the 20th of July. One flock came with us. The income of all the animals from Lahul and Spiti runs in the neighborhood of 2000 rupees a year at one quarter anna per head. There are 130,000 on the average in these two districts. The zamindars don't permit the visitors near their fields or houses, reserving this grazing for their own cattle. Youngsters under a week of age are passed free. The botanists and scoutmasters have had difficulty. They came on ahead and the animals carrying their bedding and provisions tired on the pass and they were stranded in the rest house beside a fire. They sent 13 horses from here today to succor their loads and there is war now because they won't pay for more than the five that carried goods. A Sum shepherd lost around 40 goats and sheep one day from eating poisoned grass. The Tibetans et al. are burying the carcasses at seven annas. I saw strings of the meat drying at Rahla.

July 1. We left Sum at 8:45 and reached the summit of the pass in three hours. The going was bad. The snow was melting, and in spite of the weight of the sheep herds that packed it down, the horses sank. The old horse tired below the pass, and the white one with the trunks gave out too. We tried in vain to hire two horses from the stream of travelers pouring over to the other side of the mountain, but none were to be tempted by money. Two of the horsemen undertook to replace the horses. One horse slipped in the mud and besmirched his load but hurt nothing. As we left the pass the mist clouds rolled up and it remained dark till night.

The monsoon has clearly begun and it's lucky we have got out of it without a ducking. The monsoon seldom ventures over the outer range of the Himalayas but heavy rains fall on their peaks. Clouds gather very quickly on the pass, and in the spring the people dread the storms because of the thunder. Being in a thunder cloud is described as a trying experience; heavy crashes, literally about the ears, produce extreme nausea. There were many flowers on an island rising out of the snow. Several of them I had never seen before in all my clambering around the peaks. Plant distribution in the mountains is very queer. Soil and other conditions appear uniform, and yet certain species are restricted to small areas.

We arrived in Koksar on the other side of the mountain at three. Snow continues near the town, and in several places we had to chop out paths in the ice. While crossing here once before one of the horses slipped on such a place and rolled over and over to the valley below, with no harm done. The official counter of sheep, who keeps the bridge on the Chandra over which all travelers, man and beast, must pass, says that so far 45,000 sheep and goats have gone on. They will continue to come until the end of July. One huge drove came with us and some lambs were born in the snow where we rested on the crest.

There is a disease of the sheep in which the animal's eyes grow red, the rectum voids intestine, and the liver on frying falls to pieces. If this flesh is eaten some break out near the eye or arm pit, or elsewhere, often affecting local nerves and producing distortion. Both diseases are called ཕོ་རི་ [pho ri; unknown]. On one occasion Rup Chand came near buying such a sheep. It was purchased by others and some 6 men ate of the flesh. Of these two died, one nearly died. Rest immune. The disease is confined, so far as Rup Chand's experience goes, to Gaddi sheep. The Chinese Knot [illustrated] that is pictured in rugs and paintings is known to the Lahulis as gyandüt (Chinese Knot). They can tie it. As we left the pass the mist clouds rolled after us and it remained dark in that quarter till night. These clouds do not come so far as this often though there is much more rain here than in other parts of Lahul. The clouds arise very suddenly on the pass. When there is thunder it is said to be very trying, because the clouds are about your ears and the crash is deafening. Lightening apparently seldom strikes. I saw a tree at Katran this trip and one on Chander Kam. Had a little shortness of breath approaching the pass. A few drops of nosebleed right nostril here. One of the men had a nosebleed on the pass. All pulses high on arrival. A woman came to Kyelang for an operation for an enlarged gland of the armpit and it was removed without any anesthetic. She didn't even say a word during the operation. In Tibet Ri-Choma District Tang R'ar, nice things in a monastery. At Shupke-Kyuk-Tuak, Nyang སྣུར or སྣུ [snur or snu'; unknown place name] Ri.

The shepherds tucked them in their coats and the procession pressed on in order to reach pasture before dark. Many of these herds come 200 miles and more from the parched foothills below, and are repeating an age-old trek for grass over the grim and sinister mountains that to this day hold terror even for natives who live out their days on the slopes.

It is not without reason that the peaks are feared and shrines to the terrible forces of nature are set up on the summit. Blasts of hell rise from nowhere at certain seasons, and it was not long ago that over 150 men were blown to their death from the top of the Rotang La; every year numbers of beasts are overcome and lost in some sudden squall of snow. A paisa (one-half cent) a head is collected as the grazing fee, and about 130,000 head pay the tax every year. Youngsters under a week of age go free. A troop of Indian boy scouts and college botanists are here in the resthouse. They imprudently went on ahead, and when their horses wore out in the snow on the pass they spent the night without bed and food.

A shepherd lost some forty goats and sheep by their grazing on some kind of poisonous grass. The Tibetans have bought up the carcasses and are drying the surplus meat. Every dead sheep has a sales value, but it isn't wise to buy.

September 11, 1933. The drimo came and slept in front of the tent. The horses all fled over to Ki with the new horse. She ate grass around our heads and then went off to graze before dawn. I heard an Ibidorhyncha in the night and found and got it in the morning. Also got a bluethroat. Saw three or four last night but only two today. Saw a large duck, like a pintail. Today the flocks pastured at our end of the meadow. They are tended by eight or ten children, all as usual with baskets on their back for collecting the manure. Great is the rivalry to get the precious substance. When it comes to extracting it from the little ponds a more elaborate technique is required. We bought a very nice man's earring for 10 rupees. The owner didn't want to part with it but consulted the crowd and they advised he take the four rupee profit. The children here are not importunate as at Kibor where one and all shouted "salaam zhu bakshi nang" whenever I came near. Two little manure collectors found a pile near our tent but feared to collect it on account of the dog. Finally Rup Chand persuaded them the dog wouldn't come. But no soon had they got their treasure in the folds of their kaftans (they left their baskets behind so as to be able to run the faster in case of danger) than the dog came. They hung on to the spoils till the dog caught up with them (100 yards) and then dropped it in terror. Rup Chand then gave money and explained his guarantee hadn't been kept. "But you didn't tell the dog to chase us. He came from behind of himself." They put the grain in the mill stones by hand here. We asked why they didn't fasten a hopper over the grinding but they said it wasn't the custom in former times. We parted with our drimo for a nice little donkey. They mutually pulled hair out of their animal before parting with it. Saw the W.R. Harrier twice. The day was beautiful and warm till 2. Then a squall came from below and also in above the valley and it sprinkled into the night. The birds all hid in the Eleaganus and where last night they were very numerous I

September 11. We descended perhaps 1,000 feet from Kibor (13,000) and camp was made near Rangrik below Ki, perhaps the best known village in the whole district of Spiti. The lone drimo, having lost her three companions (a drimo is the female of the yak), has taken to sleeping in front of our tent. She chewed away at the grass around our heads all night and only went off to graze at dawn. The men didn't tie up the new horse we got from Ki monastery and the animal went home last night and took our horses along.

There are numbers of blue-winged teal in the spring ponds along the river, and I saw one pintail with them. Pigeons happen to be scarce at this village, probably because all the crops are cut. Today the children brought the flocks to pasture at our end of the meadow. They are tended by eight or ten little children, all as usual, with little wicker baskets on their backs for collecting the dung. Great is the rivalry to secure the precious substance which keeps the family in fuel. Two little fellows had located a bonanza near our tent but dared not approach for fear of the dog. Finally Rup Chand persuaded them that the dog was harmless and guaranteed immunity from attack. They had no sooner got their treasure in the folds of their cloak (they had prudently left their baskets behind in order to make a quicker getaway) than the dog burst forth. They hung on to their spoils till the dreaded animal caught up with them. Rup Chand offered a cash indemnity for breaking the guarantee, but the lads said, "You didn't tell the dog to chase us. He came from behind of his own accord."

We bought a beautiful man's earring. The owner didn't want to part with it, but consulted the crowd and they advised him to take the forty per cent profit.

They feed the grain into the mill stones here by hand, a fearfully tedious process. We asked why they didn't rig up a hopper, but they said it never had been done thus in this place. We traded off our last drimo for a beautiful little

found them with difficulty. The black and white wagtail went down the valley in flocks. Rup Chand says the ibex eat till dark when a storm is coming. The Sisu lama's nephew came to camp today. He left some lunch in a cloth, some sort of mess made of the remains of the beer malt. The dog affected a great unconcern with the garment but liked the location and was finally caught eating the stuff slyly. The man has just come back from Ladakh and Rupshu. He says the Rupshu lama has finished one lakh of his prayers or readings of his Bumstok. Rabazang called snipes "shingle." Tibetans sew by pulling the needle toward them; Indians and Europeans stitch away.

donkey. Gombo said it was a fine animal because it could hee-haw fourteen times, and no other donkey he ever heard had breath enough for more than a dozen. The respective owners pulled some hair out of their animals before parting, lest their "Kismet" go along with the trade.

Gombo is going to take the donkey to Lahul tomorrow. It's a ten-days' journey, with a settlement a day on the average, and three days without any, through a country infested with charels, goblins, and spiteful beings of all sorts. Gamba is only sixteen but he comes from the most intrepid stock in Lahul. He can endure more cold than I should have believed humanly possible. His mother disappeared from the field where she was working last summer and no trace of her has ever been found. The people know well enough that Kangreta carried her off. Kangreta wears a checkered plaid and a Tam o'Shanter, and has one ear. A Lahuli lama passed through from Rupshu with the news that the Korzok lama has completed the recitation of the 100,000th prayer of his Bumstok. He has been buying sorghum meal and grain for the event.

June – July 1933

[Note: The volume of Koelz's diary transcribed here begins in the middle of his entry for June 25, 1933. The preceding volume was lost; as a result, we do not have Koelz's descriptions of the first seven months of the expedition. We do know, from excerpts published elsewhere, that Koelz's travels in India began in mid-April 1932. On June 25, Koelz was in Kulu in Himachal Pradesh, awaiting a permit to cross the mountain pass at Rotang La, and enter Zanskar.]

June 25 1933. [Tibetan characters] "Man is destroyed by this mouth; goat by his fat." Very fat goat dies in extreme cold; in great heat also. The lost letter of May 13 carried a 2 annas court fee stamp about which the District Commissioner's office had written once and dunned me one. Thereafter no word about it. If the stamp was received, the letter was received. We gave the woman who looks after this house her choice of a red glass (5¢), green glass (5¢) and a "beggar" ($1.00) necklace. She chose the red and skipped a block for joy. Shortly thereafter she brought three small squash, some green apples, mint, radish pods, danya seed for chutney. Got gun license amended to read 1000 cartridges for each "DB shotgun No 444886" and "22 bore SB rifle." No mention made of my request for legalization of .410 and .32 auxiliaries and their ammunition. The "DB shotgun" may be taken to include them as far as possession is concerned, but I don't know what an ammunition seller would do when it came to buying cartridges. A note said (date 6/23) that the crossing of the inner line was being dealt with separately. Dorje went after horses and succeeded in getting ten at 5 rupees, 2 annas to Kolung.

June 26, 1933.   Intended to start today, but rain began in the night and continued until 3. Then the horsemen had scattered and we had to stay. Dorje went with the three boxes birds and plants to Bandrole and brought back our check for 10,200 rupees and the pass to Zanskar. It is marked Pass #1, though Tucci and another person have gone across the line ahead of me. It presents rigidly where I may go and come and says I must come back on the spot if ordered. The horsemen are all Rampuris. One a boy of 28 is especially strong. Any of them walked off with one of the two maund[1] boxes. Neil Chand says he has sown some of the fine-grained rice. Transplant is women's work. The men sit calmly in the shade.

རྒན་གཏམ་མི་ལ་མི་ཟུག   རྒན་བེར་ས་ལ་མི་ཟུག

[*rgan gtam mi la mi zug ear sa la mi zug*; old people's talk does not touch people, old sticks do not pierce the earth.] Old folks talk do not affect men; old folks stick do not impress the earth.

June 27, 1933.   Left Kulu around 9. The men made camp near Raisan and collected their supplies and I went to Bandrole to fumigate the three boxes of specimens and say goodbye to the Lees. I left a registered letter to Lloyd's Bank, Lahul containing my check for 10240 ½ rupees with Mr. Lee to be posted. He is extremely careful what he says, but I suspect the Russians'[2] days are numbered and that they will be out before the year. I had some Ladakhi kumanis[3] that he has grown from seed. He says they are superior to any other. And such dahlias and hydrangeas! The former must have had 25–30 blossoms and the latter was a *solid* mass of blooms. It had rained in the night but remained neutral till about two when we got ready to start. It sprinkled then all the way and we halted at Katrain. Rup Chand had gone by motor to Bara Gaw, but came back when the rain ceased at dark. The caretaker of P.C.'s house wouldn't take four annas for sweeping it out. She said we had given her so much already! The five cent necklace and a roll of wool cloth. She wouldn't sweep today. It was bad sign to throw dirt after a departing friend. She has been very eager to get another necklace like hers for a friend who covets one. The poor soul makes her own living and can't have a cent in the world. They told me a tale of a young Englishman now fishing here. He walked down the road and, from a group sitting along the wayside, one rose to salaam. He then demanded that all salaam. They steadfastly refused. When he put down his rod, one got ready to fight and the white man withdrew. Another day he met one of the offenders and said it was

---

1. A maund is a traditional Indian measure of weight. Although weights varied historically, in 1956 these were formally standardized by the government of India to one maund = 37.324 kilograms. Other units referenced in the text are seer = .93310 kilogram, and tola = 11.66375 grams.
2. References to "the Russians" throughout Koelz's diaries appear to refer to the Roerichs and the Urusvati Institute; see chapter "Walter Koelz and His Himalayan Expedition."
3. Dried apricots, an important food for travelers in the region.

disagreeable of him not to have salaamed him. "Why didn't he have his wound dressed," he would give him something and rubbed Vaseline on it. The lad in question had had a bloody battle with a Gaddi[4] whose sheep had trespassed on his field. The boy's sister had helped to punish the Gaddi. The crops above the river floor are ripe and are rotting from the incessant rain. We met some 20 horses coming from Lahul. The wound on my heel that I got the day I went hunting across the river nearly a month ago is now healed. It was a blister from new cherogs[5] and never spread but persistently wouldn't heal. The infection hinted at spreading to the ball of the foot as it did from a similar wound my first year here. That time a grand abscess formed and laid me up with fearful discomfort for a week. The groin glands were now infected. Rinchen Gyaltsen left by motor and must have got to Koksar today. Mr. Lee has nectarines of fair flavor. Our white dog refused to leave Kulu, but the red one has taken charge of the caravan and runs from van to fore. We have 12 horses at 5 rupees 2 annas for 77 miles. The Lahulis know that maggots in the wounds of horses and sheep cause the wounds to heal quickly. The gnats around the anus [of] the maggots drill into the viscera. At Raisan, on the bare spots under a tree near the river I found 28 species of dry land caddis [caterpillars?]. One had a plain silk tube about an inch long closed at the mouth when the head was withdrawn by a soft stroking of silk. The other was one-eighth inch with a cluster of four or five leaf fragments around the mouth that made the whole look like a corsage bouquet. When walking the case fell forward and concealed the mouth [sketch]. Thus I suppose the ants never found out that there was anything edible inside.

June 28, 1933. Rained till two. Then left for Manali. Camped below town. The people are busy in the rice fields. The men plowing and digging and the women and children setting out the plants. The ground is first dug by hand, then plowed, then flooded, then smoothed with oxen, then finished with a hand instrument. Then banks repaired for ratholes, then flooded and made ready to plant. The women wade in the water nearly to their knees. The Tibetans and Rampuri are drying lingu for summer. Shot an owl near camp which we never had before, except as a nestling. Rup Chand's horses came with R.G.'s brother and are going back with us tomorrow.

June 29, 1933. Mosquitos bothersome. Some rain toward morning. Arrived at Rahla at three. Thank the Lord we shall now be free of the no-see-ums that covered us with welts at Kulu on to Manali. No pestiferous horseflies or blackflies or sand fleas or anything from here on. Several Tibetans have come for help for malaria; one a baby of four. My horseman has taken quinine daily and is now better. A three year old baby had had diarrhea for three weeks. Give a little

---
4. The Gaddi are a Hindu tribal community in Himachal Pradesh and Jammu and Kashmir; traditionally transhumant sheep and goat herders.
5. Shoes or boots.

salol.[6] The natives up to Manali have been gathering Calamus[7] for thatching. A Lahuli lad of 13 or so is going home with a dozen sheep laden with kodra.[8] We have met several Lahuli caravans going to Kulu. A group of students from Lahore came to Rahla with us. A Gundla man brought a nice very red copper gilt image and we bought it. He has the ancient iron purbu[9] that I have wanted since I first came to Lahul. The Kulu bazaar people are very kind to animals. I never heard a dog yelp all the while I was there and all the stray dogs are healthy and plump. There are two species of mushrooms that the matrons of Kulu gather in the rainy season. The *tsuntsuru* come in April and are now to be had only dried. Some smart young plainsmen came to Lahul last year and got possession of some letters that a prominent citizen had written a silly girl. Their effects were completely searched on the way home and the letters recovered. The day cleared at noon, but at night heavy fog hung over the pass from about 9000 ft. up. My malaria horseman has a red cotton string tied around each ankle and wrist. Some deota admi[10] gave them to him for malaria. The people haven't any notion that mosquitoes are the cause of the disease. Horseman's names: Tsering Wangchuck, Chotu, Sarayi, Dorje, Ram Saran, Tsering Ngomp.

June 30, 1933. A native cigarette is made by rolling some green grass in a rhododendron leaf leaving both ends of the roll open. On top is put a live coal, and on this tobacco. The smoke is then inhaled (drunk) through the palm of the second hand. Failing a pipe, you can also drink smoke by planting the tobacco and fire in a hole in the ground. Rain came in the night. The morning and until three in the afternoon was clear. We tried in vain to induce some travelers to give us two horses to Khohsar so our horses could travel better, but no inducement prevailed. Rup Chand gave up his riding horse and we thus got to Sum (Kulu for mud; Snipe = Sum; Kukri = mud hen). All the horses are weak this year because of the late spring. Besides the Lahuli horses had short rations all winter because of the drought in the summer. Our horses could hardly get us to Khohsar. All but the botany party and a few Lahuli horses stopped at Sum too. From our tent we see the camps of sheep caravans taking grain (six camps), a Tibetan lama who has been on pilgrimage below from Chang Tang, a camp of gudjers[11] with their buffalos, and two or three horse camps like our own. The alpine flowers have just begun. In the bottom of the Beas Valley among the birch trees the lavender rhododendrons are laden with bloom. The air is fragrant near them. Beside them are patches of a rose pink Androsace, or of a yellow buttercup

---

6. Phenyl salicylate crystals; used to treat intestinal disorders.
7. Genus for variety of species of cane that grows in different regions of the Himalayas (S.S. Negi, *Himalayan Forests and Forestry* [New Delhi: Indus Publishing Company, 2000], 163).
8. A variety of millet: *Paspalum scrobiculatum*.
9. *Phurpa*: three-sided knife or ritual dagger.
10. Deotas are local deities associated with the land.
11. Koelz (1935:288) elaborates on this entry, describing "gudjers" as "the curious Indian people who subsist solely on their herds of buffaloes."

or of a white Anemone, or of the glorious Primula rosea or its lavender relative. The whole landscape is magnificent with rare tones of purple, pink, rose, magenta, royal blue, yellow, that is, if you look from near. The purple dwarf iris Kumaonensis is nearly out of bloom. Karma says his baby is free of diarrhea and is very hungry. Rain followed a thick wave of fog that came from below and continued from 3:30–5:30. Rup Chand met Wang Gyel's father in Kulu and told him his son probably had mengi [?]. "I saw him out on the road and blew on him and sent him home. No matter." For mad dog bites, etc. a holy man is often summoned to blow on the patient, much as you blow dust away. Kuluese call it *mantar kama*. It's good for snakebites, boils, and for certain diseases for which there is no other cure. The same sort of practice is in vogue in Lahaul where it is called སྔགས་ [*ngag*; mantras, speech]. The carved bones of which I have a sample are useful in the performance of such rituals here.

July 1, 1933.   We left Sum at 8:45 and reached the top of the pass[12] in three hours. There is bad going now. The snow begins a little above Sum and continues to near Koksar. It has melted and the horses sink in. In a few days it will be dangerous. The old horse tired below the pass and the trunk carrier gave out too later. We tried to get two horses at two and a half rupees each (less than six miles to go) but they refused, then changed their minds, but two horsemen undertook the work meanwhile. The white horse deprived of the troublesome boxes came gaily along. One animal stepped in the mud and besmirched his load but no damage. Got a horned lark from among three or four on the pass crest. Many flowers out, several I have never seen in all my clambering around the peaks. The distribution of the plants is strange. The soil and ecological conditions otherwise appear the same and yet certain species are restricted to small areas. You simply have to keep looking. We got to Koksar at three sans other difficulty. The sheep counter says so far 45,000 sheep and goats have passed here. They will continue to come til the 20th of July. One flock came with us. The income of all the animals from Lahul and Spiti runs in the neighborhood of 2000 rupees a year at one quarter anna per head. There are 130,000 on the average in these two districts. The zamindars don't permit the visitors near their fields or houses, reserving this grazing for their own cattle. Youngsters under a week of age are passed free. The botanists and scoutmasters have had difficulty. They came on ahead and the animals carrying their bedding and provisions tired on the pass and they were stranded in the rest house beside a fire. They sent 13 horses from here today to succor their loads and there is war now because they won't pay for more than the five that carried goods. A Sum shepherd lost around 40 goats and sheep one day from eating poisoned grass. The Tibetans et al. are burying the carcasses at seven annas. I saw strings of the meat drying at Rahla. There is a disease of the sheep in which the animal's eyes grow red,

---

12. Rotang Pass.

the rectum voids intestine, and the liver on frying falls to pieces. If this flesh is eaten some break out near the eye or arm pit, or elsewhere, often affecting local nerves and producing distortion. Both diseases are called ཕོ་རི [*pho ri*; unknown]. On one occasion Rup Chand came near buying such a sheep. It was purchased by others and some 6 men ate of the flesh. Of these two died, one nearly died. Rest immune. The disease is confined, so far as Rup Chand's experience goes, to Gaddi sheep. The Chinese Knot [illustrated] that is pictured in rugs and paintings is known to the Lahulis as *gyandüt* (Chinese Knot). They can tie it. As we left the pass the mist clouds rolled after us and it remained dark in that quarter till night. These clouds do not come so far as this often though there is much more rain here than in other parts of Lahul. The clouds arise very suddenly on the pass. When there is thunder it is said to be very trying, because the clouds are about your ears and the crash is deafening. Lightening apparently seldom strikes. I saw a tree at Katran this trip and one on Chander Kam. Had a little shortness of breath approaching the pass. A few drops of nosebleed right nostril here. One of the men had a nosebleed on the pass. All pulses high on arrival. A woman came to Kyelang for an operation for an enlarged gland of the armpit and it was removed without any anesthetic. She didn't even say a word during the operation. In Tibet Ri-Choma District Tang R'ar, nice things in a monastery. At Shupke-Kyuk-Tuak, Nyang, སྣུར or སྣུའ [*snur* or *snu'*; unknown place name], Ri.

July 2, 1933. Left Koksar at ten. No horses, so got three men to carry loads. The weak white horse went back home with its owner so we are shy two horses and Rup Chand has to go on foot. The students went on to Sisu (Gua(k)ling) with us. They had a handful of grass for me to identify, all which I was able to name, but hardly to their profit, I fear. The foot wound I thought was healed burst forth again and is now finally healing. At Sisu looking down the valley is a mountain with six small peaks (Wa). When the sun reaches the second peak then it recedes and the days get warmer.

དབྱར་ཉི་ལོག་དེ་མིན་ན་དྲོ་མི་དྲོ། དགུན་ཉི་ལོག་དེ་མེན་ན་གྲང་མི་གྲང་།

[*dbyar nyi log de min na dro mi dro. dgun nyi log de men na grang mi grang*; If the summer solstice hasn't returned it isn't warm, if the winter solstice hasn't returned it isn't cold.] Approaching Zankskar Bag is Yandara(k), an alluvial plain on which are visible some 20 mounds said to be Tartar graves. Lobzang has encamped at Sisu. He treated us to tea and tsampa[13] and then brought a lot of old lama ceremonial clothes, none old enough to be interesting. He had a nice pair of Tibetan earrings but I have hopes of getting better ones in Rampur. He did have a nice image of Buddha that I bought for rupees 15. It has the lotus extending all around the base, a rare occurrence. He is going to Spiti tomorrow to buy things for me. I showed him my best tankas and he recognized the ancient ones from their style (Phuktal).

---

13. Staple Tibetan food of roasted barley flour, usually served mixed with butter and tea.

July 3, 1933. The Sisu lama came this morning and gave us a letter to the chagzot[14] of Ki Gompa who will be able to help us much in Spiti if he is so disposed. The lama will also instruct his son to stay and help us. We went on to Gundla having hired four people, two men and two pretty girls to carry the surplus load. There are no horses to be had in Sisu either—all have gone to Kulu. We looked over a large tract of beautiful land, once ancient fields with walls still standing but now barren for want of water. For 1000 rupees, Rup Chand says water could be brought from the Sisu nulla and it would fructify much of the land below. The view is lovely up and down and I am much interested in owning the land. Got a skylark[15] near Gundla. It is the only place the bird is found in Lahul and the Lahuli name, Raligang, is taken from the bird Ralgang Piatsi. There was once a plant growing on this spot that had the virtue of healing and joining dismembered bodies. A shepherd saw a lizard that had lost its tail dig up a plant and join its lost member. He noted the plant and experimented on a sheep. Satisfied with the herb's virtue, he told a companion to cut off his head but he forgot to tell his friend how to join it back again. The herb is therefore unknown to this day. We went on below Gundla and camped near the village of Qarang. A genial Rampur road workman brought a bottle of liquor and all our horsemen got gay and sang and smiled and salaamed and had a glorious time. The Gundla lama who we met at Rahla going to Kulu came today and brought an image of Buddha like Lobzang's of yesterday. We have found horses now. About 40 from this village arrived from Kulu this evening. The bakhyot[16] is in full bloom but there are few moksha because of the drought. We now have three fine images on this road and in all the time I have been here before I have got only two. The iris and the caltha[17] are still in good prime and the pink roses are beginning. They are wondrously bright and so fragrant that the odor travels far. The yellow rose is in full bloom. Such a sunny pure yellow no other rose has. The caltha grows often in a streak along the canals that lead water from the brooks. One of the horsemen had a fever last night but is all right today. Another had a felon [meaning unclear] on his finger. The malaria one is now cured, having swallowed a bottle of quinine. Dorje was ill an hour or two at Sum. Otherwise we are all right. Rup Chand has stood the enforced marching splendidly. Rup Chand's servant who is uncomfortable when traveling unless she is carrying something sent me a pair of wool sox. We have now collected a serious quantity of such things. Every Lahuli and friend sends a pair on any opportunity. The first mail of the spring brings a bundle of them, and likely the last man to leave the village for Kulu in the fall brings some more. Usually they were always too small but the people have learned that my feet are much larger than those of the

---

14. Lama who manages the administrative affairs of a monastery; one of two head lamas in a monastery.
15. Oriental skylark: *Alauda gulgula lhamarum*.
16. *Ferrula jaeschkeana*; a tall plant up to 6 feet; roots are used for medicine; an edible mushroom called *moksha* or *shamo* grows associated with it (Koelz 1979:36).
17. Marsh marigold.

Lahuli men. Pulses of coolies half hour after arrival: man of 45 = 74; two young girls of c. 20 and a boy of 25, 100 each. The people near Gundla have planted potatoes about four inches apart each way. They have the original seed that the missionaries introduced 50 years ago. A woman at Koksar (Dramfuk) saw me collecting plants among the rocks and yelled "O brother, have you seen a stray goat?" She conversed with me further and never appreciated the fact that I was a foreigner. Someone will tell her and she will be mortified to death for having been so familiar. Rup Chand saw a Tibetan Pigeon.

July 4, 1933. We sent six horseloads on to Kolung and one of our men went back with two horses. The new men are Kolung villagers coming from Kulu who have left their grain fields to earn the 12 annas per horse per stage. They will come back tomorrow. The distance is ca. 18 miles. The [illegible] is still in bloom. The young roots are also eaten, both here and in Kunoo. Shot male of a pair of black and white Capped Redstarts that apparently had a nest of young. New for Lahul or Chang Thang. We found Dr. Bhajwan Singh gone from Kyelang, but he had left a tent for us in case we couldn't find accommodation in the resthouse. We put up with Rup Chand's *choti ma*, who had shown me favor in the past and whose husband is much respected in the community. The Wazir invited me to supper and Shri Prehan Chand and Phatti Chapand were also there. It was very nice to be among such good friends again. The DSO Mahon in the Russians employ came here last year and told the villagers that they would pay the same amount as the year before and turn out and give them a reception. The villagers weren't interested. The Wazir said he couldn't command a reception. The lamas were next solicited. They gave an excuse; likewise the schoolmaster and his school. A remarkably gifted performance for an English Colonel to try to buy a reception from some mountaineers. It has been very difficult for cattle this winter. Rup Chand's servant in Gundla dug roots for four weeks to keep the animals alive. Cattle are very fond of a yellow flower thorny Astragalus (drama) and another weed that grows in fields (maybe a purmo[18]). One of the men bought five horses laden, alone through Zankskar with a troop of solders on the guard especially for him. He is ready to go with me now to Zankskar although he knows soldiers are still there. [Art is suggestion, that which is capable of arousing pleasant emotions: lewd fotos, abstract design, landscape. Artistic pleasure comes into being before humans loved object. Artistic pleasure = love thrill. All fine sensations play on the same keys and the melody is naturally the same always: love for friends, melody, beautiful view] [brackets in original].

July 5, 1933. The wanzen[19] were so bad last night that I had to take refuge on the roof, where I spread my bed. Some woman was just getting undressed in a little open shed a few feet away on the same roof, but my unwanted appearance

---

18. *Pennisetum flaccidum* (Koelz 1979:5).
19. Insects (German).

didn't disturb her. We left for Kolung and soon missed the dog. Dorje went back and found her asleep in our last night's quarters. The hosts gave us mokshe for breakfast, having heard probably that I had tried to buy some. These are mushrooms that are associated with the bakhyot and not truffles as I wrote before. An old fat man nearly blind but still working told Rup Chand he heard about our travels and believed we were very fortunate to lead such a life. Wang Gyel came along from the village. He looks ill but the doctor says his appearance is better than on his arrival. Rup Chand's mother looks healthier and happier than at any since I have known her. Dr. B.G. and Prith Chand arrived shortly and stayed to supper. We had a very pleasant chat. Z. the schoolmaster called on me in Kyelang and wanted to discuss the Russians. I have disliked the man from the very first as a mean spy. On the road along came the missionary, equally contemptible but weaker than the other. He urged me to call on the Indian botany professor at the rest house. The professor, I heard, has made two trips to Rupshu and is going again. The students 15-odd he has brought with him seem to be interested chiefly in the landscape. One below Rahla had a handful of grass, but none were seen collecting anything thereafter and I suspect the professor has come on another purpose and I shall be much surprised if Mrs. R[oerich] comes to Kyelang this summer, especially since she is wanted there! The D.C. and Mr. Gill are coming to Spiti in early August and are going to be there when I arrive.

July 6, 1933. I kept two of the Rampuri horsemen and let the other two go back, much disappointed though there was no agreement beyond Kolung. They are the pleasantest lot of horsemen I have ever had. Not a word of complaint the whole way and there was plenty to complain of. It was impossible to get the men ready to go today so we put off the departure till tomorrow. The Dr. and Prith Chand came in the morning to operate on a growth on our dog's navel. We had cut it at Koksar, since it looked like abscess and instead a sort of tumor came out. The animal suffered considerable inconvenience from the thing. We gave local anesthesia and got it out and sewn. The poor dear beast quite won our hearts with her patience in tolerating the abuse she got. Wang Gyel spent the day watching to see that the patient didn't bite out the two stitches. Both fell asleep together on the floor and amused the company. We went to Prith Chand's for supper. Nice full moon. Crops are said to be exceptionally good at Kolung. Tibetan sheep die from eating cowslip but goats thrive on the leaves. The Dr. leaves with me in the morning; he for Rupshu and I for Zanskar. Prith Chand's big black dog came home with us and slept beside my bed on the roof as last night. Our dog, whom Dorje has christened "Moti" slept there too the first night, but was too sore to scale the ladder though she wanted to badly.

July 7, 1933. We started off this morning. I went to Kolung to join the doctor. Rup Chand and Wang Gyel I left behind. Two horses of Rup Chand's and one of Rinchen Gyaltsen's are being looked after by a boy from nearby and there are two horses of another neighbor and four of the two Rampuris. All nine are lightly laden with that load growing less daily. Annas 14 per parao[20] and half for halt. That is good wages for horses these days but I give it because there are more men than usual looking after the horses and horsemen do any work around camp. There is no one to replace Wang Gyel so we go a man shy and shall have to recruit help from the horseman. The Dr. tried hard to find out all about my relations with the Roerichs, tried to persuade me to get land: there was a piece for sale in Kulu and I could have free land in Lahul, and wouldn't it be nice to go to Lhasa. Mr. Lee hard on the same tune from which I gather the government would be glad to have me settled as a spy and to go to Lhasa for them. Everyone has to spy if he is in government service, gets a government pension or is in any position (like Lee) where he can be squeezed. Wouldn't our people appreciate such a government! Poor B. doesn't want to do it but he has a family. I took pains to tell him I couldn't talk to him on matters that didn't concern him and that I understood fully that officials in Indian service were called upon to do jobs that were outside the job for which they were hired. He agreed to go to Rupshu first instead of Spiti first as he had planned, so we shall meet in Spiti. The people at Rerig say that the road to Zankskar sumdo is very bad and we shall be long in getting over it. We halted at Palden Lamo and the reconnaissance men said it was indeed very bad for a mile, but that empty horses might possibly be got over it. A villager is coming in the morning to advise on the course. The dog came along nicely today, but I didn't come so nicely. The climbs were difficult. On arrival I had a good nosebleed and felt better. I have never felt so short of breath at any time since I can remember. Dorje stayed behind to collect plants and didn't show up all night. Rup Chand has three girls for servants. We gave them a choice of necklaces: 10¢ crystal, 10¢ green glass (mate to red one [indecipherable] in Kulu) beggar's necklace, black garnet, carnelian. The first took a beggar's necklace, the next the garnet (round beads), the next carnelian. The garnet was exhausted by one selection. 10¢ diamond pieces not interesting. Rup Chand's mother much admired the 10¢ pearls. She tried them on, Dorje said, when we were gone. I intended to give that to the servant but it won't do. She asked if it were expensive. Rup Chand said no. "No matter. It's a very pretty thing." And she is quite right. I have left a 50¢ string and some earrings to match to be given her when Rup Chand leaves.

July 8, 1933. Dorje had gone on to Dozan. He arrived early. Mostly Lahulis have come there so far. Trade probably won't be very brisk. Karma bought tea in Kangra at Kulu October prices and found the April selling less than he paid. The men who went ahead to reconnoiter again came back with very faint heart. They said we would have to carry the load to Shalangdrokpo and stop there. The water was too much to cross; above that they didn't know. The old Lahuli said he could go across Bora Latza and come to Kargiak thru the Serchu-Kargiak

---

20. Distance measure.

cut. The permit that the British gave me rigidly forbids me from *going* any other way than via Shingo La and if I can't go up this way I must give up the whole of my summer plans. I decided to manage the trip according to the permit and taking two men with minimum of food and clothing we started out for Kargiak. Our horses are to join us there on the fifth day. We hope to arrive on the third. Had the permit been received when requested this route would have been easy. The snow-bridges over the torrents would have made going easy. The British must have known that and therefore arranged things accordingly. What a friendly way of doing things! Under Russian auspices I was free to go without a permit. The road turned out to be bad but passable for laden horses except for one-half furlong near the stream. The stream had however grown so much in force by noon when we arrived that we couldn't wade it, though it was less than ten feet wide. We found an old Zankskari woman and two boys camped beside the water. The old lady had the courage to cross even now but the one boy was feeble from a long illness and the other was small. She could have crossed in the morning but she feared for the boys. A woman had slipped a few days before and had been saved with difficulty by three Gudun lamas with whom she was traveling. The company joined us and we climbed up the mountain and crossed a snow-bridge high up. Rain set in at the torrent and we were forced to take the last shelter (overhanging rocks) below Z. Sumdo. The old lady and a small boy came from Ating a month ago to get the other boy who had come to work in Kulu and had been taken with dysentery. His hair all fell out and he was sick for many weeks. A few Rampuris gave him some money finally when he was able to move and he started home from Katian. He met his mother and brother at Rotang La. The Sisu lama gave them food for traveling—the one who has been so friendly to us and who is never sober. They got a little barley from Rerig which they roasted in the cave. Dr. B. said there was a remedy for sciatica: "reapar" or some such name obtainable from the Anglo French Drug Co. Bombay. It contains salycic acid and Betula lenta.[21] The grass for horses is superb on our path. Across the river there are flocks of sheep. The Gaddis' fire at night on the Lang is cheerful to see. Clusters of nice birch trees cling to the hill side. We had hoped to stop at one of these, Tagpa chen, but they say if we cross the main stream, we can't cross the branch that comes down from the Gadola nulla. My indisposition of yesterday has passed and I am able to travel without difficulty. Our poor dog tore a stitch out of her wound and I hope she arrives with the horses. We stopped in the cave all night. Rain came off and on after 3 and the mist hung low. Steady mist clouds came down the valley. One can't sit upright in the cave but it is dry.

July 9, 1933. We had breakfast at six but the clouds were still thick and shortly after a drizzle began again. About noon a breeze came up and it grew colder and

---

21. Sweet birch.

the clouds went to the peaks from which we hoped that the sky would clear. We
started off, but the rain began promptly and kept on till we halted in the cave at
Ramjak a little below the གསུམ་མདོ [*gsum mdo*; the juncture of three rivers] that
leads to the pass. Our friends came along and camped beside us in the cave. The
road to Jaskor Sumdo is perfectly passable for laden horses, except for a few
rods and these could be made so easily. The road is difficult, up and down the
hillside, in and out of the nullas which can be crossed only at their head. At the
sumdo[22] the stream from the pass is fordable but we kept on the right side and
crossed on a rutzam[23] a mile or so above. Most of the རྡ་ཟམ་ [*rud zam*; a stone/
earth bridge] have broken. At the sumdo there were four gaddis, sitting in the
rain with their heavy blankets on their shoulders. There was a cave a few rods
away. They were churning about six quarts of milk in a skin. They had tied the
liquid in the skin and then rolled it back and forth on the ground. They hadn't
got butter when I left. They said ibex were sometimes seen here but nowadays
had been driven higher up by the flocks. Rup Chand says that at least four or
five ibex are recovered yearly from the snow avalanches in Lahul. We brought
along some honeysuckle bushes, mostly root, and some juniper from the sumdo
and managed to make a fire out of the green wet stuff. There is nothing to burn
above Sumdo except manure and that is hopeless wet.

July 10, 1933. The night stayed cloudy but the clouds broke at 8:00 and at
8:45 we were off. Within a mile we met some 15 Zanskari men and women
who had come from Lakong. They must have started very early in the morning to arrive so soon. They said Ngari Tshang Rinpoche[24] was at Seni and was
going back to Tibet through India in the fall. They had expected me last year;
had I been in Naggar all winter; was the big man I was with all right (Roerich,
I suppose); it was all false that the Puhtal lamas had been fined for selling me
tankas, on the contrary many people were ready to give them to me now; how
had my two drumos[25] turned out; what price had I paid; where was Rup Chand;
I looked like a young man now (I had a beard in Zanskar before). The pass
was covered with snow. Once in a while, the whole gave way and I went down
to the hips. Dorje and the young horseman were hopelessly slow so that we
didn't get to Lakong till 4:30. I suspected they came slowly so that I wouldn't
be able to go on to Kargiak today. I said I feared the boy wouldn't be able to
stand the altitude since he had been able to do no better today than the old lady
and her infirm and young offspring. I thought he would have to go back from
Kargiak! There are a few little mustards on the bare patches near the top of the

---

22. "There are many places called Sumdo. The word literally means three directions and is applied to places where two rivers meet" (Koelz 1931:100).
23. Stone or earth bridge.
24. *Nga ri tshang*; name is always written by Koelz in Tibetan characters; likely refers to Ngari Rinpoche, the nominal head of Geluk monasteries in Ladakh and Zangskar, who was based in Lhasa (R. Linrothe, personal communications).
25. Yak hybrid.

pass on the Lahul side. A robin accentor[26] was seen below the pass on the Lahul side. The horned larks have full grown young at Lakong. Some beautiful purple primroses made a fine show in a stream above Lakong. There were some also on the Lahul side and on Rotang Pass. There is a path from Lakong to Patseo. The Shingun La would have been difficult but possible for our horses. On this side, the descent was very steep with snow to the bottom of the valley. Good pasture above Ramjak and at Lahong.

July 11, 1933. The reflections from the snow yesterday were intense and my nose was badly burned. On Rotang even the Tibetans got burned, some raw, others black or both. This morning a half dozen yaks showed up at camp. Our horses, who have never seen such animals, will be much interested. Above Kyelang a yearling yak was coming down the road and frightened the unsophisticated Rampuri ponies so that three of them threw their loads. There is no wood at Lakong and the manure was wet but we arranged a fire and ate the tin of bloaters that we have been carrying around for months. There is an abundant growth of willow (two species) about three feet high two miles below Lakong. The valley to Kargiak is green, very green in a half a dozen places where streams and springs come down from the mountains. Plant growth is rich and there are numbers of the few species of birds that frequent the country. I got all but the white breasted Dipper. There were two of them: horned larks are abundant. The natives say there are chitor and galeen in the valley. The yak herders have not got quite to Lakong. There are herds of 30 or so yaks on both sides of the river and a few horses. Snow pigeons are not rare. Shot four and saw four more. Arrived at Kargiak, the people knew me and asked after Rup Chand. I sent two men to find our horses. They think the pass (Tsa Ri Chen La) may be difficult and they will then take the horses to Tantse. The barley is less than a foot high but is beginning to head. The fields are thrifty and clean. Around them are lovely flowers: geranium, columbine, buttercup, two species pedicularis, Cicer, etc. Made up 17 skins and have six Tibetan pigeons for tomorrow. Always two pigeons and once four at a shot. From here they say you can get to Dozam in a day. On Shingun La (not Shingo) there was some red snow that reminded of Arctic red snow. This was probably due to dust though I can't see the source. There are lammergeiers[27] here *much* whiter below than those in Kulu. The little red capped linnet and the black and white finch are also here.

July 12, 1933. Last night the natives came long after dark to discuss sale of contraband things. Finally they bought two tankas, neither old. This morning we had drimo[28] milk. They bought ca. four quarts to supply our possible wants. One man bought an old gao ornamented with silver wire work and many

---
26. *Prunella rubeculoides.*
27. Bearded vulture.
28. Female yak.

designs in turquoise. It's the nicest gao I have ever seen. Another had an old scarf of good weave peshama, ornamented by a narrow border of four colors, apricot, madder red, indigo, yellow. The design was an adaption of the Persian pattern in our fine turban. The article is said to have belonged to the old Raja at Pagan. One brought a book when he heard we wanted *nam thar*[29] but it was on religion and as incomprehensible in Tibetan as it would have been in English. One had a nice man's earrings of some light metal that looked like gold with a nice turquoise set. (Tibetan men often wear only one earring [right ear].) One had a round zee-like[30] nune, but only two colors: black and white. One had an ancient lama coat of Chinese embroidery on deep apricot silk. Design of dragon, cranes [?], rocks, etc. One man asked if I could give him some gangbal; one wanted musk. They tested a gold piece by rubbing it on their thumb nail. A sample of iron pointes [?] they brought as the year before. The dressed-up Zankskaris going on pilgrimage were quite smart looking. These are mostly in rags, but a few are "sabut" and even wear jewelry of turquoise and coral and coral imitation. The place has 18 families says one. Several fields are fallow and will be left so. Some pigeons had their crops full of grain that they stole from the roofs. When beer is extracted, the grain is dried and eaten. One had its crop full of a round soft madder insect like a drakshik[31] [illustrated] that an old man said they found in the sand. An old man brought an iron arrow with grooves on each side in which poison was placed, it's said. The owner hasn't struck rock with it when he shot, as did the Renigwala. The weather is very warm in the sun. A brisk wind comes down from the pass all day. The valley is green with cultivation, a nice little river runs in the bottom, the surrounding hills of pink brown, dove grey, purple black are sprinkled with green.

[note: diary is broken here by illustrations of objects and notes, bordered by lines]

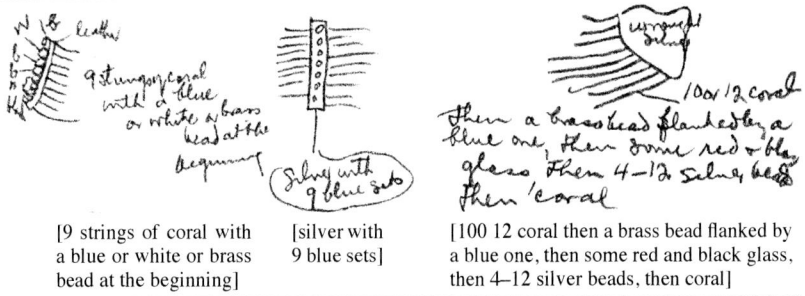

[9 strings of coral with a blue white or brass bead at the beginning]

[silver with 9 blue sets]

[100 12 coral then a brass bead flanked by a blue one, then some red and black glass, then 4–12 silver beads, then coral]

---

29. A *namthar* is a spiritual biography or hagiography that tells the story of yogis or great adepts who attained enlightenment; Koelz may have used this term to refer generically to documents.

30. *Zee* or *dzi*, bead whose name means "shine, brightness, splendor," chalcedony beads with dark geometric and symbolic patterns (eyes, stripes, swastika) on a white background. These are highly desired and many glass and other imitations were in circulation even during Koelz's time (Clarke 2004:40).

31. Bedbug or louse.

The above ornament is hung in the hair. Only one seen. The water in the streams is moderate till three. At dark it is very strong. They say fossil gastropods are very abundant in Chumurti. The houses have stone and mud walls, flat roofed, few doors, sometimes a peephole for a window. On the roofs are piled grass, willow twigs, and abundant cakes two feet long and a foot thick of sheep manure cut from sheep folds. The river here has cut only 25–30 feet into the glacial drift and the mountain slopes are well covered with soil. The cone of the mountain often forms a differentiated peak. An old man brought a zee like my ring with a brown "pupil" in the white center part. He had also some "dar yak," which is beneficial for poison. It is a white substance that looks like a dried gland that he says comes from Baltistan. It is the saliva of an ibex that has eaten a small snake (ser rul). The ibex falls into reverie and the saliva drools from his mouth. I told the old man, who is a physician, that if he would eat some of my poison and if the cure were effective I would buy the antidote. He professed readiness to take a very small dose. I asked why he needed poison antidote and he confirmed my knowledge of poisoners in Jchor. They have some poles here six inches in diameter, 20 feet long that they say came from Tchor. Some of the men have their hair cut short to the crown whorl and leave it long even to a pigtail behind. The old man had a spoon of ཆུ་ཟང་ [*chu zang*; large copper water container; mostly likely used by Koelz to refer to the metal (a copper alloy?)] that he said he found in the river at Zang La. He rubs it and distinguishes it from common copper by the smell. Tibetans value this sort of copper highly. There are fish in the river some a foot long. The people say they don't eat fish or yak, but they eat pigeons.

July 13, 1933.   Last night a nice Buddha was bought, gilded copper, lotus all around, dorje,[32] and nice incising. Off and on at intervals of 15 minutes to an hour deputations arrived from the village so that we were seldom alone. Then they repaired below and discussed probably what they had learned. A dog that comes to camp (we are housed in the ruins of the stable. It has walls on three sides high enough to shelter from the wind) didn't know that rice and milk are edible. He snuffed it and went away. Like the poor Mexican horses that were afraid of apples at Albuquerque. We sent the horse boy to the Tankse branch of the road to bring blotters. If the horses don't come by night he is to return. Pulses of seven men and ages: 80-20, 60-33, 84-30, 80-37, 76-28, 84-36, 84-29. The men of 29, 33 had their wisdom teeth. Others not examined. All apparently healthy, inbred. Gave all some chocolate. All professed to like it. Yesterday a man came at three pm and said our horses wouldn't come today, but tomorrow. He counted off on his rosary and got that result. Demand came today for provisions for the 16 soldiers at Padam. Other men 84-30, 94-20, 80-40. Last two no third molars, first had. Most pulses alright irregularity caused perhaps by excitement (?). Height of all about equal to my height. All sturdier than me.

---

32. Also called *vajra*: ritual object, representing a thunderbolt.

We wanted some more butter which, by the way, is a nice yellow and rated high by those who know it. We sent one man for it, but though he had butter he couldn't give it! Eight men drew lots. Then the man who won couldn't bring it himself. The nombordar[33] brought it. The whole performance lasted two hours. Meanwhile, the use for the butter had passed. Toward evening one of our messengers arrived and said the horses would be here late or early in the morning. Everything was all right. What a relief! Dorje first thing asked about the dog. I had forgotten her. Before dark the caravan came. They had a bad time on the pass but no other troubles. The first day they had gone to Patseo. The next day it rained and after three attempts to get started they gave up and stayed. The third day they went to above Zinzingbar. The pass was crossed on the next day and they stopped at Kuling. There was snow from pass to Kuntsi to the Yuman river valley. For about a mile they had to carry the load. The horses and men sank to the hips. Thereafter they could avoid the snow by making detours up the mountain. The horses had to fast at night at the pass crest. Feed was so scarce that they would have fled to good grass below. The sheep and other travelers waited at Zingzingbar for more favorable conditions. The next stage was to the sumdo above Serchu Sumdo. Our men met them a mile below the sumdo. The long journey to here was finished on the sixth day. Wangchuk gave out a little above the sumdo. His heart had been weakened by malaria and he got here with difficulty. Head, neck, arm pains. Others all right. No accidents. Between Serchu and ཆུ་མིག་མར་པོ་ [Chu mig mar po (Red Spring); place name] abundant grass. The Panchen La ca. three miles from here is very steep to descend. The ascent from Serchu is easy. The Zankse road leads off from ཆུ་མིག་མར་པོ་ [Chu mig mar po (Red Spring); place name] up the hillside. Our dog's ears are badly swollen, also her face. They say from the cold.

July 14, 1933. We stayed here today. The men and horses are tired and besides there is an abundance of plants. Put up some 150 from here and 10 birds. Gave the natives a phonograph concert. The whole village sans exception assembled and afterwards had their picture taken. The Indian record pleased them more than others they called Mussalman records (Negro melodies). Showed them the 10¢ store diamond pins, etc. The pearls are most admired. The polished clam shells, the snail shells, the amber were all interesting. Also the carved elefants and the ladies were very gay with the yellow pediculars in their headgear. Bought the image for one rupee, the shall [?] for ten rupees, the gao for 16. The weak man is much improved by the rest. Dorje saw a dozen ribja[34] on the hill across the rise. The men are eating sheep. They bought four quarters from a gaddi at the sumdo for eight annas. Someone told them there are lots of fish at Seni and they are counting the days to there. The thermos bottle is a wonder to the natives. They think there is a coal fire in it somewhere. None had ever heard a phonograph

---

33. Nambardar village official, revenue collector.
34. *Tetraogallus Himalayensis*: snowcocks.

before. Their delight was worth the trouble of bringing it up. They burst out laughing several times in "River stay 'way from my door." I tried to buy yak hair ropes but none to be had. Two men are weaving something or other in the village. One beside his house—the other in the river bottom. Bought a curious iron arrow head, with a knob at the tip. Have seen three lama coats so far: two blue, one apricot. One blue and apricot similar in design. All very badly worn. They say they wear them only two or three times at most in a year. There is no devil dance here. Three tankas, all apparently locally made.

July 15, 1933. This morning a deputation came to say our horses had done serious damage to the fields. The horseman had merely driven the animals to a grazing spot and the animals had improved their condition, having been left strictly to themselves. The damage was settled at nine rupees. One old man had refused the arbitration and was given one rupee more. Then two rupees were given the victims for good measure. We had tried to warn the natives vs. counterfeit coin which is so abundant now in India, but it is quite useless. They will certainly take anything that looks and sounds plausible. One at least had never seen any. We left at 9 and got to Tetha at 1:30 and stopped to camp, though we had intended to go on to Cha. There were over 50 species of new plants on way, and a number of birds to be made. Besides, Tetha is a nice place. At Tankse we found that the old Rampuri woman had died. One man took us to his house to sell us some things but the villagers promptly gathered and nothing was forthcoming. Some half dozen men and women were met between Tankse and Tetha, carrying kumanis sewn in leather with a curious red button on each. They said that was part of Ngari Tshang's [Ngari Tshang Rinpoche][35] baggage on its way to Tibet. One man who said he was in charge of moving these effects came with us. He said to put it conservatively the lama had collected 50,000 rupees in Zankskar alone. Bricks of tea he put in each zamindar's house at four rupees willy nilly. He will certainly want my money then. There are two villages across the river between Kargiak and Tankse that I didn't remember from last year. Also a wolf pit about two miles above Tankse. There is also a large patch of drama at Tankse. Roses began there and continue down the valley—the first bulbs out, fragrant blooms are on the branches. English sparrows began at the village above Tetha. There was a Hein. griffon at Tetha. A sparrow hawk at Kargiak, also white breasted swallow. This abundant at Tetha. Brown swallow also here. Gave a phonograph concert after arrival. People gathered promptly when Dorje told them we had a phonograph and became impatient that the concert didn't begin at once. They said to hurry. They had to water their fields toward evening. None had ever heard a phonograph before. The expressions of delight on all the faces would alone have repaid Strell for buying the instrument. I tried to photograph the assembly in its rapture but the sun was bad and

---

35. Identified by R. Linrothe, January 17, 2012, personal communications.

beside my movement distracted the audience. A nasty wind came down from the high mountains toward evening and the sky clouded. Bought an arrow like the grooved one of Kargiak and one with three rings. Also a coarsely executed tanka of very good design and drawing. Nothing like it seen before. There are some chickens here. A cock can be heard crowing.

Plate 34. Man on road near Phuktal monastery, Zanskar (date of photograph unknown) (Walter Koelz Collection, Bentley Historical Library, University of Michigan).

Diary: June–July 1933

July 16, 1933. སྦྲ་ཚེ་ [*scrub tse*; unknown][36] There are very interesting ornaments made of iron and copper and ornamented with brass, blue beads and green tile or china pieces. An old lady said her mother hadn't worn them. None of our men had ever seen them before. We bought five or six at Tetha. Said to have been worn on the breast. Also two beautifully executed gaos of the same materials. There was great excitement this morning during the sale. Our horses had been driven into the nulla behind Tetha and two men slept with them. The animals have learned now that fields offer the best grazing and giving the watchers the slip, came to the village. An old man sleeping on the roof heard them and gave the alarm before they executed their designs. There are some willow trees at Tetha. Below Tetha at Sale and Cha there are poplars six inches in diameter. We set out at 9:30 and at 7 arrived in front of Tchor.

Plate 35. Women's waist ornament (UMMA 17214).

The road is passable every where, but all day in and out of nullas mostly up and down them. In compensation for the bad going the way is mostly thru an exquisite flower garden. The roses are in full bloom. The nearer Tchor, the larger and taller the bushes. Some are ten feet high and some are literally a mass of flowers. The smelly purple blue mint often mats the ground around the bushes. Last trip all the roses had crimson hips and were then most attractive. We have had an almost continual rose season from April on. The Bhadwar roses; the wild white and pink ramblers, the Kangra Valley double pink wild ramblers, the Kulu musk rose in May and June, the Lahuli yellow and pink bush-roses, and now these exhilarating

---

36. May refer to a women's ornament, consisting of a pointed metal disk attached to a string of glass beads, and metal spoon. Koelz acquired several of these objects (UMMA 17214–17217).

beauties. On the face of one cliff today a gorgeous garden had formed. Moss and ferns covered the ledges and cracks. Here and there the white saxifrage hung in clusters and the purple blue poppy. Water trickled down over it all. Bushes of currant and willow flanked the display. Tamarisk bushes six to eight feet high are in bloom in places along the river bed. The game jamadar[37] and his cook, both Mussalmans, overtook us at Pangpepal and came to Tchor. They said they had been sent to Kargiak and had to be back to Padam tomorrow. We met them three miles below Kargiak yesterday. At Tangkse a man told us they had said they were going to Kargiak to meet a sahib. The government has effective communications in this country. Last year we beat the game watcher in time and he traveled to Hanle after we had left. I shouldn't be surprised if the British found some pretense for turning me back from Padam. The two I met in Tsomanong said no one had collected birds in Zankskar and they certainly don't want me to be the first. We met 14 people and two yaks laded with Ngari Tshang Rinpoche's things. Wangchuk's heart is behaving better. He got here on this difficult journey with no undue fatigue. My heart, which gave me much concern all the time in Kargiak: shoulder and neck pains and short breath, forgot its troubles today. Dorje complained today of fatiguing easily on this journey (to Zankskar). Saw a sisson wagtail (blackbacked) and a WC redstart and a sparrow hawk at Cha.

July 17, 1933. Sent a note to Rup Chand today by some men taking goats to Lahul (Lahulis buy them for milk). Have heard galien here and at Tetha in the morning before 6. Jamspa Tsondras, soldier of Ngari Tshang Rinpoche was met at his camp above Reru. We had intended to go to Bardun but the Tibetans who are going to take Ngari Tshang Rinpoche's goods to Lhasa were interesting and when we wouldn't stay with them came to Mune. One youth of good stock was apparently in charge. He had an older assistant who knew some Urdu. We bought sheep at Mune and invited them to supper. Four of them came; the other two went back to camp. They didn't come and sit till supper, but had to be summoned from the monastery when things were ready. Our men gave them meat, onions, and [indecipherable] rice; afterwards honey. The Tibetans were by all accounts well pleased. The youth ate all he could and handed the plate back saying he had eaten a great deal. They returned to camp but came back in the morning to say goodbye. First thing they told me that the big lama was just about to write me a letter saying he was going to Lhasa and that he was soliciting permission for me to accompany his suite from the Dalai Lama. I should ask the British for permit to cross the frontier. The youth is in charge of the message to the Dalai Lama. I could get as many paintings and objects as I ever dreamed of he said. Things here were few and expensive. There are iron, gold, copper, brass images up to 18" high. Paintings he could guarantee to procure. I would be permitted to shoot also, he said. He had a Winchester rifle model 1895. My guns were very interesting and he shot two birds with the little rifle,

---

37. Military officer.

with as much glee as a 14 year old boy. He had a letter for me to read from an American girl and her mother whom Ngari Tshang Rinpoche met last year at Mulbek and I translated it for him. No one else had been able to read it he said. He presented me with a shin of butter on my arrival, two-plus pounds, and when I later gave them one of my very nice American knives he was for giving me a sapphire or a zee. The thomsonites and chlorasts. were much admired. The big shell had few equals in all Tibet. He said the phonograph was listened to with rapture. ཆུ་ཟངས་ [*chu zang*; large copper water container] he said was worth ca. eight rupees for four pounds. He gave me a piece from Zangla. From Kargiak he would take the lama's goods on 30 mules. Until then the Zankskaris assume transport. We passed a dozen men and women with leather-encased loads today. Mune is a very nice place. We camped on the crest of a hill looking down the river valley on two sides and in front of the monastery. To the one side the valley is barren, flanked by huge mountains plentifully streaked with green. To the other, far below are the fields and houses of Mune. The present monastery is new. The old one is in ruins but stood near. It was protected on all sides but one by the nature of the slope.

July 18, 1933. We left at 9. The Tibetans came to say goodbye and to get paper money for silver. Out of 300 rupees, they had only seven counterfeit. Notes they wanted new. Notes are current in Lhasa. A sheep in Lhasa costs two rupees. Other things are equally cheap. My Austrian metal ruler was much admired. Saw a blue rock thrush male, the first blue pigeons, another sparrow hawk male. They told us we couldn't get to Padam on account of a bad stream at Trakkor. The stream had a bridge but the bridge was weak. We didn't find it any weaker than the rest of the bridges. They are all implausible—three or four poles laid on two pillars of rubble and overlaid by huge rock slabs that are ten times the weight of anything that will pass over, sagging too. Every one said that the stretch from Mune to Badun was very bad so we would need help. They told us that when we went up the valley. It wasn't so either. None of the road is so bad as that between Cha and Tchor—up and down, in and out of the deep valleys that the currents have cut. The men made gyuma out of the sheep. You fill the intestines with spiced ata[38] and boil it. The heat was intense till after Padam when a stiff breeze was spilt out of Seni Nulla. The broad valley at Padam is capable of enormous development. There is water in abundance and the crops grow well, but the greater part of the plain is grazed. Some barley is well headed. At Padam we met a man who recognized me from the year before last and accompanied us on the way to Seni. The soldiers are in the houses he said, and have several dogs. The natives here at Padam are half Mussalman and half Buddhist and some of the Buddhists are clearly half Mussalman due to their mothers' carelessness possibly. A C. mongolus[39] with young was on the Padam

---

38. Wheat.
39. Mongolian plover.

plain. Got a nice young one that could barely fly. Ravens here as at every town. One pair the natives say. Skylarks numerous at Padam. One man said there is a spring at Seni and many fish come to it, some 18 inches long. The old lady and two sons who crossed Shingo La with us came to camp in the evening. They will reach Seni tomorrow. We pitched camp near dark above Seni. The horses have to be tied up tonight because of the wolves. The view from Padam with the white houses of Kursha and the abundant patches of green, the snowcapped brown, purple and pink mountains, and the long vista toward Pens La is the finest we have seen in Zankskar. All told, the country has been most pleasant to behold. We will rest tomorrow and meet the big lama. The Bardum gompa[40] is a beautiful located old building, completely impregnable from all sides but one. The building is ancient, but there are ruins of a still older building up on a crag in front of the present structure. The gate was barred by a huge log and a violent dog barked out of a hole in the wall above. Finally, a lama came to let us in. He turned out to be one I had seen in Kadung Gompa two years before. The monastery has nothing of interest that is apparent, except for a huge copper prayer cylinder, nicely written. There was a piece of very old cotton print, apparently Indian with peacock that I should have liked, but it would have involved the day, and much bad treatment probably. The lama showed us two tankas made out of pieces of embroidery. I had never seen small ones of the sort before. The cloth pieces are beautiful and the effect pleasant. There were some ten silver tankas hanging on the walls, with a central male and numerous small male figures around. The silver background is very dull and the whole not particularly attractive. In Lahul I have seen these with gold background. There was a bundle of some 25 tankas that the priest said he couldn't show, and we departed. He did however, show the palms of his hands when we left, to no purpose.

July 19, 1933. We spent all morning trying to dry the plants. We have moved so fast and through such varied country that taking only one sheet of each species we have our blotters full and the stems of the dried plants haven't had a chance to dry. A sprinkle came out of the nulla at dawn and lasted ca. one hour. It was quiet and warm thereafter, even hot until noon when the brisk breeze began that was blowing yesterday. It lasted till dark. Natives say such winds are the rule here. Various natives in full dress passed to Seni throughout the day, because tomorrow there is a sort of a fast that they are celebrating with the big lama. After lunch I sent Rinchen Gyaltsen with the dynamo, a calendar, and gastr. [?] to the lama. The servant was well received, offered tea, and a red rag was tied around his neck. I think Rinchen Gyaltsen gave a rupee. One doesn't go into such [indecipherable] sans money. I went with Dorje. The boy who accompanied the priest met me. He had grown a half foot since I last saw him and was as tall or taller than I. I played all the records to the appreciation of the audience. Such nice sound they

---

40. Monastic fortification.

had never heard, nor seen such a small instrument. The Lhasa machines must be very old-style ones. The big man thought the Negro songs were better than the Indian ones but the rest didn't think so. The priest then brought the subject of tankas, saying he had sent for five for me from Lhasa. What should he do with them? I told him he needn't have bothered. I probably wouldn't like them and wouldn't want them as a present anyway. I expressed hope he would give me the Likir tankas. Much circumlocution from which I got the impression he much feared I wanted them in return for my gifts. He finally asked the meaning and price of my gifts. He said he could use, would pay for three or four other gifts, but I said couldn't be supplied. He pressed me to stay tomorrow and said go to Karsha and take out what you want. That was agreed to. Not a word about my going to Lhasa until I broached the matter. Then he said it wasn't difficult but permission had to be got from the English, not from the Chinese, with whom they were fighting, but from the English with whom they were very friendly. There was a telephone to Lhasa and many English came. After a servant brought a huge pan of dry grown rice which one of our men identified as from Pango, a cake of the HR tea that the lama forcibly bought and a large slab of native sugar. The servant inquired whether I would want the letter to Karsha Gompa in the morning. He protested at the rupee tip, apparently sincerely, but took it, asking Dorje if he might keep it. There are huge springs that issue below the village forming a little lake. There were several species of sandpiper; one glottas, rest unidentified, except for six to eight Mong plovers with young. Shot a Tibetan tern. The priest wanted to know how much I had paid Phuktal. I couldn't remember. He said 100 rupees for two tankas and a tea table. The clever chaggot.

July 20, 1933. Saw a sparrow hawk and Ibidoryncha at Seni. Also several Tibetan terns. The plain at Padam is a mile wide with water in abundance, fertile soil and room for hundreds of people. If nothing better they could plant it to trees. Poplars grow well but there are no willows such as one sees everywhere in Lahul. The cattle have grazing amid pasture on the hills. Early the lama sent a man to ask if we weren't going to Karsha today. I said yes, to give us a man along instead of a letter. This was done. The monastery was an impregnable fortress with houses for 100 or more people. An older building in ruins is the highest of the series. The village is the lowest of all. Construction is clearly very old. The entry is through a long dark passage and then the way winds around among other buildings and up to the monastery at the top. A whole side of one of the rooms fell off last winter and some 30 lamas were repairing it. The men were friendly, two of them were fond of Rup Chand's family. The tankas on hand were ordinary Tibetan-made of some age, but nothing that I cared to spend money for, though I should gladly have bought any a year ago. They said there weren't any older ones but I insisted, having seen the ruins of three in a mane[41]

---

41. Room or chamber?

at the gate. Finally a dirty pile of rubbishy clothes was bought and among them, a beautiful old thing in repairable condition. I spurred on the search and three more of the very oldest type came up. They were willing to let me have all at my price, so I said rupees 100 for the five. Then they said the best and largest couldn't be given or ill would befall them. I suggested that the dirt of the ages on his face couldn't have pleased the demon it represented and anyhow how much more could happen—a wall had fallen already. The things were wrapped up and then the lamas with much excitement got their picture taken. The big lama's servant said he had never put in such a hard day in his life. He had to walk seven to eight hours with us. He objected now to my having the big demon tanka[42] but his master laughed and said I should take them. Such things he could give me in Lhasa by the donkey load. It was decided all were to be photographed again in the morning.

July 21, 1933. There were 16 birds to make before we could start but we got off at 9:30. I went to the lamas and photographed the establishment. Then I asked him to give me permission to buy the Likir tankas. I said I would give 600 rupees for 15, 100 rupees bonus, and him a goat. He wrote a letter to the head lama at Likir ordering him to let me take any 15 tankas I wanted. Rup Chand says there are 17 of the loveliest old things, the likes of which there are not in this country. I then asked to have a man to accompany me to be sure that his orders were fulfilled. He said he couldn't spare one of his men, as he was going up the valley but he would give a man from Rangdum. He showed me six seals of various sort. One of wood with a beautifully carved grip (Chinese). The horses had to wait at Togring[43] bridge because the officers of all sort were coming from Kargil to hold court. Their things were coming across the bridge. We got things carried over finally, including the dog, then one of the horses swam back and couldn't be induced to enter the water again. It ran hither and far and was only finally dragged across by a rope. The Zankskarians got five annas for hauling him over. Some 50 or 60 people were assembled at the bridge, all men. A soldier said in spite of the guards at Padam the thieves come for Kut. Some came through Lingshol and Trakkar nulla. Last year 300 came armed. The Zankskaris stole it too and sold it to the Kampas (one man wanted to sell me some). The big lama has collected 8000 rupees cash. One of the officers stopped Dorje and Rinchen Gyaltsen separately and asked what we were hunting, we mustn't hunt. Then one rode back to me and told me the same. I said we weren't interested in anything but small birds, etc. and gave him three rupees which he wanted in the first place. Saw another Ibidoryncha at Hamiling and two mergansers. Black swifts about 12 at Seni, one white rumped a little green sandpiper at Phe. We had intended to go to Hamiling but it got dark and we camped below. It was two pm before we left the rope bridge at Tsgring.

---

42. UMMA 17465.
43. Or Tsgring?

The crops are good, with barley and wheat headed and the fields a magnificent flower garden of yellow and rose pediculars, geraniums, yellow [indecipherable], forget-me-nots, to mention only the abundant and conspicuous ones. The big lama said the Tibetans like American silk, figured on red and yellow, large knives, pistols, flashlights, binoculars, diamonds. There are frequent flashes of lightning on the mountain peaks since we have been in the country, only at night, like heat lightning. Rinchen Gyaltsen says the women of the village where we got fuel were all drunk and gay. Their husbands have gone to the big doings at Seni and the ladies are having a good time on their own.

Plate 36. Thangka from Karsha monastery depicting (*clockwise from upper left*) Shakyamuni Buddha, Tsong kha pa, Green Tara, and Sadaksari Avalokiteshvara; seventeenth or eighteenth century (UMMA 17467).

July 22, 1933. There are numbers of galiens on the rocks just above the pang where we camped. The people of this village keep a Tibetan turned Mussalman to watch their fields. If one of the village cattle strays it is fined four annas, strangers double. The money goes to the watcher! At Abring shot seven pigeons (Tibetan) at one shot. Shot from 12 a young Mong plover there and a Lahuli gugti. They were cutting mudbricks there. The Zankskar houses are made of such and stone. The Abring fields are wondrously thrifty. None of us could help but comment on the very favorable place for humans in this valley. Compared with Lahul existence is easy. The old glacial plain is broad, of very fine silt, no retaining walls are needed, water abundant, trees thrive, for cattle grass is abundant on the hills and the Raja's kut[44] forests are across the way. One man said he had six horse loads to sell. Above Abring birds were very numerous. We got to a place two miles below Bok at dark. The horsemen were so lazy that I had to scold them and threaten to give them no more sheep. The camp place was splendid—a clearing beside a stream surrounded by willow bushes. With 25 bird skins to make up after dark I got to bed at 11:30.

July 23, 1933. Today I kept Rinchen Gyaltsen with me and let Dorje see what he could do with the caravan. He moved off fast and furiously and soon left us behind. Anyhow the plants are so abundant on Pensi La—we got 117 nos. The Chagzot of Rangdum soon overtook us on the way back from Seni whither he had gone with the crowd of officials. The big lama he said had sent him back to see that we were properly looked after at Rangdum. Day after tomorrow he must come back to Bok to meet the Tchsoldar. The roses have finished—rather there aren't many above Seni, but the road today on the Rangdum side of the pass was through meadows of rose and yellow pedicularis, rose purple onions, pink primroses, asters. The little primroses six to eight inches high often cover several square yards. It was growing dark as we approached Trashi Tongde [Tashi Tonde] and two lamas told us our horse had come to the monastery. Arrived at a stream after crossing a half dozen small ones and wading in the marsh for the last two hours. The meadows are a mass of springs—we came to a little torrent a rod wide but now become too deep and swift to wade. It was now dark. *Qel faire*. No light from the camp. Shot gunfire—no response. Built a fire but the tamarisk wood wet and not much use. Bitter chill when the cold water touched my legs, so that there was no joy in going further for the willows that are abundant. We had passed two miles back a corral of horses where a man was

---

44. *Saussurea lapa*; root is collected for incense in temples, sold in Indian market. Koelz (1979:47) noted: "the plant is especially common in some of the Kashmir forests and extensive and dangerous smuggling has been carried on against the state monopoly. Of late years, Lahulis have taken to cultivating the plant and in my last visit in 1936 hardly a cultivator was without a patch. New land that was never considered desirable for ordinary crops was taken up and rose bushes that in spring beautify the slopes were almost extirpated for fencing. Latterly the price has declined and the boom is fading. The root is used to some extent as incense in the temples and the leaves are sometimes smoked."

camped. His fire now evident. Went back. Scared the poor man to death with our mechanical flashlights. Surely, thought it was the devil. Served sattu[45] and tea. Said water probably too strong for horses. I could have a bag and could sleep in a little stone sheep pen. Had I had some wool clothing should have done it. We had five pigeons we could have roasted and anyway dead tired and not sure of crossing water. On three horses went forward. Old man reconnoiters stream. O.K. A little colt got across too. Arrived at camp at 11:30. Fee two rupees. The wolves eat colts, so have to corral the horses at night. Our men had crossed below, where the water is not so swift. A woman had brought wood and milk for us. Tankas soaked.

July 24, 1933. The mosquitos after sunrise were so fearful that it kept us slapping. Got the 22 birds made up and plants labeled by noon, mosquitos attacking every second. We had fed them yesterday from below the pass. When ready to go, a whirlwind, such as are seen daily somewhere in Zanskar picked up our blotters and frightened a horse so that it crossed the stream and ran three miles, a piece of paper whirling gaily after it when it took off. Lama Ishi Gering who is accompanying us to Likir came in the morning and said he couldn't go conveniently today but if we insisted ok. I went to the monastery where tea and kumanis were served and I got a cake of tea for a present. Asked if I wanted anything from the monastery. Couldn't find the tanka I had admired before but bought the old tea table I had wanted. Had three of the same design. There were many monks reading in full dress in the main hall so I couldn't go in looking for tankas, though I was urged to. The lamas will read another seven days. This is the fifth. Many are from Likir. The big lama's cook or steward whom I had photographed with a sword was at the monastery. He said they intended to take me back to Lhasa with them. We left at 2:30, our horses having gone on. The road is along the spring-fed meadows, grazing endless and flowers beggar description. A lovely deep rose purple orchid appeared for the first time and mingled on the ground with the yellow pediculars that Zankskaris are so fond of (men and women wear it in their hair, head, dress, or elsewhere). The pink primrose around Rangdum may cover a half acre. There is a village of a few houses within three miles of the monastery and another, Shakkar, ca. five miles below that. The streambed is wide and the stream slow till some two miles above a now ruined village: Gyama Tongdze. There, on a broad green dry plain scattered sparsely with huge boulders, we halted for the night. Such luxurious grass has nowhere like been seen. At Shakkar an old woman accompanied by three or four others offered me a dish of sattu. I was to take a pinch and give baksheesh. The idea was so novel and the people so wretched that I did. They have large herds of yaks and a few fields. They say it is too cold for crops here. The valley at Ramgdum at its

---

45. *Sattu* ("seven cereals") is a mixture of ground pulses and grains and is often served with fruit, sugar or milk; Koelz (1935:290) describes it as "roast barley flour."

widest must be five or six miles across and half that the other way. The road to Wakka, they said, was very bad and horses couldn't go. Some dozen men just came back from mending it so the tehsildar can go. Some four of our horses have for several days had a broad swelling on the belly where the ropes come. Puss is now forming. The phea[46] which since Abring have been numerous are still with us. The dog caught one at Bok. Mostly they whistle from their holes' edge or from a big rock and then disappear. The redshank[47] is common in the pools that are formed on the marshy swamp. Saw a rock creeper at Abring and five or six snow pigeons at Rangdum. Ephedra since Abring has been very abundant, often an acre matted with it. Dorje pulled the strap fastener out of the camera case and the horse with the old tankas stumbled into the stream bed, so I had that to greet me when I got home at midnight yesterday. We camped last night at dark.

July 25, 1933. The horses last night left the splendid pasture and went far below. At Gyama Tongdze we met seven [indecipherable] soldiers going to relieve the Padam watchers. Three or four more came later, one escorting the baggage. They had come from Parkachen. The road now is up and down over old glacial drift from the now narrowed valley walls. The stream becomes a torrent. The mountainsides have been everywhere since Rangdum bare, except for the lower slopes, but on today's march the peaks were lower and instead of bare rock peaks with a green talus, have now mostly rubble below. The dry plain patches may be carpeted with flowers too. On one broad patch of an acre or so the whole was scattered with pinks. At another ruined village below Gyama Tongdze the old fields were full of purple blue geraniums and yellow and rose pedicularis. On the hilly drift are the last of the roses, the short one to five foot thicket-forming variety. Parkachen is a large village with thrifty fields at the bend of the river. The place is very pretty. The valley broadens; the hills are green; and the fields on the new plain and on an old one above. The lama went on ahead and then two or three sophisticated looking Mussalmans came out, one offering me a plate of sattu. I didn't take any. The men wanted to stop on a nice green place behind the village but our stop Tangola is four miles below and on we went. There are some powerful springs below the village. Flowers are abundant in the fields: geraniums, purple aconite, purple fleabane, yellow composite geraniums. There are small beds of a sort of wild lettuce that they probably use for vegetables. The bird trunk got wet in mountain torrent today and I shouldn't have known it except that I decided to shift the skins to a larger trunk. We camped at sundown a half mile above Tangola. There has been no village for some 20 miles above Parkachen. The mosquitos weren't quite so bad this morning. At dark it's too cold for them but they come at once with the sun. A half grown calf stopped our dog on the road and gave her a thorough licking, to the dog's great pleasure.

---

46. Himalayan marmot (*Marmota Himalayana*).
47. A wading bird of the Scolopacidae family.

July 26, 1933. Burdocks are used to catch rats in Lahul and Kunawar. The Kunawari name is *pir zumba* (rat catcher). You fasten the burs near the rat hole. We left Tangola at 8:15 and arrived at Sangu at 6:15. There were three species of pigeon at Tangola: a W.C. redstart below Yuljuk, also six green sandpipers and two Ibidorhyncha. Shot one. Men at Sangu say plenty near here. Hoopoes at Sangu. My cherogs from Kulu developed a hole in one heel and had to be discarded. The women all along route today were much afraid of us. They fled from the path at our approach. There is an old fortress now totally ruined two miles above Sangu. At Sangu two nullas come in at an 'X,' one the Poo nulla they say has big ibex. The other leads to Dras. Our dog got terribly tired on the march today and often toward last remained behind to rest. All the men drank heavily. The road today was for the most part through heavy cultivation. Villages of stone, partly or not at all mud-plastered houses were one quarter to one mile apart. The crops: barley, wheat, peas, buckwheat, and tobacco were the nicest one could want to see. The mountains to Sangu have heavy glaciers on their peaks such that we had many streams to cross. There were bridges on all. The road was well kept. Crossed the Z.[48] River at Tangola and stayed on this side since. Made up 13 bird skins before bed and then a cat stole one promptly the light was out. Got a cuckoo at Yuljuk. Men complain grass expensive at Sangu. Camped inside a sort of corral with serai accompanying magpies from Tangola, also black swifts, brown swallows, and WR Swallows.

July 27 1933. We left at 8:15 again and got to Kargil after sunset. We camped in the resthouse. The road was less beautiful than yesterday. We left the great range with its glaciers at Sangu and the walls were now bare, the valley not so broad and villages fewer. Except for Tsaliko at a broad sumdo where water had been brought from Sangu eight miles or more. Here trees and crops were spending. Houses (said to be 1000 families on both sides of the river) were scattered among the land and seemed new for the most part. Willow, Lombardy poplar, and another poplar common from here on. Crops: two kinds peas (white and red flowered), wheat (2 kinds), barley, and tobacco, changtan, tzetze, kana, kertze, buckwheat, two species. In one place only a small patch of potatoes. The barley from Sangu on is ripe. In places the second crop (buckwheat) has been sown and is up. Guntring, a little above Kargil, is another luxurious spot. Here a few apricot trees (no fruit seen) and mulberries. The latter very small. Saw an oriole here. An accepiter[49] medium size at Tsaliko. Dorje said another at Guntung. Crows from Zaliks down. Black swifts abundant. The men were incensed that the people here plow with milk cows. They have cows and churries.[50] Saw few animals today. Say they keep them inside except for sheep and goats which go to the hills. Goat and sheep run heavily to black and brown. The heat was less intense than yesterday because cloudy

---

48. Suru River.
49. Sparrow hawk.
50. Donkeys?

Plate 37. Women's waist ornament, *dolcha* (UMMA 17217).

all day. Though the heat on the high altitudes is strong one doesn't take off his coat. There is always the lurking of cold in it. There is by way of crops also a little sarson,[51] a few stalks of corn in one place, a small bed of onions and of potatoes and three or four apple trees with fruit at Tsaliko. The children from Tangola down have been splashing naked in the water. At Sangu and at Kargil the nights were very warm. The women in Purig wear black, many have the brass ornament that the Ladakhi women wear with strings of shells attached. Red opaque glass beads and the silver gao beads are common. None of the people asked in lower Purig or Wakka had seen iron arrows. At Lonchen they had a broken zee and said there were others in the village. Children from Tangola down splashing in the icewater naked.

July 28, 1933.   Rinchen Gyaltsen went for fresh apricots and didn't get back till nearly ten, so we started at 10:15 and got to Mulbek at 8:00. We camped on a pang beside the river. The country is now different in character. The mountains are nearly bare of vegetation and have no snowcaps. In the river bottom for the first ten miles or so is abundant cultivation with trees. Then follows a long barren stretch then a narrow gorge til Shergola near Mulbek. All day we wound in and out of the little river that comes down from Wakka. You would think that now we will get a view, only to find another elbow cutting off what is ahead. We met four or five caravans carrying numdas[52] down. They said the

---

51. Mustard.
52. Rugs.

goods were from last year. The Yarkandis haven't come yet. We overtook two nice Wakka men going home. They said horses can go from Wakka to Rangdum easily in two days. From Mulbek to Wakka you have to cross the little stream 100 times they said. Wakka has 100 houses. There is much wood in the nulla, birches, they said, and willow. We had difficulty in getting grass. The tekidar said his grass was only for government people. There are two small monasteries on a huge high rock overlooking the town of Mulbek. One is Hemis, one Ngari Tshang Rimpoche's. At Shengdu there is a pretty little white one posted on a cliff. They say there is a big monastery above Likir belonging to the Yura Monastery—Sankar. The valley at the monastery is open and the hills of the pass rounded, but in front of the town sharply cut and worn. Last night in the blue and pink of the sunset the sharp sulfur and pink tinted peaks were beautiful when in the light. The heat today was intense. At times a cool breeze reached us. All day, even within a mile of Mulbek they told us we had two daks to go (dak = four miles). The Purig people in general had a good idea of miles and usually gave a correct mileage.

July 29, 1933. The houses since Kargil have been large and very attractive. A little above Mulbek carved on a huge boulder in bas relief is a beautiful Buddha 30 feet high. A house has been built against it. We left the Nakka River shortly and went across Namika La to Bodkarbu. Left at 8:45 and arrived at 3. It's the first day of travel that we have camped before evening. The men have been fearfully worked, but not at all cross. They have to get up at 4 and look after the animals and don't go to bed before 11 or 12 when we arrive late. The people at Bodkarbu are very friendly. A crowd promptly gathered and we played the phonograph. Delighted and tsa-tsa, as usual. Then the women came and sang and danced for us. The performance was remarkable, rather for the good nature of the performers. They brought a four-cornered arrow. It seems they use such here now in their arrow-shooting matches. Later they had many new ones of the sort. One man brought an old brass gao with silver plate over it. One man from Chiktan says in the old chief's house there, there are carved boards over the door ways. The Mussalmans villages have a common building with a little cupola which I take to be a mosque. They said the old monastery at Mulbek had been destroyed in the wars. The ruins can be seen. There are also old stronghold ruins on the crags at Takhtse and at Bodkharbu.

July 30, 1933. Left Kharbu at 8:15 and got to Yuru at 2:30. Day partly overcast so heat wasn't so bad. The people at Yuru were very friendly. The chaukidar treated the men to chang and the villagers invited the lama to food. Bought two gaos such as I had never seen and an old door seal. These pieces of carved wood are used in this district to seal a jar of grain or a door before locks came into use. The head lama has gone to Sheshukul and has been there nine months. Gawan, Shan, Kar, Yuru, Kukshu, Skirbuchan, Sheshukul are all

Plate 38. Wooden door seal (UMMA 17030).

one group. The men complain that provisions are higher here than elsewhere. Chikor[53] plentiful. Say no galeen here, but many toward Linshot. There was a heavy rain above Kangi so that the flood carried off a jomo.[54] The river today seemed no worse than when we managed it in September, and we might have come that way except we would have been detained by the rain and flood.

July 31, 1933. One of the men went back and forth half the night to my tent bringing tankas. He never brought over three at a time. Kept me awake during the process. He always wanted to be sure I wouldn't say a word or the head lama would skin him. I found among his bringings two very old specimens, one beautiful, one interesting. If I stayed a day many things would come he said, but they don't know the difference between a good and bad tanka and it would have been a big chance. They are going to an old monastery in the nulla and bring things to Nurla tonight. We left at 8:45 and got to Nurla at 4:45. We stopped at Khaltse to buy some apricots and post a letter to Rup Chand, telling him to be at Tso Tskuntsi on August 25 and what food, etc., to bring. There are many trees of apricots, some well laden walnut, and mulberries at Khaltse. The apricots were tiny and not well flavored but the man said the better class had been picked and these are now in season. In our haste we believed him. It wasn't

---

53. Partridge.
54. A female yak-cow hybrid.

Diary: June–July 1933

Plate 39. Copper container for monk's alms bowl (UMMA 17245).

so as we found a few paces farther. The sky became overcast shortly after dawn and stayed so, so that travel was pleasant, else the deep gorge from Yuru to Sengge Nangpo would have been a bake oven. I found the rattle snake plaintain of which I had got a single specimen before and the lovely pink apocynum in the gorge. Arrived below Nurla we found some 20 men and women returning from salvaging wood from the big river. For a mile big willow trees, timbers, etc., were floating down the river. They said there had been a cloudburst above and the water had carried off trees, [indecipherable], etc. An old lady with a nice clean face, but with a very few stones in her *ferag* asked if I were all alone, no men, where going, and invited me to stay at her house. The people are very friendly and soon brought all sorts of things for sale. I bought a beautiful gao of copper, silver, and turquoise, a copper cup carrier and a native copper little plate. One man brought some very nice apricots. There are many trees here. The planting is luxurious, a strong contrast to the apparently bare but purple, madder, pink, brown mountains. The mulberries here are medium, but stick to the tree, black and white sorts. Walnuts well laden. Some apples. Bought some small ones. Red, soft flesh, but insipid and very wormy.

August 1933

August 1, 1933. We set out at 8:45 and got to Likir at sunset. Bought some more old things this morning: a nice carved sattu box and a silverwood teacup. There were several old selchoks[1] but not original designs. Some very fine apricots bought. We came by the Saspul road. The upper path, they say, is only a footpath. Shortly out of Nurla on the "one way" stretch we met some 25 soldiers and their baggage. Two convicts were handcuffed together and two Ladakhi women on horseback were also in the procession. The Indus was still more turbulent than yesterday. Saw a chikor with two new hatched young. Latter hid under rocks. Big flock of blue pigeons at Saspul. The barley to Saspul has been cut or is being cut. At Alchi it seems to be ripening. Came the pass road to Likir. At the crest of the pass a magnificent view toward Alchi of the range that separates Ladakh and Zankskar. It is very green in the high valleys of the range and snow patches rest on the peaks. Toward Likir, a splendid view of the fields in the valley with the monastery on a knoll at the top. Toward all other directions peak after peak. We hastened to the monastery and found they had heard word of our coming. A room had been fitted up for me with a sort of a cradle for a bed and everything covered with rugs. A shrine was at one side and a balcony came off it. The monastery was very clean and the tankas abundant with richly ornamented cloth. Images few, practically all of earth. Many rooms in the rambling old building. Found most tankas new. In one room some 30 old ones of the oldest type and their copies. Horses arrived at sunset and camped in the grove below the monastery. I made my bed on the balcony of the room. Gave Ngari Tshang's letter.

---

1. Textiles (Central Asian?).

August 2, 1933. Last night two of the monks came in to see that I was well in bed. The view from the balcony is grand—toward the Alchi Mountain wall with the new snow on the peaks and the melting mist clouds, and the scattered green fields nearby. A musical performance on two big metal trumpets of a minute or two heralded the sunrise. Then later there was intermittent trumpeting, musically pleasant, from the assembly hall where the monks worship morning, noon, and night. The monk read his letter this morning and said his abbot ordered him to give me 15 tankas of my choosing. I now again went through the monastery and came out with some 15 very old and very beautiful ones—so old that they are not registered in the monastery list. Of these 15, five were large and in fair condition. The others were damaged badly, two large and eight small. These I was willing to consider as three and to take the seven neden chudruk[2] for the lot and make 600 rupees and 100 rupees tip and shell in payment. It was clear now my choice wasn't pleasant and I was urged and made to make a third inspection of the collections but not much more could be gleaned. There were many new tankas, many with beautiful brocade, but I remained firm in my choice. They said now they couldn't give the seven. Very well then, give me a letter saying they wouldn't obey the superior's command. That they wouldn't do. Finally I said I would take the 15 old things at 250 rupees and 30 rupees baksheesh more tips and a letter saying they wouldn't give me the seven. This they finally did and after 20 rupees more tips I left at 4 pm with my tankas. All day of negotiation. The monastery presented me with a huge basin of each rice, flowers, kumanis, and ten eggs. I took the eggs and a meal of rice. They treated us well: gave wood and pasture for the horses. There is a huge rosebush still in bloom 12 feet high walled in beside the monastery and a large thrifty cedar. There is only one crop here. It is higher and colder than the places along the Leh highway. One man came with the monastery chagzot and claimed damages for our horses in his field. Everyone was skeptical that our horses did the damage, it seemed old work, but finally I paid the damage money. The monk from Rangdum whom I brought along, foreseeing the monastery's refusal to obey orders signed the letter as a witness. The monks didn't want to sign it, but two finally did.

August 3, 1933. It has been cloudy with light rain squall for the last two days. Today at two it cleared. We left at nine and got to Nimu at two. The monks came again this morning and asked for more tip: I had given to the monastery in general, to the monks in general, to the caretaker, to the building of new houses, and this was now for Rimpoche. We made it clear that by their refusal to obey orders that had cost the monastery 420 rupees plus a valuable present. Dorje on gathering a specimen of the large rose bush was told it belonged to the "nether folk" and harm would come to him. We bought a basket of chilies at Basgo—the last we will get in situ. Saw a spotted backed dove at Basgo and shot one. Reception at Nimu such as we had seen in Punig and at Likir. Our 20

---

2. *neten chudrug*: the 16 Arhats, disciples of the Buddha.

lantern glass that has been cracked many days went to pieces yesterday. Four of our six men could lift the rajas stone near Basgo today. One man at Basgo recognized me as the sahib who had lifted it last year. A horse on coming to Likir had broken out in a heavy sweat after drinking water abundantly and the men had to carry the load on their backs. Men bought a goat and made gyuma again. I tasted it, but it had no seasoning. The monks served us with a plate of boiled rice flavored with sugar, butter, and several kumanis at noon yesterday. It was very agreeable. The phonograph made a hit as usual. One or two of the lamas had heard one at Leh. No audience has failed to laugh when the record comes to "you're just a lovely little river." Yum was the least appreciative audience I have had. One or two men walked away there after the first record and none there had ever heard a phonograph. Saw a kite at Likir. An old lady at Kurla who had the choice of a green opaque glass necklace, a glass crystal and a beggar choker was for taking the latter but was induced by others to take the glass.

Plate 40. Phonograph performance (date and location of photograph unknown) (Walter Koelz Collection, Bentley Historical Library, University of Michigan).

August 4, 1933. We left Nimu at 8:15 and got to our old camp beyond Patog at 2:15. There were four female shapu[3] in the gorge above Nimu. I got two more Accipiters at Nimu—they have nested in the "forest garden," and three falcons on the way to Patog. The snipes and Ibidari are at Patog as before. It sprinkled a bit in the night and most of the day remained overcast. A man carrying apricots from Kholake to Gawan to be relayed on thence to Kushog at Shurkukul went back from Gawan. Our Basgo apricots were a mess at Patog. What his are? The old female chaukidar at Nimu had a very attractive gao-like thing of silver and turquoise, which was worn around the neck. A double string of coral and turquoise and silver beads attached at the sides. She wanted double price. They said it was Balti. Dorje and Wangchuk went to Leh. Dorje met at Nimu a man who had gone to Poo with young Peter when he went to fetch the two Xians that remained of the missionizing there. The man had seen the beating a Tibetan gave Peter.

August 5, 1933. Wangchuk came back from Leh at 9 and all the rest of the men went up. Rung Mo Shakh came in the afternoon with a horse-load of all the rubbish you could imagine: pieces of brass, wood, glass, papier-mâché, stone, many nameless, some old but wretched tankas, two new images of brass (these he said he could sell for the same price as old). Of the whole there was nothing worth purchasing or having for nothing except a brass gao and an old copper seal. These I bought when the old fox brought out the loveliest murti[4] I ever saw. He said it cost 100 and he would take 120. I finally got it for 75 after he had locked it up twice and said several times he never lied. It stayed cloudy and sometimes sprinkled a little during the day. The full moon rose tonight through a rift in a blue black mass of clouds small illuminated fragments floated sinister. To the south, edging in front of the bank and fading off the southern horizon a huge blotch of diffuse yellow light. In the west, faint afterglow of the sunset through heavy cloud masses. On the northern mountain rims, ragged clouds rested, lighter in color than the darkened peaks. The sky overhead was clear. There is no bridge at Nurla so that the Zanskaris can't come out till the Indus is low. At Likir, when I went into the rooms where shrines had water dishes, a man quickly emptied the water into a pitcher. This has been done under similar circumstances before. In my bedroom however the water was left before the images. There are a few mosquitos here at night but if the wind blows they are soon dispersed. At Yuru, they say also at Khaltse, there are the tiny sandflies that leave such a nasty welt. Ringmo yesterday had an eye shade of woven horsehair screen that fitted over the eyes like spectacles: the men all came back from Leh.

August 6, 1933. Dorje went to Nimu to get the Balti gao, since there are none to be had here. If I had the wit of a chipmunk I should have known there

---

3. Tibetan sheep or Urial (*Ovis orientalis vignei*).
4. Image.

weren't any here. Lompo went back to Leh to see about tankas. He asked me the usual questions: would I go to Lhasa, where going from here. He omitted the land questioning. Rinchen Gyaltesn and I went to Spitug monastery. The chagzot is a clever and very likable man. He said there were no old tankas (we didn't see any) because the monastery had been sacked by the Sikhs. He said he would send the old Namgyeltsemo[5] table in the morning. Out of a recently made hand-written note book he gave me the ages of the various monasteries: Spitok 600, Nubra 60, Hanle 800, Hemis 870, Likir 800, Namgyeltsemo 900, Gawan 600, Marsho 800, She 500, Tog 200, Basgo 500, Tiktse 600. They are getting food ready. Tomorrow for one and a half months, all their branch monastery lamas will be assembled. The chagzot asked twice if I weren't connected with the Roerichs. The Pangchen R[oerich] portraits and the two R[oerich] prints are on the shrine in the main hall. We saw some 25 tankas, none interesting, of the ordinary type 100 years or so. Ringmo Shakh came again with the bag of rubbish. He had three tankas, one printed on paper and hand-colored such as many of Timoo monastery. One was rather new but well designed and drawn and one was the circle sort as we got from Likir. This was the oldest and best. The image Dorje and Lonpo had so much admired turned out to be a brand new crude thing. We bought a little Balti gao and let him go back. They said the Ladakhi raja got 2500 rupees for his daughter. No one knows her; she is unhappy. The Marsho raja is said be crazy. [drawing] shoklok square perag ornament = jubjur. Balti gao = tumar, sattu box = zhipor (h.), pepor. Round perag ornament of onyx and turquoise is shoklok. The children of Nurla had put with apricot gum some seeds of shoklok on their faces.

Plate 41. Round *perag* ornament from Nirmu, Ladkah (UMMA 17140).

---

5. Nyamgal Tsemo: monastery in Leh District.

August 7, 1933. The wolves last night attacked the horses within 10 rods of camp and tore the hindquarters of Rup Chand's nice horse. She fell and sprained her shoulder, so that she is unfit to walk with us. Lonpo will feed her for a month and send her home with some Lahuli. Dorje has gone to be nurse for several days. The first Lahuli arrived yesterday. His horses have not come. We got a new brand of Wans,[6] white but peppery. I can't taste anything but pepper but the Ladakhis all say it is better than the red or another kind full of garlic. We still laugh when we think of Mrs. Lee eating the cakes with bread. Chulis are good at 10 pounds per rupee. They said we should have bought them at Kargil. Cheaper and better. The men bought four medium cauliflowers. Many fresh apricots daily. We went on to She and camped on a little dry spot in the marsh below the old palace. A little lad recognized me from before. A horse stumbled in one of the endless pools along the river bottom and soaked all the kumanis. An old man brought some turnips, beets, and lettuce. He had sold me some boxes in Leh before. There were two Hobbes, several sparrow hawks, two big Accipeter nepalensis, a Pallas Fish Eagle between Spitug and She! The Ibidorli weren't so numerous as last year, but the children say they are out on the shingle. A flock of small bw teal at Spitug. We got some goat-hair bags at the same price as the wretched cotton ones last year. Rinchen Gyaltsen and I went to Tektse monastery. The chagzot was away but his father and two lama boys showed us the old tankas. They have at least 50 of the very old ones, uncared for as usual. Twenty-five can be made into first class exhibits. I am delighted beyond words. There is good reason to believe the acquisitions can be engineered. There are some good Lhasa tankas but none worth buying. They showed things freely. The monastery faces up the river and is exquisitely located. The view up and down the valley is surely the best in Ladakh. The road between She and Tiktse is sinking. There are springs below and horses can't walk in several places. Some Baltis had a bad time getting theirs out of the mire. The field walls are about to slide. We were able to replace the lantern glass that gave out completely at Likir. The well-to-do women wear a deeply embroidered Kashmiri pesham shawl instead of a goatskin reversed as the other women. Most men have on goat skin when carrying their little baskets. The weather continued cloudy with more sprinkles than usual today. After dark the sky was completely overcast and it rained a bit off and on.

August 8, 1933. A man on horseback came at bed time last night with two gao necklaces and a beautiful yubjur. The turquoises were perfect. He asked such exorbitant prices that I told him it would be impossible for me to deal with him—I'd be ashamed to offer what the things were worth. At 3 o'clock I was awakened by the chanting of several men nearby. One took a pulsating vibrato at C and wove back and forth with very slight interesting variations. Several others accompanied with a bassorial hum. There have been a few mosquitoes every

---

6. Meaning unclear.

night at Spitung, but when a breeze came up as usual with the shifting storm they were quickly dispersed. They say these men were washing and shrinking cloth. Why it had to be done at 3 AM is not explained. Went to Tektse again. A man came early and said the chagzot had arrived. The chagzot is a friendly timid person who is ready to give the tankas but says five people must be consulted and two won't be there till night. He has a watch, two flashlights and a small thermos. He knew that all other monasteries had given us things. One man bought a piece of linoleum apparently of the Old Stone Age. One must be very careful in buying things; seemingly old appearing; gaos with what one would say archaic pattern—turn out to be the poor work of some extant blacksmith. The Ladakhis call burnt སྨན་ [sman; beneficial, helpful, also used for batteries] batteries "masala." Some men were singing all afternoon on the repair of some of the forest of chortens beside She castle. We went to the monastery at sundown and was told no tankas of any sort could be given a sahib. It wasn't their custom. I then asked to see the committee and five men came into the chagzot's room. One stupid looking (bullheaded–I) old man was the leader and I saw there was nothing to be done with the group. Besides, there is an old man in the abbot's shoes with whom these do not agree and he would have to be consulted too. I offered 500 rupees for 30 torn old things and a dung but they didn't know what the latter was. The C. says they always at war with each other. Ngari Tshang put out some dozen married monks and some Englishman and reinstated them and they are warring over that just now. I might have known better than to try to do business with seven men. The Tibetan proverb says to put out the third if there are three to do business. The iris that grows along the field borders in Purig are not pulled up when they spread their leaves into the crops but are tied back. In Zankskar yellow clover was tied up in the same way when it encroached on the crops. There was a woman at Tiktse gathering the silkia grass (Lahul name) Omo, Ladakhi. Lahulis get the grass from here to stir sattu into soup so it won't clot. We came home from the monastery at 10 in the moonlight. What we should have done without the nearly full moon traversing the quackmire that road has become. Today clear for the first time since Lamayuru.

August 9, 1933. We left for Tog. Got two species of cuckoo at She. Saw the big Nepal eagles again. The bridge on the Tog side has been so damaged by high water that we had to unload the horses and take things over on men. The water that had to be waded afterwards was above the knees in places. There is a ford lower down but we didn't know where it was. Met a Tog man who said the raja had received a letter from She announcing our arrival and he had watched for us with field glasses yesterday. A man met us near Tog and showed us into the nice garden in which we camped last year. The Jispa Jomo came promptly to welcome us. A little later the Rani's sister. Toward evening we went to the palace. They had sent beer and tea for me and the men before. The Raja and Rani and latter's sister, another old woman with a beautiful perag [headdress] said to have just come back from Spiti, a middle aged man of good family made up the company. Rinchen Gyaltsen had a place opposite the raja.

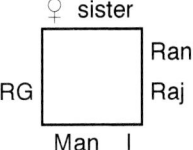

The Rani gave me the nice biscuits I had first eaten in her house. She wanted to know what to do with the perfume I gave her. I gave a beggar choker from Rup Chand; the Raja—a pearl handled pen knife. She showed me a bottle of 3-in-1 oil and wanted to know if it was hair oil. The sister's little boy whose appetite was so phenomenal died in one month of stomach trouble. The poor sister is very lonely here though she says they treat her well. The Lahulis were all glad to see their own people. On leaving I asked the Raja for a few words in private and he came to my tent. He said he couldn't give the two tankas I wanted; it would be disgraceful, such things weren't to be had nowadays, money was too ephemeral, etc. I gathered he wanted more money and let the matter drop, promising to call in the morning. Food cooked was sent me and the men. Two of the men, Rampuris, wouldn't eat it, suspecting yak flesh. Everything possible was done for our comfort. The horses were grazed in the garden and spent the night among my tent ropes.

August 10, 1933. The two Lahul women came to camp early and had a visit. The Jomo sent letters home. I went up to say goodbye but I didn't sit. Rinchen Gyaltsen gave the Raja 10 rupees for the servants. The Jomo says they won't see a copper of it. Two came with biscuits from the Rani and tarried. Rinchen Gyaltsen explained their baksheesh had been given the Raja. One wanted a picture of Ngari Tshang. The Jomo escorted us out onto the dry plain. We camped below Masho monastery beside a nice spring. Got a big hare. Weather clear and hot. One man came to see us and said some people, male and female, had gone to Takognath from here. We should tell them to hurry home. He would give us a little wood this evening if we would do that. Saw five shapu on the plain below the monastery apparently feeding in the fields. Raja said he didn't have any shawls like the Kargiak one. The two Lahuli women hadn't seen any either. A bedbug village had located itself in the little tea table bought yesterday. When I put it in the sun they all came out and soon died. Went to the monastery toward evening. The Kushog was down below helping build a chorten. He asked much for Rup Chand. The music pleased him and delighted a little waiter who looked at the machine on all sides to see what was making the noise. Dorje arrived at sundown with some cauliflower, Chinese cabbage, turnip tops, and a basket of white-seeded delicious apricots. The hare had four embryos nearly ready to be born. Our horse is being looked after by a government attendant who says

Plate 42. Mashro lama at Talognath (Walter Koelz Collection, Bentley Historical Library, University of Michigan).

in one and a half months it can go home. There is another horse and a bungu [?] that have been torn by wolves too. We have to supply food. A man walked over from Tog last night with two little old tankas and an image. He told us in Tog he had them but wouldn't show them there. He sat beside the stream in the dark and wouldn't come to the tent. He says he has others which he will send to Hemis tomorrow.

August 11, 1933. When I got up, the little Mussulman from Leh who last year forcibly presented me with vegetables had been waiting for hours at the camp. He had two bags of fresh vegetables: rutabagas, kohlrabi, carrots, turnips, radishes, new potatoes (egg size), chard, peas and some sort of cabbage tops.

A stiff sprinkle came out of Masho nulla at 2:30 and kept up for two hours. It cleared then until we got to Hemis. Then a fresh shower came out of the nulla and continued an hour. The men got well soaked but no matter. Their bedding stayed on the horses during the rain, under the saddle. An old woman whom I remembered from before was coming up from the river with a basket of green twigs. She took shelter with us beside a chorten and chatted a bit. When the men went for grass and wood she came back with another man, both with a basket of green grass. Our men invited them to tea and sattu which they both enjoyed tremendously. The old lady, it came out gradually in the conversation, has nothing at all. She went to Chemire to beg yesterday. The Kushog gives her a dress every year but he isn't there this year and the chagzot isn't so kind. She has a daughter whose husband died a month ago. She said if only she could die, but death doesn't come. And with it all not a word of complaint or begging. Our men gave her some dry tea, a spoon of yellow butter which she much admired and carried home wrapped in a cauliflower leaf. They invited her to breakfast. It is custom to give all the deceased's personal property, including jewelry, to the monastery on death of any member of family. This property the lamas divide. From such stock of clothing, the Kushog gives such as our old friend. Our dog eats fresh and cooked apricots and cracks the seeds. We have had lizards since Karsha. The orange collared ones were there and until here. Here there is a flatter one, like the Tsokar species. The Leh vegetables are all insipid. The foliage is so tough that it can't be cooked easily. The old woman took home the turnip tops.

August 12, 1933. My little Mussalman was on hand first thing this morning. He brought an old lovely tanka which he said the owners swore was painted with 100 rupees of gold. I bought it for 15 rupees. My friend appeared at the bridge at M. as agreed but brought only three tankas. I bought one very old one. They will have more for me next year. Rinchen Gyaltsen and I went to Hemis monastery and the horses went on to Upshi. We saw some beautiful old frescoes in three rooms, mostly ruined by repair to plaster. In one room were six huge chortens silver with gold and stone ornamentation. The stones were carnelian, lapis, turquoise, coral, jasper, etc. In one room was a bundle of tankas carelessly dumped in a corner. Almost all were beauties, Tibetan, none of the very old style. Several were by the same artist to the one the Kushog gave me before. They don't know when the Kushog is coming back. No word received. Chagzot gone to Hanle. The young monks were approachable, but the old ones nasty. The whole monastery is carelessly kept. Dust an inch thick on the precious chortens and rubble in the corners of the rooms. We saw only three rooms. There must be many more tankas than we saw. Arrived at Upshi; an European and wife were camped. Some Lahulis from Kolung brought them to Upshi but had to go back because wolves killed a horse and wounded two more. Spell of rain toward sunset. Upshi in bloom. Heavy fragrance. Had a little cold from Masho and the Hemis rain. Uncomfortable till supper time. Our old female friend came for breakfast as invited. She-Ghe road is mostly barren to Upshi.

There is nothing after Marshalang on this side of the river. The traders cross at Yugu. The grass is better. We met a family of Tibetans on Nekor. A huge flock of laden sheep was moving above Yugu.

August 13, 1933. The night was clear but there were evidences of heavy rain above Upshi. The road to Gya is all day in the narrow valley along the Gya River. At Miru there are several houses, also at Latho. At Gya the valley opens and there is considerable cultivation. There are small rims of green along most of the river and on the opposite side a narrow fringe of roses, tamarisk. There is a monastery at Miru in the ruins of what probably was a fort. A man said they had some 40 tankas. The monastery at Gya is across the river from the town. Both are under Hemis. The flora becomes Tibetan at Gya as at Nirmamud. We left at 8:15 and got to Gya at 3:30. Sky overcast all day. Decidedly cool at Gya. Saw WB Dipper at Latho. The little Mussalman was on hand this morning early with 3 [indecipherable] tankas, a lovely brass incense burner sans top and a new brass belpo gao. I bought none. The Rampuri servant of the Sahib who photographed wild game says he fell in the water at the Leh bridge with the sahib's mail; a family with whom he stayed at Miru stole half the sahib's kumanis and four batti of sattu and said the dog took them; the Lahuli horsemen went back with his salt and blanket; the old sahib and his young wife quarrel like cats and dogs. There are red cows at Gya, no trees. Large trees at Miru.

August 14, 1933. A dog, not starved looking either, came last night and ate all the shucks of my apricot seeds; also sorry troop of horses, four wounded, came into camp during the night. From their shoes, the men said they were Lahuli and sure enough as we were ready to start, a Lahuli came for them. His party was camped at Tiarnak on this side of the pass and the wolves attacked the horses, in the night. One horse was seriously torn, two others fled somewhere, these nine he traced to this village. A large wolf and small one followed them to Urntsi, the last village. The four are only slightly torn. The other two were recovered uninjured. Three or four Rampuris have been camping at Tiarnak just below the pass for several days but they tied their horses beside camp at night. The Taklung Pass has two paths—one follows the stream that comes down from its crest and is a straight and steady ascent. The other winds up the steep hill at once, and then runs along the ridge crest. The latter is followed by laden animals. I came the མགྱོགས་ལམ་ [*mgyog lam*; shortcut, literally quick path] and had a good stiff climb. My head ached for an hour after we got to camp. There are numerous plants all along the way from Gya to the pass crest. The road to the foot of the past is along drama[7]-strewn pangs. On the Rupshu side, drama is very abundant and the stream is soon green-bordered, but from the crest for the steep 1000 foot descent there is very little vegetation. The horsemen saw ribja.

---

7. Shrubby plant, *Caragana polyacnatha*; important source of fuel for travelers; sometimes used as horse feed in spring when fodder is scarce (Koelz 1979:29).

All we needed to make the journey as stiff as possible was rain and that began at the crest and continued into the night. Everything was soaked and everyone. I took off all my clothes and got into bed. The rain turned to snow at dark and covered the ground. The four horses stood perfectly still beside the tents all night. The other horses last night were so frightened that they didn't stop to eat in the fields and never left our sides all night. Toward sundown two women came along with two donkeys, four small goats, and two sheep. The poor things were miserable but said they had to cross the pass. They had no tent and on this side there are no rock shelters. Several men with donkeys came along later but they camped near us. They had come from Lahul and said rain came every day. Such bad weather they hadn't seen. The Lahuli horseman says the two British officers have gone to Spiti. The M.C., A.C. and the Wazir accompanied them from Kohsar to Dozam.

August 15, 1933. It sprinkled briskly off and on all day with intervals of sun. Our things were so wet that we had to sit and dry them in these intervals. In the late afternoon a protracted squall came accompanied by occasional distant thunder rumbles from Tso Kar way. The small horned lark is abundant here. It hasn't been seen before on our trip. Phea[8] are here and hares. Our men saved the bitter apricot seeds from Khaltse. They ground them and soaked them in water, then strained. The milky liquid they boiled. By stirring, the bitter taste leaves. Eat with salt, [indecipherable], and butter. Very good. The men are packed in the tent like sardines and the dog is squeezed in between. My tent has plenty of room, but the dog never comes into it unless I am eating or unless shelter is badly needed and isn't to be had elsewhere. Dorje found the lovely big single bloom gentian on the slope opposite camp. I have never seen it before except for a few specimens on Chang La. I bought a pound of barley from Gya to send home for seed. The season must be short there. The growth was splendid. At She men and women, adults and children, 20–30 strong came to camp before breakfast with their little kiltas to carry off horse manure. One enterprising man came before dawn. He of course filled his basket. Some came from a mile away. At 3 it turned cloudy again and sprinkled till after dark. The dog was restless in the night and the men went out once and yelled, suspecting wolves.

August 16, 1933. We left at 10. I told the men to be ready early and consequently got off two hours late. It stayed cloudy till two when we camped just above the ridge that shuts off the view of Tso Kar. Then the sun came out and it was quiet and lovely. There has been heavier rain this way than we have known. Large pools have just dried. The road is in a broad green valley all the way. All the men are delighted with Rupshu. The smooth, jumbled, pastel-toned hills with the wide walking spaces and no people. Even the grass is a pastel green and

---

8. Himalayan marmot (*Marmota Himalayana*).

the pinks and purples of the hills melt into it. Kiang[9] are scarce. We saw three scattered ones and two together. In a large flock of twites, horned larks and sand larks were two hoopoes. Ravens two for today. Met a large caravan of mules accompanied by some ten Indians, presumably with tea. The stream that comes down from Taklungh Leh soon dries into the sand. There would be a lake in the valley before Rogchen if there were drainage enough. As it is, in the bottom of the valley in about half acre extent are large mounds and corresponding cavities as if a big bubble had come up from below. The mounds must be 10–15 feet high. All dry. The mountains on all sides are capped with fresh snow. The dog this morning cracked and ate shucks and all about 50 apricot seeds from my breakfast. Nowadays, stewed apricots and sattu for breakfast. For lunch I had hot milk with butter, salt, pepper, and spices and a cracker. Grand supper onions and rice. The Amritsar onions will last about seven more meals. They are still firm but hard and strong. The Kulu onions were very perishable. Met two men with two donkeys from Khaltse coming back from Karzok. The Rupshuans are at Pangchen they said. They had met much rain. There were four cranes on Tsokar but they saw no young ones. Have still a slight headache; Wangchuk says he had a bad headache on the pass. Rest said no effects of the altitude. Today two or three pairs Tibetan pigeons. Pulse at 6 pm after four hours rest before supper: Rinchen Gyaltsen 80, Buddha 86, Wanchuk 94, Dorje 90, mine 92.

August 17, 1933. Last night was perfect. The sky was clear. The campfire of drama burned bright and illuminated the two tents and a broad circle from which the dog barked at fanciful things outside the range. A light breeze carried the scent of the burning shrub. The horse bells tinkled in the kiang grass beds across the whispering little rill. And we knew we were alone for miles. At 11:30 a band of horses fled down from above. The owner in the morning said they had come from Pangchen. He didn't know of wolves frightening them. When ready to leave, a Korzok man, boy, and woman and a Padam native returning from Lhasa stopped at our camp and made tea on our fire. I gave each male a pencil and the women three safety pins and got three zhus. The Lhasa pilgrim had a very nice old tanka of the good workmanship but uninspired school. He said he took it to protect him vs. ཆུ་རུད་ [chu rud; flood] of which his people had so much to fear. He didn't want to sell it. It could be bought for three to 15 rupees in Lhasa depending on the owner. While we were visiting, with much noise of horse bells, the band variety, a troop of 15 mules and horses came into view. The animals had long red yak hair tassels from their throats and looked very fine. A Tibetan was conducting them and a Mussulman, who they said made report of the traders traveling over the various routes, was riding along. The caravan was mostly empty and was going to get kumanis for Lhasa. We headed straight down the plain and soon ran into the upper arm of the lake. I made a turn toward

---

9. Tibetan wild ass.

Thagge Naz now but too late. The mounds had been softened by the rain and were like dough. By making a right hand turn and skirting a marsh, the horse arrived at the monastery and then came down the Puga road to a pang around two lovely little pools where the horses had such grass they never will see again. Nothing had pastured there and the sedges were all in bloom. The plain at the upper end is level and firm. It is scattered in places with small curiously reticulated pebbles, black in color and of hardness 5. As the lake is approached, the plain is thrown up into mounds and within 100 rods of water are craters as if from explosions, and deep steep banked channels. In some water is running. The lake is bordered in many places by cliffs 10–15 feet high. Apparently the water that comes down from surrounding hills gets too boisterous at times and tears up the plain. Sirkabs with young were numerous and B.H. gulls by 2–300. Mong plovers exceedingly common, mostly adults and very fat. Redhawks, a few very small sandpipers, hoopoes, and sand larks and wheatears. In one flock, 12 hoopoes, afterwards two and three together. Saw a male sparrow hawk and a duck hawk. No kiang but hoof marks. Fox tracks common. No wolves or cranes. A man in Thugge Gompa. He called a direction to our horses. Said to be a shepherd on the fresh water lake.

August 18, 1933. Saw two stilts and a flock of 30–40 teal. A Rudok trader from Horzam [?] is camped on the fresh lake. A strong river comes down from the lake. There are two flocks of sheep visible beyond the trader's camp. We crossed the river between the two lakes on the sheep path and headed toward the Tahsumba path. We camped on an old Rupshu campground beside some green bordered pools. The Rudok traders said one crane came to the camp morning and evening. I went in the evening and the crane was walking around the tent. At sight of me he promptly betook himself off and by the time I got through trying to persuade one of the Tibetans to shoot him for me, he had got almost to our camp one quarter mile away; I called to Dorje to bring a horse. Dorje rode up to him, dismounted and bagged him, a very thin male. Had remains of small rodent in stomach, probably one of the innumerable little voles. The Rupshuans said there are usually four or five, but today only one. They have come down from the heights (two families) to give their sheep salt and water. This grass is for the winter. Saw three of the big Nepal eagles feeding on a sheep. Also a kurral and falcon like a duck hawk. Shot a very emaciated brown headed harrier. The men to dry fresh grass put it in a pan and covered it, steaming it. Saw 12 geese on the freshwater lake. Saw two foxes: one red, one cross. Terns a few, but no grebes. One old sirkab had eight young. In the lake—beside the [indecipherable—perhaps gammarus?] are often streaks and clouds of fairy shrimps that make bloodstained patches a rod square or a rod long in the blue. Heavy frost at night. The gulls sound like a flock of crows. Sirkabs (2) and teals all females. Unlabeled.

August 19, 1933. Last night just as the men were getting ready for bed with a violent commotion the horses burst their ropes and made off at a gallop for the fresh water lake. Our men went after them but lost trace of them at the Tibetan encampment. At 3 am they returned. One white horse was too firmly bound and couldn't get away. At dawn, the men went off again sans breakfast confident the animals must be near. At 1:30 they brought them. They had fled up across the pass and had wandered from one nulla to the other. The poor men had the devil's time tracing them among the kiang and native horse tracks. The kiang have showed up today: six to eight were in sight most of the day. Four natives came this morning. One knew Dorje from Gartok, Richen Gyaltsen from Lahul, and me from the year before. Two got empty tins and no one else has ever appreciated cans as they deserve to be. Both men were delighted. The phonograph was yantsempo.[10] The cranes leave here the fourth month and tomorrow is the last of the sixth. Our crane had got left here by the Ku-yuk in punishment. Birds' social arrangement is still unspecified. Our horse had got frightened from drinking some unknown water. It often happens at this season. Water unknown. I however saw a fresh wolf track near camp and have my own opinion. Sand grouse are common at times. Their blood at once heals a sort of boil that breaks out on sheep. Sand grouse are here in flocks. Got eight tonight. There is a sort of water in Tibet that causes all the hair to fall out of horses that drink it. Our dog won't touch duck bones. Sand grouse ok. Two of the men came to draw a petition against their brethren who they say intend to rob them. The chief had lost a yak and these two are suspected of stealing it. Hence the intended punishment. Our men persuaded them that as long as they hadn't been robbed they really had no cause for complaint. The old horseman went along with Rinchen Gyaltsen to the Tibetan camp to buy butter. He had walked 15 hours sans food but started off after lunch. They didn't get butter. It was four rupees a batti[11] and they wouldn't pay it. The yak disease didn't reach these people. Wangchuk was totally exhausted so we couldn't leave today. The Kurrat, the other eagle, and a white-headed harrier came to camp today. They took water indiscriminately out of a neighboring pond. My rice was as yellow as turmeric and as bitter as wash soda. Later the natives showed them where good water was flowing. The natives say they go over Polokouha to the third month in spring returning to the lake where we found them.

August 20, 1933. We got ready to leave and sent Wangchuk on ahead. He made the trip splendidly, in spite of the fact that Tsabogorka must be as high as Takpung La. Two cranes showed up as we were leaving and we got a fine male weighing 18 pounds. Fresh tracks of two goa[12] on the plain. Natives say they stay

---

10. "Melodious"; an edited diary entry for this day was published in Koelz 1935, where he substitutes "miraculous" for *yantsempo*.
11. In Koelz 1935:292, he writes "35 cents a pound."
12. Tibetan gazelle (*Protapra picitcaudata*).

Plate 43. Man with yak (Walter Koelz Collection, Bentley Historical Library, University of Michigan).

at the Rogchin side. Nice big grasshoppers with blue legs and wings common on the plain and over on the Nidr Plain. The dog caught some of the nice little lizards that scurry over the place looking for one of their several holes and missing one or two by six inches in the panic. Saw a lark that undoubtedly was the Hanle lark. Met one of the men we had met just before going to Tso Kar. He had some 10 small maggots in his eyes, says Rinchen Gyaltsen, and they gave him some of the black residue from their pipes to kill them. Shot a yellow bellied finch on this side of the pass. Was alone. Got one at Cha the first year. We camped in the nulla above Kiling Chu where we put in before. All that water has dried up and we had to go up into the nulla to get it. A large pool from the recent rains is bone dry. A colony of cony looking things is located on the path on the Tsoka side of the pass. He [the man mentioned above with the maggots in his eye] wanted some raw meat to bait the worms, but of course on the march we had none to give. Then he wanted Rinchen Gyaltsen to spit tobacco juice in his eyes. A squall came out of the nulla toward sundown but not much inconvenient. The horsemen wanted to turn the horses loose tonight. Night before last the wolves routed their horses 10 miles away. Finished the Kahaling:[13] three female, five male. Crops of all packed with flowers and seeds of the barrenplain blue purple astragalus. The dog won't eat crane. *Tsabo gongka zhidak sō lo so lo so lo so / Ki Ki so so hlar gyel ho.*[14] The first is the Lahuli cry on crossing a pass. The name of the pass is given first. The second is Changpa and also by some Lahulis.

August 21, 1933. On leaving the nulla this morning two red wolves were digging near a mane. Dug out two coneys and shot one and two hares. Saw a little owl big as the little Kuli owl but pale like a cochen. It sat on the protruding rocks of a broken strata line and bobbled up and down on its legs. Saw such a one in Tog nulla before. Very shy. Shot two adults and young of M. blanfordi.[15] Saw several hoopoes. Got one. Camped below Pang at the junction of the rivers. A nasty squall came down the river at two and threatened to carry off the tent. Saw a laminergeier. Weather cloudy and windy after two pm. Morning calm and bright. Two of the men took the horses down the river where the grass is said to be good and stayed all night. The fanciful cities carved in the cliffs of the river in front of Bong La were more splendid than last year. The dog for some unknown reason slept with me last night. Saw a dozen tiny maggots circling in the eyeball of one of the horses. Dorje says he has seen cases in men at Gartok. Mercurochrome put in the eye cleared them out promptly.

August 22, 1933. It stayed cloudy all night but the day was fair with more or less heavy clouds till dark when rain sprinkled. We left at 8 and camped at 5:30 at Giam.[16] The one horseman wanted to camp somewhere above here where there isn't a blade of grass but we drove the horses across the river. Someone will have to swim for them tomorrow. Met a [indecipherable] Lahuli a little above Pang, saluted and passed. He asked the horsemen where I was. I had gone on ahead. He said he had met two servants but not me. At Tai Shunggam Shung saw two rock creepers and got one, also a rock coney, some Rupshu hedge sparrows and a dramon warbler. At sumdo two She traders were camped. Some 25 mules of Indian ownership were met coming to Sumdo. They had grass which they had collected at Rachogha and dried for the bare halts. One of the mules had a monkey

---

13. University of Michigan Museum of Zoology logs show that Koelz recorded killing eight Tibetan sandgrouse on August 18, five male and three female.
14. These are two different phrases that people yell at the top of a pass. *Tsabo gonga* is the name of a place, *zhidak* (*shi dag*) means "protector deity" and *so lo so lo so lo* is an invocation to a local non-Buddhist deity. The second phrase *Ki ki so so hlar gyel ho* means "victory for the deity."
15. *Montifringilla blanfordi*: plain-backed snowfinch.
16. Or Gram?

fast asleep. A Kampa whom I met last year at Rangdum and who recognized me at a distance with a four month beard was met a little beyond Sumdo. He had a letter from Rup Chand carefully tied up in his holy books, etc. Not a spot on it. Rup Chand says the sky hasn't been clear in Lahul since we left. The Russians didn't come to Kyelang this year. Everyone well. He has 30 species of Lahul birds and over 200 plants. The little pang by Gonajil with its crystal-clear stream and the view on the city of spires in the bank of the river below charmed me as fully as at first. There is a shrine among the marble fragments that broke off Gonajil and all our men added a stone to it with the *so lo so lo* cry.

August 23, 1933. The rain spattered into the night. We could hear the bell of our one horse till we went to sleep so we knew there was grass. The morning dawned cloudy and the animals visible in the morning where we had driven them. After a while the sun came out and Dorje and Rinchen Gyaltsen went to swim the river for them. They soon came back half frozen and said the water was too swift. I told them to get warm and I would show them how to cross. About 10 we all went down to the river and I discovered it was too turbulent for an inexperienced swimmer so I went in myself. The horses came to the bank to watch the performance so I didn't have to travel far in my original costume, but the beasts didn't want to go into the torrent. I got four started finally but they came back. Then I drove them up where the river washed a cliff and they saw the joke was on them and willingly went in. With great difficulty I managed to grasp the tail of one and got pulled across. A dry sand flat saved me from perishing of cold. I buried myself in the warm sand until the men brought my clothes from downstream. We got off at 12:30 and got to Serchu at sundown. There is abundant pasture across the river all the way to Serchu, but not many people are able to take advantage of it. The wretched marble fragments and the three hills one crawls up and down on the way to the boundary vexed me extraordinarily today. At the boundary is a stone marked 126-7 which I take is the mileage to Kulu. Birds exceedingly shy on this stretch. The horned lark is the Lahuli form. At the tsamp we met some Changpas on the way from Rawalsor. An old lady was apparently suffering from heat trouble and a baby from malaria. They had two donkeys: two men, four women and a child. Heaven knows how they will move across the days of desert with such a layout. The tsamp was up to our knees. At Serchu a big camp of Lahulis and a couple Changpas whose sheep we had met on the way. They lost a horse here and are trying to find it. I never noticed the powerful springs that issue from the mountain wall in front of the Tsemp mountain. They call it the 100 springs. The water at Serchu is deep and swift. Horses have to swim. Day clear with a few cumulus clouds. Night clear. We have used 16 pounds of kerosene so far from Kulu, including what got spilled out of the lantern and can in transit. Saw Pallas fish eagle at Serchu.

August 24, 1933. The day was clear and warm till about 3 when it grew cloudy and squally. Birds plentiful but of few kinds: two Phoenicurus, Wheatear,

Drama Warbler, Redwinged finch, and a few Kauan finches. The vast plain above Serchu must surely be fit for crops. The road is on the old glacial plain except where it dips into drainage gulches three or four times. Gaphan's grain piles stretch about a mile from the opening of the gorge below Kunlung. We camped below the serai. Toward evening a man brought a letter from Rup Chand saying that on August 17 he had got our registered letter of Khaltse July 31, and would be at Zingzingbar tomorrow. Our men's provisions are low. They got some flour from the men at Serchu and yesterday a gaddi gave them a goat that had been wounded by a falling stone. They gave back the pelt and couple handfuls of kumanis. The gaddis are all gone from here. This one crosses tomorrow. Made up 18 birds. We will have 300 skins and 75 species by tomorrow. Plants 1704 from Kulu.

August 25, 1933. Rain began after dark for an hour or two then became perfectly clear. About 4 am began again and turned to snow, covering the ground with slush in the morning. A flock of Phoenicurus stayed around the camp till the slush melted. It stopped at 9, but stayed half overcast and threatening toward the pass so we didn't move. Above there is no fuel and we had none to take along. If we got wet on the way it would be a pretty pickle. Rup Chand will find at shelter at Chorten Ranjpha if he gets to the appointed place today. The gaddis went on across today. They will be able to give him our news. Two Gya men came from the pass crest today with a horse which they recovered from the Lahuli whose horses the wolves had demolished at Gya. They got 16 Rupees and the horse back. The men went into a side nulla for wood and brought some tamarisk. It cleared over the pass in the late afternoon. Another gaddi with a flock of goats mostly white went toward the pass in the afternoon.

August 26, 1933. It cleared reasonably this morning and we loaded on our wet green tamarisk wood and started off. The road is a revelation after the snow-enshrouded silence over which we passed on our first trip. At the bottom of the valley above Kinlung is a detached mountain of schist fragment that came from no visible source. Tso Yumnan is mostly dry with a few Mong Plover and Sol. and Green sandpipers. A little above we were attracted by a yellow meadow before a diminishing snow patch and found it a garden of yellow composites, fragrant and so bright that the natives call *nyima gangshra* "thither the sun turns." A few blue aconites and purple asters enhanced the gorgeous beauty. Near the crest of the pass many more flowers appeared in the freshness of spring. We camped beside the crest on a stony meadow just free of snow and covered with tiny plants. A yellow Saxifrage and a white Arenaria make the meadow a garden. Many plants grow near and above. The pass is given 16,300 feet but there are plants up another 100 ft. We collected over 100 numbers; some we have never seen before. We sent a man ahead to find Rup Chand but no news was received by night. Travelers coming from below had not met him. Rain set in again after noon and sprinkled into the night. The weather is not cold; snow melts promptly; butterflies and bees flit about but we are all half frozen from the chill wind that comes up from below (south).

August 27, 1933. Rain continued all through the day. We had no hope of finding our friends. Toward 3 pm our man came from Zingzingbar where he had spent the night with Rup Chand and Wang Gyel. He brought two batches of letters. Mother and Shulz are well, Rhea got her necklaces; of our birds no word. Bartlett writes our Lahul collections have new species of grasses and mosses and undoubtedly other things. Mrs. P. is leaving the house September 15 and is terribly upset, poor dear. A little later Rup Chand and a lad of 16 with four drimos[17] arrived. He looks fine and says he has had 30 injections of the medicine Dr. B and I talked about. The Deputy Commissioner didn't come as planned nor did the Russians. Mr. G. promised to give back the cartridges borrowed from me last summer, but made no mention of the gun promised Rup Chand. He went to Spiti on the 29th with the missionary. Rup Chand bought only a nice old purpu[18] in Lahul. Has 37 species birds, 115 specimens. Galeen are numerous

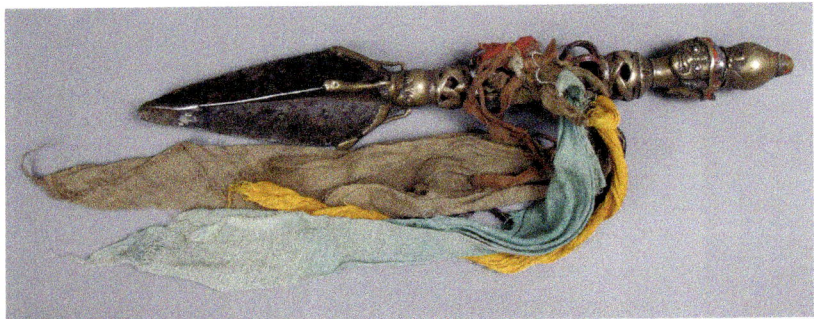

Plate 44. *Purba* or ritual peg or blade (UMMA 17264).

on the surrounding peaks. They chuckle and whistle when the sky clears a bit in the morning. I gathered some dandelions for supper. The natives don't know they are edible. Rain continued into the night and there is no prospect of leaving tomorrow. Paid the horsemen 40 rupees baksheesh. Half to be divided among the four drivers and half to be prorated among the nine horses. Those returning to Lahul will have 45 paraos and 13 days of rest at 14 annas and 7 annas. Ram Shom gave Rup Chand all the news in the time he was with him. I gave the lad a shirt last night and he said the Balti nono had given it to him.

August 28, 1933. Continued cloudy and squally as before growing colder and less rainy in the pm. Wang Gyaltsen had taken again with malaria. Four men went back to the Kinlung nulla for wood and bought each a huge load mostly of tamarisk ཤོམ་བུག་ [unknown] and some honeysuckle ཅན་ཅན་ [unknown]. The latter burns readily green. All looked like hay bundles but we had nothing left to cook

---
17. Female yaks.
18. *Purba*: ritual object; three-edged peg made from metal or wood.

with and there won't be wood at Dopo Gongma, our next stop. Rup Chand and I went over the tankas and packed the things that go to Kolung. The question of boxes is acute. We shall need them for bird-skins. Wang Gyel brought in four Kahalin and saw a flock of 15 galeen. In another place he saw six young, full grown. Rup Chand saw a pair of Kyak at Kyelang and two cuckoos (hepatic) at Jispa and at Dartse. Wang Gyel says he shot a crested sparrow in Zilling Nulla. A rose pastor he got at Kolung. A golden eagle stayed all summer between Kyelang and Kolung. The lad who is with Rup Chand lost his mother last summer in an unaccountable fashion—they were working in the fields cutting barley till dark. She told him to go home and prepare supper—she was going to look after the horse. They found her sickle where she had laid it but no other trace of her. She was habitually cheerful of extraordinary strong constitution. And there is nothing to account for her disappearance except that some devil carried her off. Rinchen Gyaltsen reports 19 stilts at Tso Yumnan. If it doesn't clear tomorrow the men are without food. The nearest place where anything is obtainable is Dortze 20 mi away. Wang Gyeltsen has food till Lobar but all the help will have to eat of it once we get under way. We have four loads of wood and nothing to cook. Our own food is at a minimum, flour and butter will be exhausted at Losar.

August 29, 1933. It didn't actually rain though it stayed cloudy, and we got ready to go. I paid the men for all horses to Kolung or to two paraos beyond Bara Latza La. It came to 45 rupees, 1 anna per horse. Twenty rupees I gave the four horsemen, 20 rupees to be divided among the nine horses. The total came to something over 430 rupees. Wangchuk was very ill with his malaria and heat and went down with Rinchen Gyaltsen and Sonam Naolis, taking his two horses and Rinchen Gyaltsen's one. The dried plants, all the ornaments, birds, art objects went on these horses. We loaded two of our churries with tamarisk and other odds and ends much to their disgust. They threw the loads off a couple times each, lying down, as did a couple of the remaining five horses. We finally got off and got to Drokpo Gongma about 4. There is a nasty stream from a big glacier that has to be crossed on the Lahul side of the pass and then another (the headwaters of the Chandra) that is worse. The cattle and the dog know how to cross streams. They head against the current. The foolish horses go down with the current. The road all the way to Drokpo Gongma is like the pass, bare stone, gravelly soil scatter with a yellow saxifrage, a white armaria? [in original] and a lavender aster. At Drokpo Gongma there are springs and wide pangs on both sides of the river. Many sheep were grazing on both banks. From one gaddi our men bought dried meat, of course of sheep that died of accident or poison grass during the summer, or worse. One boy came to camp and said the wolves had eaten three or four sheep out of his fold at night during the summer. There are no leopards or nabo[19] or galeen here. No one has passed here in two months but

---

19. Himalayan blue sheep (*Pseudois nayaur*) or *bharal*. *Nabo* is the Spitian name.

the SDO. They are leaving tomorrow and will go up to Kohsar. Hamta is nearer but the Kuluese below Hamta are quarrelsome. There are no villages along the Chandra until Koksar. The lad said there's no use sitting on the pass waiting for the weather. It being a pass there was bound to be rain. Wang Gyel went to Tso Yumnan but the memsahibs have all gone. Good plants on the pangs. Birds: larks and rosy rumped finches. Somewhat clear and warm in the afternoon but heavy clouds at night. Rinchen Gyaltsen is to start for Kulu on Sept 20.

August 30, 1933. We left for Drokpo Yokma (the lower little river). The first wasn't so difficult but one of the cattle wet its load. The road is (there isn't a road). You go over and between rocks and stone, up and down, in and out of nullas. Near the lower river there are pangs again. We camped this side of the river, it having swollen so that passing will not be safe till morning. A gaddi who lost two sheep came across this morning to look for them. He will cross on the glacier tonight. Our men invited him to tea and tobacco. Our dog got a phea. The poor phea made a hideously painful squealing, but got finished. It stayed pretty clear and warm all day and at night cleared completely for the first time in days. A gaddi came to camp in the evening and said one of his horses with load was lost. We passed a half mile back and it is strange where it could have gone, leaving another horse on the path. The lad washed his coat today and wanted to wear it wet. Last night they gave him a blanket and asked if it was warm. He said he couldn't tell till morning. In the morning he said he didn't recall whether it was warm or not. Ordinarily he sleeps with a little cotton sheet or nothing. The hunters at Bragnak slept with wet grass on their backs and said it was warm.

August 31, 1933. The gaddi's horse had left its companion and gone back to the old camp. Some of its load was not recovered. Morning dawned overcast but cleared soon and stayed with some clouds all day. Two ravens came in the morning, kept the dog busy chasing them away. She assumes they are after the phea. She barked all night at real and imaginary thieves. We left at 9:30 and got to Chandra Tal at 3:30. The way is as yesterday up and down, in and out. The Drokpo Yokma was to our knees. Four flocks of sheep crossed the stream with us. The Chandra has a wide bed on which three or four streamlets flow. It looks as though Ibidorhynchers might be found along it but none heard or seen. The valley walls all the way have been glacial drift furrowed copiously by torrents from the side. There is abundant pasture in many places, along streamlets or on pangs. The peaks are barren with patches of snow. Here there are numerous glaciers on the peaks. The Chandra Valley is perhaps a mile long of clear water and grassy shores. Bottom near shore gravelly, suddenly deep. A species of green black gamnarks occasionally. One Glottis shot. No other birds seen on [indecipherable]. Saw the large yellow Phyllocopus but couldn't collect it. There are three gaddi camps beside or near us. In one, a Rampuri lad made over to a gaddi. It is apparent from his features that he is not [indecipherable—asli?]. He gave our men milk night and morning.

September 1933

September 1, 1933. It was cloudy and sprinkled in early morning but cleared by the time we got ready to go. The bottom of the lake the right side of the river was white with clouds. We ascended the mountain chain and ran for four or five miles along the rest, diving into one nulla after another and crossing the flanks again. Rain began with the descent and a sleety drizzle kept up all the way to the pass. Sometimes it turned to snow. You couldn't see more than 500 feet ahead and with each crest we hoped for the pass but only to drop down into another nulla. The crests are all marked with buyes,[1] of course. By the time we got to Kunzam La I was soaked and ready to stop, but just beyond the pass it cleared and the spring wind dried me out. We came on down to the sumdo above Losar. The country beyond the pass is very green and swarming with gaddis. A camp of Tibetans told us Gill was at Losar today. At night a young Ki lama with a Juggatsuk man came to camp. The lama is carrying Gill's letters and will go to Phuti Rundi. Dr. B.S. hasn't shown up in Spiti yet. Tucci got nothing from Ki monastery, but many old tankas from the people. He also bought old silks. We camped at 3:30 a little above the sumdo on a broad green meadow. We got a lone BW teal, saw the solitary sandpiper and wounded a green sandpiper.

---

1. Stone cairns.

Plate 45. Member of Koelz's team on mountain top by stone cairn (date and location of photograph unknown) (Walter Koelz Collection, Bentley Historical Library, University of Michigan).

September 2, 1933. This morning a number of Spitians came along to repair the road. They asked if I were a "mikser," the local name for sahib. Lahulis also refer to sahibs as "yellow eyes" as well as "moksha" from the mushroom hats. The natives say it is cold here this year because so many sahibs have come. The lama carrying Gill's letters was summoned back by a night messenger and didn't return. He says he was told the Kulu man could carry the letters. We went plant hunting in the morning and on our return found Gill and the padre in camp. They had been sitting in the mud waiting for us for over an hour. They were most cordial and stayed long. Mr. Gill wants to write a paper with me on the Lahul and Spiti birds. We found the Losar people friendly to the point of being a nuisance.

They brought some old things of which we bought a copper gao. We hear on all sides that Tucci gave the most generous prices and bought many things. The three Spiti school masters came for a visit. They are on their way home for one and a half months leave. The one is Tantama Nath's brother whom I met the first year in Lahul. The women are gathering fuel from the hills. They came down with huge kiltis of the Saxifraga pennsq. The Cotoneaster and Honeysuckle are also gathered. A Spiti boy with black curly hair Rup Chand says some in Lahul dig porlok[2] roots to dye black. Boil cloth (wool) in a mineral from mountains (alum?) in which it becomes yellow. Then the porlok makes it a fast black. The purple musk-scented aconite is used to kill sheep lice in Rupshu and Lahul. Here they use sarson bil. Crops: barley three kinds, red flower pea, sarson. Women don't cut hair; some men all cut, some have back part of head unshaven and pig tails. One little girl like porcupine. Here and there, dirt clotted bunches of it and her mother pulled it out so that it is like a porcupine with a few broken quills.

September 3, 1933. We stayed all day. I collected birds of which there was an abundance among the fields and houses. The people brought many things but not much of interest. Their tankas are never exceptional. Rup Chand made a tour of the houses. The image 18" high of which we had heard is new and not exceptional. The people are extraordinarily friendly. A little girl saw me eating the husk barley and ran back and got me some huskless. The crops are very nice. One seed of barley gave 12 stalks. Children are free to eat peas in anyone's field. The children followed me around in droves, not at all shy and bright. The houses are all whitewashed. Much fuel collected. Women apparently do all the work. They are off before daylight for fuel in the hills or are weaving before the house at dawn, or pulling weeds (the fields are very clean). The peas are now being harvested. The cattle go to the hills at daybreak, a man or child taking them. There is a grove of melons around a building they say is a monastery. Fences of eleagnus, the small sort of some roses. There is more land that can be watered (water indefinite apparently) but they complain they don't raise enough to feed the village.

September 4, 1933. We left for Hanse in the morning. It froze well last night but dawned clear. Late in the afternoon a devilish cold wind came down the valley and made the temperature very low. We stopped at Kiomo on the way. There were some interesting tankas but all were 100 Rupees each. A very nice prayer wheel that we couldn't buy. We crossed the river with the district Munshi and a coolie he got from Kiomo. The Hanse people were friendly as above. Several Irish faces such as I have seen before among the west Tibetans. A broken phonograph record was much appreciated and sought for. Empty cartridges are a desideratum of all sexes and ages. One mill had a wheel with propeller blades partly of wood and partly of yak shoulder blades. Pigeons

---

2. *Geranium Pratense*; Porlo (Koelz 1979:30).

Plate 46. Woman gathering fuel (Walter Koelz Collection, Bentley Historical Library, University of Michigan).

(Tibetan only) common at all villages, i.e., a flock of 20 or so, so we eat well. A woman here said she was 73 and a man 79. The valley below Losar closes and the walls are steep and mostly bare. There are five or six scattered houses between Losar and Kiomo. The people have almost clean faces, much cleaner than Lahulis. No trees or bushes (willow) at Kiomo. One or two at Hanse. One man who came to camp at night said he had to hurry back as he would be suspected of thievery. At this season when the crops are out folks may not wander about after dark. The horsemen threatened to eat our lad because he couldn't pay his food bill and had no relations to pay for him. He was ready to cry. The new

pieces will probably be kept in the house monastery. One at Losar had galeen eggs. In Lahul they likewise keep eggs. The first of the crop is also entered.

September 5, 1933. There is a small slim black rock lizard in Spiti to here. We left Hanse and came to La Dorse. The nice prayer wheel came from Kiomo. They bargained at Hanse for a cow but wanted to give only nine rupees. One man and his wife came in early morning and tried to pick a quarrel. They said the horses had eaten their barley. There was a handful of heads picked off and some horse manure thrown on the ground there but no horse hoof marks. They said we had camped on their ground then. Finally they left and we heard from them no more. At Kioto an old lama brought out a nice wood seal and a book which we bought. The book is in characters we can't read so we don't know what it is. It cost only 6 annas though. Beyond Kioto the road descends to the river and then begins a long ascent to La Darse. There wasn't a bird along the river. The wide river bed now has only some anastamosing branches of clear water below Kioto. Beside the river some huge stacks with roof-like flat tops are clustered, giving the appearance of a village. There were five or six plants in the dry slope I have never seen before. At La Dorsa there is a broad amphitheater one-quarter mile in diameter flanked on all sides but the one toward above with high rolling hills. Parang La's snow peaked flank is to the below side. It is exactly a piece of Rupshu. Meadows of drama and abundant grass and springs. The full moon rose over the snow peak and we didn't use a lantern. In Lahul on the day of this full moon the people above Tinoo assemble at noon for a dance. Three men get inside a lion mask, etc. one in a monkey suit, one represents a gaddi, and some six others dance as men. They eat and drink and the next day grass cutting begins. The festivity takes place at Pratap Chand's house. Until this festival no grass may be cut except at Karing where the springs dry up and they are free to reap and sow at season. Other villages will be fined if they don't act according to custom. If this custom isn't kept the season will be cold. The lion is represented because once an old lady saw one in the glacier in front of Kolung. Sowing takes place in the spring at a time the lamas say is fit, not before. On the sowing days they pop barley and on this they put popped buckwheat. The buckwheat is so light that it flies and is blown onto the animals. The barley is eaten and a big bowl of ཆང་ [*chang*; barley beer] with butter dabs in it is served. In some places by noon they are so drunk that sowing stops. The seed is thrown with recitation.

September 6, 1933. We started off for Kibor, but a mile on the way we saw Wang Gyel's galeen place where he said there were countless galeen and stopped. There were three kestrels feeding on a black and white horntailed crichel. We got two of two species. There were four or five of the brown sand larks. A family of nomads came to our old camp as we left it. Day clear and warm till the afternoon when a wind came from above. Rup Chand and Wang Gyel went after galeen but couldn't get near them. They were too scared from the morning. Saw a big eagle, probably golden, hunting over the mountain, probably for galeen.

September 7, 1933. Descended to a village of two houses in the side of the rock called Dumle. Chased five or six chikor. Then continued descent to the stream that comes down from Parang La. The stream has cut a very deep gorge and probably got help from an earthquake: 300 feet deep and very narrow. After scrambling out of this nulla we arrive at the Kibor fields and shortly the village. The village is picturesquely located, but the village across the stream (Chikim) is white-washed and clustered like a Lahuli village and more attractive. The crops haven't been frozen and they are cutting the half ripe barley. The fields are full of women and children also men. They are cutting what little grass there is on the field borders. In the winter they say they can drive the animals to the hills. When passing by three groups of people in their fields we found a blanket folded with gravels and flowers in the path [drawing inserted]. This was for a tip. Rup Chand says when passing from Dilbung to Gundha pilgrims find children's blankets in the path and leave their food on it. The children are up there watching the sheep. Dorje went after galeen promising a number and got nothing but a hare. Wang Gyel saw a rock creeper near Kibor. A boy says there are several geese with young on some lakes near Kibor. The brown sand lark is here also. Saw it first below Losar. Very erratic and haven't gotten a shot yet. Saw two harriers here. Got one. Weather got cloudy and disagreeable after noon.

September 8, 1933. Today no one was in the fields. There is a rule that everyone must do the same work on a certain day, or nothing. The fine for disobedience is a sheep. They also appoint days of rest. All the way to here, the children run naked where the temperature isn't exactly tropical. Wang Gyel went back and got the rock creeper today. Rup Chand got two sirkaps and saw a gallinago. We shot a queer [indecipherable] lark in the field; this probably was the thing I took for the Hanle Lark at Tso Kar. I went up to the three or four pools formed above the small village beyond Kibor and got a Tringa glareola and two solitary sandpipers. Also got a queer piper with toes smaller than the common field piper. There were three high up on a pang. It turned cloudy toward late afternoon and stayed so all night. The people here don't put the dūng on their head, [indecipherable] too eagerly the image. At Losar almost every thing including a piece of copper they rubbed to their forehead. The women strip to the waist to work in the fields and are not shy. Chortens and manes are few in this country. There is a large one with a cluster of six to eight small ones a little above camp and here and there feeble ones on hillsides. The cattle are mostly cows here. Donkeys are common. Not so above here. None seen before. There are no pigeons in the fields here, though a few Tibetan Pigeons come to sit on the rocks.

September 9, 1933. A beautiful sunset tonight. The sky to the west was banked with cumulus clouds tinted strong tones of purple to pink. The valley beside the village was filled with a purple pink haze and the sky overhead was overcast with blue purple clouds. I went hunting in the fields early. Saw what I took to be a female and two young of Propasser nepalensis but missed at close range. The brown sand larks eluded me all day though I saw one to seven five or six times. Shot another rock creeper beside the village. The cattle couldn't be found this morning till 11. Two had strayed over to the little lakes. Disposed of three to the Ki Gompa for an old wood teapot and a four year old horse. The big [indecipherable—zho/gho?][3] is so weak that I fear he would die if we took him farther. We now have a sterile drimo left. Bought a nice wood press from Ki and a very old wood plate of nice grain. Rup Chand says wood plates haven't been in use in Lahul for many years and this is probably over 100 years old. A man went after a three year old horse to trade for our cows and suffocated it on the way down with the rope around its nose. Rup Chand says a strangled animal must be given full lungs of air at once or it is fatal. A golden eagle flew over the corpse site, clearly identified.

September 10, 1933. Left Kibor and descended to the river and crossed to Rangrig. We didn't go to Ki Gompa. The chagzot is a stupid inflated person who affects Tibetan clothes and wears glasses. Tucci gave him such a price for his things that it is hopeless to look at anything. The old Lhasa teapot and the carved wood are enough. Someone put a new stone on the Kibor mane engraved with a '+' and these words

དཀོན་མཆོག་ནི་བྱམས་པ་ཉིད་ལགས།

[*dkon mchog ni byams pa nyid lags*; Loving kindness is the greatest rarity]. Rup Chand suspected asboe. Finally got one of the brown larks. There was a flock of 50 or more but five settled and I got one. Saw a white rumped Harrier over the fields at Kibor. At Ki the people invited us to stay. They didn't recognize Rup Chand and spoke Urdu to us and talked about us among themselves. They have wheat at Ki. The descent to Ki is sharp. Then one descends 100 feet more to the river plain. The river is in several branches, none deep. We camped on a broad pang with several strong spring streams, below Rangrig. I went up to Rangrig to look for pigeons but there weren't any. All the crops are cut. There are a couple dozen poplar trees a foot in circumference. The village is perhaps slightly smaller than Kibor. On this side above here are three more villages. We got a small rail and a Gallinula, a Gallinago, a brush warbler, a couple small snipes, a glottis [?, not clear], a black and white wheatear and saw several bluethroats. The children say that the little rails are very secretive, so probably they are common. The Gallinula had a sitting place scooped out of the mud and had crept into it when I shot him. A companion had been eaten within five feet of the place. The valley is broad, three-quarter mile, and has a broad strip of pang with streams—ideal for Ibidorhynchus but I couldn't find traces of it. It was very warm and pleasant for the first time in Spiti. The wheat here is exhausted and our men have to buy barley flour which is mixed with pea-meal. The dog won't eat it, nor the horses. There is a large Eleagnus[4] tree at Ki, a foot in diameter.

---

3. Phonetically *dzo*, or *mdzo* in Tibetan transcription; *zho* is closer to the actual pronunciation; a dzo is a yak-cow hybrid (Joe Leach, personal communications).
4. Silverberry or oleaster.

The Ki Gompa is situated above the town on a hump on a steep slope and is not easily photographed. It is not as large as the main Zankskar and Ladakh gompas.

September 11, 1933. The drimo came and slept in front of the tent. The horses all fled over to Ki with the new horse. She ate grass around our heads and then went off to graze before dawn. I heard an Ibidorhyncha in the night and found and got it in the morning. Also got a bluethroat. Saw three or four last night but only two today. Saw a large duck, like a pintail. Today the flocks pastured at our end of the meadow. They are tended by eight or ten children, all as usual with baskets on their back for collecting the manure. Great is the rivalry to get the precious substance. When it comes to extracting it from the little ponds a more elaborate technique is required. We bought a very nice man's earring for 10 rupees. The owner didn't want to part with it but consulted the crowd and they advised he take the four rupee profit. The children here are not importunate as at Kibor where one and all shouted "salaam zhu bakshi nang" whenever I came near. Two little manure collectors found a pile near our tent but feared to collect it on account of the dog. Finally Rup Chand persuaded

Plate 47. Sunkyil, the expedition's dog, near tent (Walter Koelz Collection, Bentley Historical Library, University of Michigan).

them the dog wouldn't come. But no soon had they got their treasure in the folds of their kaftans (they left their baskets behind so as to be able to run the faster in case of danger) than the dog came. They hung on to the spoils till the dog caught up with them (100 yards) and then dropped it in terror. Rup Chand then gave money and explained his guarantee hadn't been kept. "But you didn't tell the dog to chase us. He came from behind of himself." They put the grain in the mill stones by hand here. We asked why they didn't fasten a hopper over the grinding but they said it wasn't the custom in former times. We parted with our drimo for a nice little donkey. They mutually pulled hair out of their animal before parting with it. Saw the W.R. Harrier twice. The day was beautiful and warm till 2. Then a squall came from below and also in above the valley and it sprinkled into the night. The birds all hid in the Eleaganus and where last night they were very numerous I found them with difficulty. The black and white wagtail went down the valley in flocks. Rup Chand says the ibex eat till dark when a storm is coming. The Sisu lama's nephew came to camp today. He left some lunch in a cloth, some sort of mess made of the remains of the beer malt. The dog affected a great unconcern with the garment but liked the location and was finally caught eating the stuff slyly. The man has just come back from Ladakh and Rupshu. He says the Rupshu lama has finished one lakh of his prayers or readings of his Bumstok. Rabazang called snipes "shingle." Tibetans sew by pulling the needle toward them; Indians and Europeans stitch away.

September 12, 1933. We left for Kahze after seeing Gombo off with the donkey. He is to go to Sisu with the Sisu lama's nephew. They will reach Sisu in four days from Chikim. He had to be supplied with shoes and a blanket for the journey and ten days' food. Our nice little donkey is well gone or we should certainly not have bothered with it after it had traveled with us for a while. The men didn't like to stay at Kahze because grass would have to be bought so we came on to Shelgo, a spring watered thorn strewn pang, some half mile down. Kahze has nice level fields and the crops are almost cut. They are plowing for spring. Saw the W.R. Harrier again and our men saw another. At Kahze there were many brown sand larks, in one flock saw several hundred. I got 27 at one shot of no. 8: 24 female, 3 male. The blue pigeon is at Kahze and I thought I saw a sky lark. One Ibidorhyncha near the bridge. The queer grey bird was seen again at the bridge. Have seen it every day since coming to Rangrik. The swallows attacked it on two days. The first day saw two. It is probably the same bird we got at Kibor. Saw the block rock thrush above K. Saw a Gallinago at Shelpo. We passed the nono's village on the other side of the river, Kuling. It is a small village but his house is rather larger than the others we have seen.

September 13, 1933. We stayed in camp making up skins: 36 today. Got a ruff. The Chikor are abundant. Saw a flock of chicks not half grown, also two hares. The little warblers (Phylloscopus, five species) are all common. Dr. Bhagwan Singh and Tsering Tache came from the village a half mile down.

They have come from Pin and are on their way to Kuling and then up the river. They buy wheat and grind it themselves. Our men went to the nono's for wheat and flour but he had none to spare. They couldn't buy a sheep except for double price (8 rupees). They got some barley sattu that he said was specially made for him. The men say it is better than the other sattu: that has sand, straw, beards. Dorje went up the river to look for the queer grey bird, the harrier and a couple others. He brought back six skylarks, a batu, and four blue pigeons. We have 60 species of skins from Spiti now. Three Tibetans came today from Purang with the wooden cups out of which the silver-lined things are made. One man with an animal went by in the dark, muttering our name. He yelled for us to look out for the dog and the men told him to stop his shunker which he did for a few minutes.

September 14, 1933. Hoopoes, Wang Gyel decides, are not fit to eat. Dombu, Domthru, Sengthruk—dog's names. Forged ahead along river, green and [indeciperhable] rust. [Indecipherable], the latter commonest, people say common. Dorje went hunting again and got nil. We came on up to Larh where the doctor is camped. He gave me a book of Vivekananda's lectures. This is his fifth. Saw an Ibidorhyncha on the river below Lora. One family was threshing with three donkeys or three cattle tied to a revolving stick. They tramped on the grain, a man singing accompaniment. Afterwards, a man and three women tossed the chaff into the air, the man whistling a few bars of a ditty and repeating ad infinitum. Most of the crops here are still in the field. We bought 16 pounds of fine wheat and will grind it ourselves. There is no flour to be had. Doctor says people suffer mostly from stomach disorders and ear trouble. The children have whooping cough and measles sometimes. Has seen no cancer or tuberculosis. Got a hare.

September 15, 1933. Bought a nice teapot that is the mate to the one we have. The cat caught a weasel, and the doctor saved it for me. A little boy was cutting its head off. The weasel heads are of use in certain lama ceremonials. The nombardar is called *gatpo* "the old man" regardless of his age in Spiti villages. The grain is being hauled home on donkeys and on men. The women are cutting. We departed from the Dr. and came on to Lidang. Two Tibetans camped there too with a flock of 30–40 sheep that they have come to trade for grain. The doctor's man went up and got us three galeens and a hare. They are feeding on the dry hillside, eating Ephedra berries, the wooly "burtzi," the jye-like grass. Mercifully they weren't fat. Saw the WK Harrier again. The people call me *acho* (achi, fem.) which is fit for anyone older than you, a stranger, or a brother. An old man is *mimi*, an old woman *abi*, regardless of who they may be, your grandmother. Ram Shom has taken to eating hares. The women often wear their hair in a dozen small braids down their back reinforced with wool, so that it reaches below the waist. Each braid is tipped with a blue porcelain bead and a brass bell. I saw a man fondle and snookle a baby just as one of our old ladies might. House sparrows here and in the last two villages very abundant. There are some six to eight old willow trees here and at Lara—a foot or more in diameter. On the upper slope are clumps of birch shrubs now gold. The blackflies are in swarms along the streams here.

September 16, 1933. We stayed at Lidang while Dorje went over to Giungul to see about a teapot we heard was in the possession of the Chota Nono, a relative of the nono. In the morning, we gave a concert and such an appreciative audience we hadn't had in Spiti. They dropped their work at the excited calls to come of the listeners and all attended. The donkeys carrying the grain were left to themselves. The Tibetans selling sheep came too. They left at evening for the next village. We got a sheep for 4 rupees, 4 annas and gave back the pelt. They asked ten last night and didn't give in till they were ready to go. They sold one other here. Our men killed ours by cutting out its heart. "O dear I am afraid" said the Tibetan and left. "We strangle ours." I gave him two Chiclets: he ate one and said he was going to take the other to his companion. Dorje said the people of his village had all gone to cut grass. There is much dry grass above. A man would bring the two teapots in the morning. Rup Chand asked what sort of wood was giving off such a pleasant fragrance. One said it was burtze but Ram Shon said he had put a bone in the fire.

September 17, 1933. The man bought two old Spiti-made teapots that were crudely worked. Two lamas (Tibetan) that have been here two years came and [indecipherable] so this morning. One says he's the companion pupil of the Muling lama—the one from whom Rup Chand bought the purpa. The men have been able to get wheat flour here but they are grinding the weeds in with the grain and the bread looks as if it were made in ashes. In Kulu there is a weed that grows with the grain called *ubur*. The seed is round and grain colored and when ground with the grain produces a sort of intoxication. When one buys wood in Spiti a little is given from each house and the money prorated. A little girl three and a half feet high did the collecting. She is a bastard. Her mother is an attractive woman. Went along river to a camp a couple furlongs below the new bridge. Saw a small tern in the river and chased him up and down but he always flew so he would fall in the torrent. No crossing. Finally flew up the river and settled on my side. Shot. Flew across and dropped dead just on the edge of the swift water. Found a ford above and got him. Arrived at camp, a gale came from up the river, promptly reversed and blew down the tent. Sprinkled for a couple of hours, the wind coming first this way then every way. Natives say bridge (atirgu) is no use to public, except for the few people in the Pin Valley. A man brought a yak and horse cover such as we had bought at Ranguk. It had the same red, green, blue, yellow, that kept their color and which I took for vegetable dyes. He said it couldn't be over 30 years old, as he remembered when his father brought it. It had some stitching of lavender thread which had lost color. A pair of gugti flew out of a thicket at Lidang and we got one. The W.R. Harrier came and sat within 25 feet of us at camp but Dorje had the big gun above.

September 18, 1933. The night was clear but the morning was overcast and in the late afternoon a drizzle set on in that continued into the night. We crossed the river by camp and came up into the Pin nulla to a village called Guling. The path now is along the flood plain but in the summer there is much water and a path fit for ibex is cut into the slope. The valley is steep walled and narrow and for ca. three miles there is no cultivation. Then two or three villages appear on the opposite side and three houses on this. A little farther on a cave are three or four little buildings, Pin Gompa, and then Guling. There are some trees here, one large poplar over two feet in diameter. The houses are lofty, clustered and some have the Kuluese cut doors and windows. The crops are not all cut. There is much bakhyot on the way and probably the accompanying mushrooms. The gatpo didn't show up and one very poor appearing man gave us a little wood. We gave him a rupee for it and he promptly came back with a load of nice cedar roots. Cedar trees have been in evidence since our yesterday's camp, sometimes a small tree 20–25 feet high, but more often a bushy thing with great underground development. I saw a number of such, mostly stumps at a house beyond Losar and it may be the trees grow there too. After a certain day in this village, the responsibility of watching crops is with the owner. We don't have to watch the horses now.

September 19, 1933. The rain continued nonstop till 10:30 when it stopped for an hour and continued again into the night. In the clear weather I got two small creepers and a wryneck. Rup Chand got a W.R. Harrier and wounded another. Five or six hawks came out of five species. Wang Gyel got a wryneck too. We bought our first tanka, once fairly nice and remediable. We ground our wheat in the local mill. Turned out well and makes good bread. Got a coney. A weasel yesterday. There is a group of four to five houses above this and another beyond. Mikim (two houses) is visible about two miles above. Khar and Sangnam, the largest villages in this valley are in the nulla opposite it. Muth above is about 50 houses they say. From there one crosses into Rampur and comes out at Wangtu. Ranguk is said to be the largest village in Spiti. The children here are zealous manure gatherers. A flock with baskets follows the grazing horses. If one raises his tail some alert youngster spys him, points and yells "mine, mine."

September 20, 1933. It cleared in the night and we set out for Khar. We went the path along the river and the horses went up over the hill. At the sumdo Rup Chand got two Ibidorhynchus at one shot. Got a little dove from a flock of two. The streams from both nullas are strong and we had to choose a crossing. The stones are covered with green algae and fearfully slippery. The large village of Sangnam is a picturesque cluster of high houses and above it rise the terraced mountains such as we first saw in the sumdo above Losar. The strata are horizontal and the sides furrowed by erosion. The strata clusters vary in tone to the peak. Up the nulla above Khar the peaks are snow covered to half way down. The view is delightful and we were all glad we came. I shot a fox at Khar. The people say they are common. The faces are unattractive here as everywhere in the Pin. The people are friendly enough. We got two dried *moksha* (they call them *shamo* here, and with the Ibidorhynchus and eight pigeons I shot at Khar we had a fine supper. There are three more small villages visible beyond Khar before the nulla bends to go to Muth, the last village. There are four willows (large) at Khar. A wind blows down from above every day after noon, they say. It was not unpleasant today.

September 21, 1933. We got another fox today. We took two horses to the last village Shidang and then came on to a plateau between the nulla mouth and Mani. An old lady who lives at the mouth wanted us to camp with her. Her husband had previously invited me. Our camp has a magnificent view to a row of snow peaks behind Mani. The terraced fields of Dankhar and the cluster of Dankhar houses out between some pinnacled crags in front and up the river. Behind the rise, birch, juniper, and cedar stub scattered slopes to peaks with perpendicular strata exposed at the crest. There is no water. Grass enough. The cedar is more abundant here than anywhere we have seen but all shrubs that have sprung from the stumps. There were once good forest growths of cedar. There are plenty of ibex behind us and nabo in front. We heard galeen at dusk. Got a snow pigeon, the only one since Chandra. Some of the ephedra plants are totally covered with coral sweet edible berries. The [indecipherable] Chongpas are feeding on them. The rose bushes are laden with their lovely crimson red fruit. We gathered a lot of a purple flower with spearmint flavor. The Lahulis used it with pepper as a condiment and it is medicinal also. The bird trunk must have been under the tent drip. Almost all the birds were soaked and had to be taken out, dried and rewrapped. Khar is the first Spiti village we camped at and bought nil.

September 22, 1933. The ephedra here has an abundance of yellowish juice in its leaves. Rubbed on the teeth it promptly removes tartar. Round balls of manure burned to a glow and placed on a flat stone wet with saliva and struck with another likewise wet gives a noise like a gun report. Lahuli boys play now with such weapons. To stimulate saliva they eat the Eleagnus berries. The red wood of the cedar, rubbed two sticks together makes smoke. There is a cedar at Mani nearly five feet in diameter. They say 50 years ago there were many cedars in this country. Today some five little boys ten years old or less went by on horseback, bareback at a gallop driving a horse herd ahead. Five little girls came on foot behind with some dozen donkeys. The children usually know the year of their birth. The men went hunting behind camp but got only two rosefinches. Nabo signs abundant. We went on then to Mani. The road descends at once to the river, then ascends again. Mani is in a sort of ampitheatre on the old glacial plain, but by a stream from the mountain wall. A low ridge circles the other side. There are houses on its slope. On the opposite side is a long grove of poplars and along the gulch border some large Eleagnus trees 18" in diameter. There are springs below them which form the water supply of the town. It is carried on backs in closed wooden buckets. There is much dry wood on these trees, and on the ground but it belongs to the lu and must not be touched. In Lahul at Christmas everyone

Plate 48. Team on mountain top; Rup Chand in foreground (date and location of photograph unknown) (Walter Koelz Collection, Bentley Historical Library, University of Michigan).

makes small dough figures of domestic animals and elephants and fries them in fat. These are about six inches high and are kept a week in the window of the sitting room. Jewelry and tankas etc. are also hung there. One man near Kolung had an excrescence come out of his figures one year and the matter spread far but wasn't satisfactorily explained. Rup Chand says it was due to uncooked dough fermenting inside. The Manians eat pigeons and other fowl flesh. The military officer who came to Lahul this year gave seven rupees tip to the resthouse chaukidar for a seven day stay. The chaukidar was glad but considered the dohar a fool.

September 23, 1933. Saw a wryneck, four crows, and a "queer bird" like Tabo, probably. The people here keep many of their yaks and sheep in Chummurti, with the natives there. They then get butter and wool from there. There is no grass here for them in winter. More Irish faces here. The children say you can find skulls among the rocks here. Either now or formerly they hid their dead thus. One woman here furnished supplying wheat flower from washed weight. The grinding is by hand as everywhere. A lama here just back from Lhasa says there is a rumor that the Nashi Lama has returned. The children here are close attendants, struggling largely for the empty cartridges. Their families have difficulty in getting them to do work. One who had been sitting long in front of the tent told another who was going to the village "if my people ask where I am, tell them I have gone to look for the horses. See, I am ready to go now." A woman here says she is a relative of Rup Chand's but she couldn't be sure which of us was Rup Chand. Unmarried women in Spiti (from childhood) wear a single turquoise in the part of the hair above the forehead. There seems to be an excess of females. In Lahul, Rup Chand says the excess is worse. A Ladakhi, son of the She lompo who has married here invited Rup Chand to supper. Rup Chand went to his house and had a visit with one of the Hemis chagzots. He says the Rimpoche may come back in the spring. Two men have gone to invite him back. He told Rup Chand the tale of how he got Padre Joseph's property away from him and about the Marsho lama. The lamas of the monastery took the money and ordered new tankas from Lhasa. Our landlady gave us dried sarson and it was very good. At dark a group of men came to our tent and said we had to give four annas a horse for tying our horses last night in this woman's field. I sent the horseman to acquaint Rup Chand of the singular demand and he came back and argued with them. The gatpo said we'd probably have to give it. If we did, we'd be plundered in every town in Spiti. At last I became exasperated and dragged the ringleader to the light and got his name. Kundra. He had previously said that if we tied the horses in the field again he should stone us. Yes and get shot said another. Whereupon he bravely remarked that if we fired, our tents would be demolished. They now dropped their demand for the rental of the fields that didn't belong to them and said our horses had eaten their crops—a complete falsehood. They had been tied all night and in the daytime the fields were full of laborers. I saw a chance now to get the better of them. I summoned the gatpo and told him I had been threatened with a stoning. Secondly, that his people had come at night and attacked me. Whether they had a grievance against the horseman or not had nothing to do with the case. They were guilty of serious offences that I intended to report in Kulu. He begged Rup Chand to intercede. They invited Rup Chand to dine, etc. And there was peace.

September 24, 1933. The gatpo left early on horseback and wouldn't be back for a couple of days. The ringleader and all his associates stayed out of their fields and there was no sign of any hostility. The Gatpo said he hoped we would hold nothing against them as they had nothing against us, etc. He refused baksheesh. His wife sent our laundry and then came herself. We left without more ado and came to the rope bridges and then along the big brushy pang Shichiling below Dankhar to Dankhar. Got a fly catcher, two of the dark Kansu finches, two gallinulas, a shrike, an althea, a Zanskar Y.B. sparrow, and saw an accipiter. It grew cloudy in the morning and in the afternoon it sprinkled and kept on till dark. The Mane Ladakhi came to camp and asked about our quarrel. He said we needn't have crossed on the rope bridge. In front of Dankhar that water isn't to the horses stirrups. The natives come to Shelgo in summer to settle difficulties. They also come to Dankhar. There are two fields of *Tsitsi* [*Chini* (Kulu)] at Dankhar. It is sown below Mane, also. Our landlady at Mane is considered a witch because she is a good farmer and manager. They brought her to trial before a lama once but he dismissed the charge. She and her husband came to camp in the night highly pleased at the treatment Kundra had got. Mane is the lumber center of Spiti. Nowhere else are there so many trees. They sell the timbers for two to three rupees each. Rup Chand says house sparrows do not stay in Lahul in winter. Gave some eight year old Manian some gum. Our men call it *puk-pul* which is also their name for the big goatsucker of Kangra. The children called it "*getshoarshovra*" "laughing stuff." They call "funny" anything interestingly queer by that name, even people. If a man should persist in his opinions after a clear exposition, for example.

September 25, 1933. All people in summer who die are put up for the vultures. They are taken to a definite spot. In winter they are usually burned, but may be decapitated and thrown in the river or buried according as the lamas dictate. Saw a bater, gugtis common. No rain but sky overcast all day and cold. The town of Dankhar (Drankhar their pronunciation) is entirely located on the summit of a conglomerate remnant that rises almost sheer 2000 feet or more above the river. It is steep on all sides and its summit is cut with stacks and crags. The houses are pasted on the peaks like an eagle's nest. A path leads up straight from the river among the stacky pinnacles at the top and through a gate under a house to the summit. Another path comes up from a stream in a gulch on the east side. There are some fields to the east and one 600 feet from the houses and more large fields are 2000 feet below in a thicket of Eleagnus bushes. There was formerly a closed staircase descending 300 feet from the summit to the little stream just mentioned, not now it is chiefly in ruins. The crops are now being cut, barley, wheat and tsitsi. W.R. and W.H. harriers here daily. The nombardar (gatpo) gets all the price of the sattu we buy from the town. The lake has no birds now. Owls are commonly heard nowadays, say the natives. Pigeons, mostly Tibetan, are seen in flocks of 100 or so.

September 26, 1933. Saw two wall creepers at Dankhar. The rose finches are common halfway between Dankhar and Po. Saw an accipiter and a large blackish hawk mottled lighter at Po. Sky overcast all day with a light sprinkle for two or three hours. Heavier clouds up the river. At Dankhar is the court of Spiti. Over the door is written in English "Divisional Provincial Durbar," "welcome." In Urdu, "Khushandi Div. Prov. Dur. Dar-ul-Khlafan Spiti." In Tibetan something else. Inside they said was an ugly dog. The welcome to the court was probably inspired by the thirst for fines that might result. There is plenty of dirt and manure inside a small monastery in which the only worshipping lama is allowed. A special room is sealed with manure and the little hand seals we have bought everywhere. There is no jail in Spiti. If it were, it would be full if food were free. It couldn't be more unattractive in jail than in their houses and food couldn't be plainer than they are accustomed to. Our men tell of a crafty old man who bought an old donkey for a young one. The vendor had filed its teeth to suitable shape and had sewn a tail in place of one it had lost in battle. He discovered his fraud when the children pulled off the tail. *Jabtaik tali men bahat* (boiled rice), *tabtak teva memsath*. Lahulis call gooseflesh "chikor flesh." In Lahul the wild roses occasionally bloom in November. These are considered the devil's flowers. In Gundla they hang old shoes in the blooming apple trees. The gods see the old shoes and don't consider them fit for anything. Otherwise they would come for the blossoms and there would be no fruit set.

September 27, 1933. An old Lahuli who has lived here many years says the brown dipper and blue whistling thrush are here but he hasn't seen Bonaparte's thrush, winter wren, or oriole. A pair of Golden Eagles lives here and has eaten more than a dozen lambs this year. The eagles stay all winter here and in Lahul. There is one-eighth mile of thicket along the hill that bounds the Po fields. In this thicket are many cedars, mostly dying and Eleagnus. The whole belongs to the gods and not even the dry branches may be burned. Travelers have, however, made some inroads on the dry store. There are some 20–30 tall poplars and two willows near the town. The town has 30 houses, they say, and there is relatively good land. Buckwheat, chini peas, wheat, and barley are the crops. The houses have the little notched windows [sketch] and one at least has a lot of carving around the opening. The notching has been evidence since Rangrik or Lidang. The Lahulis say nabo are certainly always to be met with. Gill told the nono he was not to overstep his bounds as magistrate. He is fit to decide only ordinary quarrels. In a lawsuit each side deposits 30 rupees to start with. Then like as not both are fined, one more, one less. Beer, sheep, etc. are consigned to the litigants' expense. Sky partly overcast but mostly warm and pleasant. Went across to the fields (two to three houses) just across the river but got nothing new.

September 28, 1933. We left Po at noon. Rup Chand went to look at tankas and Wang Gyel went looking for dippers. He saw a W.C. redstart only. Rup Chand's tankas were good but not Class A and price rupees 70, so they stayed where they were. Gave the gatpo a little mirror and comb as at Lidang and he was tremendously pleased. A mile or so below we had to unload the horse to go along a platform fastened on the cliff where the river crowds the shore. We met through a large flock of salt-laden sheep coming from Tibet. There is a five-house town on the opposite side of the river a little below the balcony construction and then two houses on this side and then Tabo. At the edge of the Tabo we found a stone cutter at work. He had been in Spiti seven years make stone for mane walls. He has an order to cut 10,000 *Om mane p h s*[5] for a Tabo man. He lives

---

5. *Om mane padme hum*: Sanskrit mantra particularly associated with Avalokiteshvara, the bodhisattva of compassion; it is commonly carved into rocks or written on papers that are inserted into prayer wheels.

in a tent with a woman and a small child. He has a much swollen neck gland and the child boils which he says have been infected by the Lu. There were three camps of salt bearing Tibetans on the Tabo fields. One strapping boy had a silver disc on a string of beads fastened to his hair. He wouldn't sell it. Such an act he hadn't yet been guilty of, though poor. When dead, the string must be perforce removed, but until then no. The Tabo village is on the hill border. The monastery is a group of wretched hovels on the good fields. It is not made by human hands, they say. The gods have much to learn from the Spitians in the way of building. There are several ponds among the fields filled with long weeds in which water is probably stored. There are five or six apricot trees here. There were a couple opposite Po. The fruit is said to be excellent. The Lahuli got some kumani seeds from us to plant. He planted Kut last year and it is thriving. There are at least three nice litzi[6] trees at Po. The fruit is crimson, small and persistent. Tree two feet diameter, much and irreg. branches, attractive. Flowers said to be white. There are a dozen or mud chortens, 10 feet high or so round the Tabo monastery. No such accumulation has been seen in all Spiti. The monastery was padlocked so we couldn't see inside.

September 29, 1933. Went out to Lari, ca. three miles and camped in a field. There is a cluster of ca. five houses and two isolated ones that constitute the last Spiti village. Fields are extensive relatively. Pigeons, rock predominating, are feeding in the fields in a flock of 150. All crops out of sight here. Our men say if a seven month or so pregnant mare is fed chini or wheat straw in November or December, she will drop her foal that night. The Poo Zehldar arrived and sent for Dorje. How he knew who we were is not known. There is one apricot tree here and some dozen poplar and willows. Got a little rail in the streamlet that waters the field. Day clear and warm. Said to be many nabo in this area. Saw c. 100 pairs of horns on a mane at Tabo.

September 30, 1933. Left Lari and came on to Huling just across the Spiti border in Tibet. The road most of the day was on a path pressed into the almost perpendicular slope of decayed shale 200 to 300 feet above the river. The day was warm and bright. The Tibetan cup sellers who camped with us stripped to the waist. Saw two nabo opposite camp, very near. Grass non-existent. Tsepat[7] three feet high and abundant but no fruit. We passed the town of Sumdra in Rampur on the opposite river bank. It has some dozen apricot trees and a few poplars. Wounded a brown dipper in the river that marks the Spiti-Tibet border. Stone Martins here. Several droves of Rampuris with sheep, goat, donkeys and horses laden with chulis going to Spiti. One had 13 walnuts and some of the little bitter green apples. For 16 of the latter, he got a wood cup from the Tibetans. Our Tibetan tea finished today. Two bricks lasted a few days over a month.

---

6. *Malus baccata*: Siberian crab apple, eaten raw (Koelz 1979:25).

7. *Ephedra gerardiana*: branches are used as toothbrush, ash from plant mixed with tobacco to make snuff, also medicinal (Koelz 1979:5).

October 1933

October 1, 1933. Air early morning filled with small aphids. On the mane hill behind the tent is a stone inscribed with the information that two men gave 30 Rupees for its construction in order that they may be reborn in the Eastern Paradise. All who would have this happen to them should do likewise. Rup Chand thought the man might better have given a few rupees and removed the obstruction in the road a few miles back. The horses came through an arch that doesn't allow the load to pass. The hills opposite have put on fall colors. Some small plant has turned crimson and large patches and [indecipherable] of that color adorn the upper slope. There are a few birch groves of gold, purple, and green and black green clumps of juniper. The roses that in summer must have been a dream are great with red berries and the leaves are also turning red. Roses have been common till below Lari, apparently the same two sorts that grow in Zankskar. There was yesterday a nice cut showing glacial stratification of fine sediments. The sandy silt was almost a foot wide, three feet from the top and the clay layer two inches. At the top of the stratification the sand was less than an inch wide. The road today was more stable but up and down in and out of huge nullas. The road was in Tibetan territory until we crossed the stream that came down from Tibet to form the Spiti River. There were three hair-raising bridges to cross today. The boundary river smells strongly of sulfur. They say there are hot springs at the camp place Sangzam. A group of Tibetans on the way to Rawalsa are resting there. The tall coragana is above. Today it is used for baskets in Rampur. There was also some cedar and ash, the latter up to ten feet. Cedars are very abundant on the Rampur slope. We met more Rampuris. Got a basket of green grapes of fair flavor from one. They also had loads of chulis for the Tibetans and sacks of a species of juniper that is used for incense (gyeshuk). Saw a Phyllora Tytler, two blue rock thrushes, a white wagtail, and shot two tsut? sparrows in Tibet. We camped about four miles above the river beside a deep gorge. Place called Chala. The cup sellers had got there ahead of us. Yesterday and today are the two most uninteresting camps I remember ever to have made. View of dry earthen mountains that shut in the view on all sides and only ephedra and a few roses and artemesia for vegetation. The ephedra here is also luxuriant—three feet high often, but almost no fruit. One Tibetan sheep that I examined today carrying 30 to 40 pounds of salt and carried it easily. One doesn't sleep with his feet toward an image or tanka. The Lahulis don't sleep with their feet toward Tar.

October 2, 1933. We climbed out of the gorge and then saw a still higher pass ahead: Chanko La. A golden eagle was on a crag within gunshot but I didn't have a gun. From the last pass the descent is straight down for 1000 feet. When you are down and think the town is near, there is another perpendicular descent of 2000 feet through a dry nulla. The town of Chanko is in a large plain, watered by two streams that bound its two sides. Eleagnus bushes are abundant on the field borders, and there are plenty of willows and poplars and some healthy apricot trees. The crops are all in and some people are ploughing. There are some green pumpkin like squash that taste like potatoes, and also turnips. One man from Khap said they had cabbage, garlic and onions in his town. We bought a cake of Tibetan tea but it had no taste and only discolored the water. Our Lipton tea has taken on a camphor ball taste, though we haven't any of these articles. Saw a whistling thrush and got a large Redstart male. Saw another male. The hills on all sides are dry. There is some small vegetation on the hills across the river. They say not much snow falls and it promptly melts. The houses are scattered around the edge of the plain. Chikor abundant. Blue and Tibetan pigeons. The men lost a leg of their sheep and the dog picked it up. She ate some and then brought the rest to Wang Gyel who had gone with the horses.

October 3, 1933. Day clear and pleasant. Erratic strong breeze in the afternoon. Got a flycatcher not heretofore seen in India. Another big Redstart and saw a third, both male. Saw also rock creeper, blue rock thrush, Griffin wagtail, whist thrush. Bought an old tanka that shows the resources of the original artists and an old Nepali image. Tucci offered 600 rupees for a lamp in the monastery here that Rup Chand says is as ornamental as a cooking fan. He also buys plants the natives say and jewelry. The aphids were most annoying before the breeze came up. The people here seem to be very poor. Some of the small naked children have such pot bellies that it seems they must explode. Gave a starved puppy a crow carcass which he greedily seized. A native chased him and got it away from him, for medicine, he said. Two huge donkeys larger than any I have ever seen were here yesterday. Said to be Tibetan donkeys. Easily two times Spiti donkeys.

October 4, 1933. Left after finishing bird skins for Nago. They said there was no pass today and only "manuli charhar," but we ascended to the peaks by the time and we got halfway to Nago. The road just before Nago was particularly bad. At one place on a perpendicular slope of shifting earth the path had been cut into the slope in front of a huge boulder and the horses had to be pushed and pulled over the way. One slipped and nearly went into the abyss. A few rods further the descent was perpendicular to the streamlet that has cut a tremendous gorge. Then a perpendicular ascent out. There is a village just before Nago pasted on the slope called Nintling. The people were having a festival. The women coming home from the nulla in the gorge told us to join them. They would give us ཆང་ [*chang*; barley beer] and ཤར་ [liquor]. By the time we got to the village, the males, some eight to ten adults and as many children with a drum and a hideous voiced trumpet were escorting a huge dust mop, 15 feet high from which streamers of cloth were hung up to the top of the hill. They jiggled the mop as they went and accompanied themselves with the metal drum beat. Occasionally only the terrible trumpet noise issued forth. Our youngest horse had never seen or heard such a sight and bolted. Arrived at a certain stone, they rested the mop, then poured something on the rock, blew and beat and then sang a bit. The mop was then brought down with the accompaniment. There are in many places between Nago and Chango at 1000 feet above the river huge boulders six to ten feet in diameter with potholes cut into them. The nulla near Nago was once filled with glacial drift after having been formed. The village of Nago is a cluster of rough huts of stone and mud with the usual heaps of brush along the roof edge. There is a small pond 150 paces long with willow trees and poplars and rose bushes on the slope toward Drashigong. One poplar is six feet in diameter but completely hollow. The children gather the rotten wood of all the old trees. The fields of these villages descend nearly to the river and rise above the towns. They have been constructed with great labor. Almost all the walls are six feet or more in height and none of the fields are over eight rods wide. Saw two Golden Eagles. Li is a beautiful picture of green beside the river far far below.

October 5, 1933. A Tibetan trader got us out of bed to show us his goods. He had some very nice things and we bought over 100 rupees worth. He knew values and didn't take us for fools so the price was settled in less than ten minutes. After that people came all day bringing us things. One brought one of the old style tankas. The women toward evening had lost their diffidence and left their round wooden closed-topped water buckets (such as Spitians used at Mane and elsewhere) and gathered to discuss our purchases and prices. Many tsa-tsa-tsa of wonder. Usual with the west Tibetans. One man saw I gathered rose seeds, etc., and gave me some grape seeds. He said grapes got here originally by a man finding the seeds in bear dung. There aren't any bears here now. Two hawks (accipiter) one large and one small were hunting together, one on one side of the tree one on the other. Bought an ancient table of Rampuri workmanship for three rupees to the great interest of the populace. They were very appreciative of the pieces of record we played. So were two cows. Both listened attentively and one joined in twice with Indian songs, to the improvement of the company's enjoyment. The plow of the Spiti, Gundlu, to Koksar is the same [sketch with notes "ox rope fastened here, iron point"]. Rest of Lahul same sort of plow, except at Dortse both sorts [second sketch, with labels "ox rope here, iron, handle"]. All parts wood here shown except the iron points. The nights are very

cool here. They get only one crop a year, two at Chamgo. A road officer, Beli Ram from Poo, came today and says Tucci will arrive at Poo on Oct 19. He gave the Dzong 100 rupees bribe and got permission to travel at liberty in Tibet. The Spiti Lahuli who lived at Nago a couple of years ago says there are Hodgson's partridge above Nago. The natives don't know it. The women here wear one or two large zinc bracelets, size and shape of Kuluese. The men here wear hair to shoulders and not Tibetan pigtails. Here as in Spiti there are many monuments of three stones six inches in diameter or so, whitewashed often and placed one on top of the other on a big rock usually. These are put up at the time of bumkyor.[1]

October 6, 1933. Left Nago and the nice pond that is blue when fair weather and green when the clouds come from the Sumnan mountains and descended straight 2000 feet to the river. I thought it must be 1000 feet down but surely it is 2000. Two Tibetans whom we last met at Hanle came before our departure with some things. Bought an old monk's snuff bottle and tried to buy a nice pair of turquoise earrings. Two rupees difference in price and bargain fell through. The passage across the river is on a wire of braided iron. The span is ca. 300 feet and the river runs 60–75 feet above the torrent. Some men from Li were ready to take us across but it took three or four before everything was ferried, including the horses. None of the men could remember when the wire was put up. A well balanced piece of apricot wood, a foot long and four inches in diameter, grooved lengthwise, slides across with the bundle (each nearby village has its own piece of wood). They used pulleys at first but one broke in midstream and threw two men it was carrying into the foam. A yak tossed another in while they were tying him for transport. The horses are bound around the belly in two places and then pushed into the abyss. They were a little surprised at first but only the colt appeared at all concerned. Some looked down into the torrent once or twice and then hung motionless till they were landed. The dog got sewn into a sack with her head sticking out and she enjoyed the view below. It cost one and a quarter

---

1. '*Bum skyor* (hundred thousand): refers to the recitation of 100,000 verses of a text (Joe Leach, personal communications).

rupees per horse. Chunregog drank so much ཆང་ [*chang*; barley beer] last night he wanted to die this morning and was in a stupor all day. Got a pair of BW teal on the Nago pond. The people say that a little later yellow ducks come. Two ravens also at Nago and huge flocks of 300 Blue pigeons. No Tibetan pigeons. Two young Tibetans were camped at the bridge and were trading salt for grain with the Li people. Equal trade. The whole town of Li came out to meet us and swarmed close to the tent. Our Tibetan trader of yesterday met us drunk with a plate of grapes and we had a hard time to get rid of him, so good natured and genial. His wife has a huge silver gao in place of the turquoise studded one we bought. A Nago man found a rupee I dropped in the sand and returned it.

Plate 49. Walter Koelz crossing a river (Walter Koelz Collection, Bentley Historical Library, University of Michigan).

October 7, 1933. Li, they say, has 60 houses, 15 one side of the river and 45 on the other. There is a high rock on which are ruins of the former village in front of the large stream. The land is abundant terraced, but the drop from top to bottom is not over 50–75 feet. There is an abundance of trees, mostly apricot, that improve the productivity of the land. All the fields are full of buckwheat (gyandre) (our buckwheat and some few fields of the bitpet sort). We went to the monastery and saw the piece of carved wood that Franche illustrates in his "Art of W Tibet" and which Tucci, they say, offered 200 rupees. The arrangement of the figures is very good, but the execution is only fair. It is in hardwood and is in my opinion, not Tibetan. There were some dozen ancient torn tankas of the sort I bought a sample or two of in Nago. I scolded the lama for not selling the thing. The people continued to be present in droves through the day and usually stunk us out of the tent every ten minutes. Our Tibetan trader friend Sonam Tsering came sober in the morning and hilariously drunk in the afternoon. Rinchen Gyaltsen arrived. The season has been so bad that the Lahulis can't harvest their hay. It rained heavily in Kulu at the end of September. He came from Kulu to six miles above Banjar then Rampur, Panygel, near Nago, here. We bought the start of a [indecipherable] necklace and wanted to get a nice fox skin from the same man in exchange for a [indecipherable]. He said he was a trader and it was a sin to catch foxes, but he had three to sell. He took back a dirty rag (holy) that was tied on the back of his necklace. There are no books in this place. They loaded them all on a donkey and took them to Nago and sold them to Tucci. He apparently bought them one and all, complete or otherwise. Goiters are in evidence from Nago. Don't remember one in Spiti. We are called "lama" here as in Spiti, if not "achho." A Tibetan had a two-eyed peacock feather, three-eyed would be still better, he said. White or yellow seed pods of any shape are valued here. Other colors not wanted. Yellow best. Circa 10 rupees for tolla. The sun comes to us late and the moon which was full several days ago is evident for an hour or two in the light summer clouds that hang over the eastern peaks. The water that comes out of the nulla is led thru a tunnel of 100 years in the hill. No one knows how the tunnel happened. The gatpo told us a log fell down from the rock where the old village was and killed a man. After that the old village was deserted.

October 8, 1933. We left in the morning for Hanko. There is a steady ascent of 1000 feet to a ridge and not a pass, as have been all our passes since Kunzam La. The pass to Summan Hang La is a real pass. There is no great descent to Hanko. A Golden Eagle came almost within gunshot, and we saw a large black vulture with a great streak on his rump. The Kushog of Li, a young man who is Ngari Tshang's pupil came this spring and said he was the reincarnation of their deceased Kushog. He went to the cell and pressed his hand and foot into the stone and left a print like that of his previous incarnation. He also cured a man whom he met on the way of leprosy. He left Li today and came to Hanko with a large procession. The people at Li went to the monastery after dark last night, and again this morning. Arriving at Hanko a crowd of women, one bearing some smoke, met him and his entourage. They beat a couple of drums and blew trumpets for five minutes while the Kushog blessed the crowd and then he went on to the village where lamas were gathered to receive him. The people say that there is an unknown poison grass here at Hanko on the hills that kills horses and donkeys. An old lady at Li brought a string of mussel beads all one half inch diameter, dull, dirty, white. Why wouldn't we buy them! A Sahib at Poo (she knew his name) bought a string from her mother that reached from the neck to the floor. It must have weighed 10 pounds. Gave dozen pieces of chocolate to as many people, children, men, women; none could eat it. One said it was for vomiting, one like dye.

October 9, 1933. The people came most of the day with things, but in general they had nothing good. They have many of the big round wire earrings and none had even a passable necklace or a good stone in one. They are ploughing. One crop a year. No grapes. Saw trees and these willow, no apricots. Nights were cool. Dorje went to Sumnam to get the silversmith started on some work. Birds few in species and numbers except for blue pigeons and small sparrows. Sent a clamshell to the lama and he sent back a plate of grapes. He is a subject lama of Ngari Tshang and is going to ask permission to stay here a couple of years. Garlic here excellent. Bought a small yellow pumpkin and some dried moksha. There are some at Li also, they say.

October 10, 1933. All night at intervals the town sang. A man came from Li, walking in the night with moksha. This morning the people were mostly busy carrying manure to the fields. The sacks on all sorts of animals but sheep or goats and in baskets. An old man brought the square silver gao he wouldn't sell for seven rupees yesterday and took six rupees today. The ascent to Hang La is stiff. The pass must be 15,000 feet! The descent is abrupt. Met a man with a few sheep and goats carrying walnuts to Chango. Sumnam is very pleasant. The pass was very cold. If the sun hadn't shone it would have been bitter. Got two horses to the pass crest but so lazy that one man had to pull, one beat from behind and one thumping in the middle. A little Tibetan lad brought two donkeys with the lame white horse's load. One horse slipped near S. and nearly went in the river and goodbye our things. Got four chikor one shot. Got a N. Kestrel from a flock of ten on the pass, also a raven from three there. Same a shrike at Hango. A child to whom I had given grapes told its mother མི་མི་ had given them (grandfather). མི་མངས་ལ་རོ་མང་ [*mi mang la ro mang*; many people/many corpses] They say this when there are many workers and no work is done.

October 11, 1933. The Tibetans have many tales about people who rose from the dead རོ་ལངས་ [*ro lang*; zombie, literally risen corpse]. If these touch another, he becomes the same, etc. Much dreaded. Tale of X's[2] resurrection is therefore not particularly pleasant to them. The rings were brought at noon. The men had been working one and a half days on each and ask wages. The work is much more elaborate than any seen. Bought several pieces of silverwork and the workmen stayed until dark watching the bargaining. None of the silversmiths have any silver and only one man has any for sale. He says he will give it in return for an equal weight of silver coin! The current price is eight to nine annas a tolla. They brought a very nicely executed teapot of copper of Mongol design and wanted to sell it as an antique. They confessed later to have made it here. A dog last night carried off two chikor skins and a lark skin from under the tent. I recovered the latter but found no sign of the others. The crows last night carried on a considerable conversation in the dark. They are very numerous here. Ram Shom went to Ropak to visit the seat of his ancestors and the remaining descendants who have faced starvation and remained. In Kulu they have abundance but the fearful malaria. Three of the boys who took our luggage to Lahul had had it. A friend of Rup Chand's father brought a plate of rice, a bowl of milk, a plate of flour, and some kumanis as a present. Another man brought a green pumpkin. The kumanis have little flesh but are sweeter than most. The seeds, though sweet, are insipid compared with Ladakhis. A very nice looking Tibetan lama said our image was remarkably fine, worth 200 to 300 rupees. If it had a certain mark it was special. A raja once said most of the images were no good and struck them. Those that spoke he saved and the rest he threw in the river. The silversmiths found a mark of such a beating on the images. A goat at Nago managed to get up 12 feet in a willow by walking on a branch that came near the ground. One man with some broken metatarsals as result of a log falling on his instep brought a green pumpkin before asking for medicine.

October 12, 1933.   Ram Shom's relatives sent me a seer of rice. They could have had no hope of getting anything back. Gan Rane is the name of the chorten which they have in this country. There is one above Sungnam. We had fewer sellers today, but bought some digra in the morning (pins for holding the shawl). A man brought a nice horse bell of nice tone made here. They also make bells above here of fair quality. One man sold a digra by force for three rupees. He said it was སྤུ་རྒྱུར་ [*pu rgyur*; unknown] — when the Tibetan government or lama puts any object in your possession and names the price, you are to pay. No silver has been forthcoming til this morning when a man said he couldn't give it possibly for 13 a tolla. I berated him and then he said all right, I'll give it for 12 provided you take all I have. Our wounded man said if we needed silver he would give us his wife's bracelets. An old man brought two such bracelets for sale at 13 rupees for tolla. When we weighed them he found they had shrunk with age probably and wouldn't sell. One man sold us some ན་བོ་ [*na bo*; unknown]. Ram Shom came back laden with presents — a seer of rice for me. The donor couldn't

---

2. Christ.

Diary: October 1933

have hoped for a return since we expected to leave here today. He also sent our men some food. The good chulis here are of Ladakhi seeds. Though not equal to their parents, they are superior by far to the local native. A nice Chinese lama who came yesterday brought a very nice old image. He first ascertained whether we would tell where we had bought our other images. Another friend of Rup Chand's brought a present of 50 cents worth of rice. The Chinese lama said rats had eaten all his things while he was in Li, couldn't we give. Rup Chand said མཐུ་རྒྱབ་ [*mthu rgyab*; black magic/sorcery] them. This is a magic rite in which a large bomb containing axes and other destructive instruments is sent skyward by gas with instructions to destroy a definite enemy. Meteors are supposed to be such bombs. If certain lands are inclined to slip they say མཐུ་ [*mthu*; black magic, curse] struck them at some time or other. Languages: above Surynang and below Pano; Summmang Chiassu, Kaman Labrang, Pilu; Lipe Asrng, Jangi; Pang like, Telingi, Rahang, Kuangi, Birlingi, Chini, Dunis, Rolg to Tarand wth some variations; then below. In front of Kanam is Nessang with a different language, Tsarang, Kuntrang, Chitkul, Raksam in front of Chim. This does not take into account other valleys than the Sutlej. Hindus don't use a donkey's name early in the morning, so fooling the animal. It's the only thing you mustn't give a lama. The people here call aluminum "silver." Their treasures here and below at least to Darkali they keep in a small windowless house on top of their residence, or even beside it. Why, heaven knows. Here the grain is fed the mule by hand. The Chinese lama says in his country the nether stone revolves in the upper as here.

October 13, 1933. Stayed in camp and got bed mended. Bought a sheet with a gay border and cut it up for lining to my eider robe. Several rings were made and the marble snuff bottle was nearly finished. We were delighted with the ཆུ་ཟངས་ [*chu zang*; large copper water container] rings with green sets. The woman of Kunauar are favored for the fine thread they can spin. Even the little girls are constantly spinning.

October 14, 1933. Melted 27 rupees of silver for the wood plate. Watched the man work most of the day. He hit each coin with a hammer before smelting. If bad it will break. The George rupees he wouldn't take. Edward is better

Plate 50. Wooden plate with silver work added (UMMA 17253).

and Queen Victoria 1840 is still better, but folks won't take the latter and it isn't in circulation. The languages here are related to Gundla and Patan. Below Kanam they speak the Ropap language. Dorje went to Kanam. He says it is nasty and gives one a headache like the Gya-Rogchun pass. We paid off and dismissed Dorje, but he didn't leave. The new oil we bought at Li won't burn in the lanterns so we have neoza[3] wood for light. It is soaked with pitch and looks greasy and burns with a bright flame and much smoke. The children here plea for empty cartridges.

October 15, 1933. Dorje left after breakfast. We spent most of the day watching the silversmiths make the lions, etc. The design for this work is first drawn in pencil, then incised. The plate is then reversed in the wax and the space inside the line driven out with a blunt punch. Cats in Lahul have one to two young. There are no chickens to here except I heard a cock crow at changes. The children hold their chin in one hand by way of accentuating the appeal of begging. The walnuts here are often ground and put in tea by way of oil. The Rampuris and Lahulis do not beg in travel, except for whey. Spitians, Zanksaris, Baltis and Tibetans beg anything and everything. The children call the empty cartridges shui-shui from the whistling noise they can make with them. The women are busy gathering twigs of the poplars with leaves for fodder. If the trees were cut off last year, only the leaves are stripped. The buckwheat (our sort) has been since we have been here. Two Englishmen got all the Poo villagers to dance and then didn't give them anything. One man sold Tucci his monastery complete for 1200 rupees. Tucci wanted to buy a door and frame in Poo for 1000 rupees.

October 17, 1933. Spent all day again with the silversmiths. The work will not be finished tomorrow. A new man came to work and made 12 buckshot balls we need. The other old silversmith won't work with Puntsok, but agrees to work separately tomorrow. We hope them to finish with the plate day after tomorrow. Dorje's wife sent presents with a messenger. H. Namgyel sent cabbage, potatoes, kerosene, and rice from Poo. A boy with gonorrhea and malaria responded to one day's treatment of quinine. Yesterday two men made the six medallions (lion and fleur di lis) and the centerpiece, having done half a day's work on them previously.

October 18, 1933. Snuff here is made half of ephedra ashes and half tobacco. Some people add a little perfume: white chomdan, laichi, many users. There are scorpions and snakes here but people say they don't bite. Two of the former were under my blanket. At the Tibet-Rampur Bridge one of our men saw a three foot snake and at Tabo they say there are such. The silversmiths are almost finished work, but it will take all day tomorrow to finish it off. Wang Gyel went to Chim with a letter for Mother and to get silver for paper. Gyeshi

---

3. A pine, produces edible nuts.

Rimpoche is going back to Drong next month. The Tashi Lama is coming back to Lhasa. The Gyeshi Rimpoche was consulted by the Rampur Raj on what to do to get posterity and the lama said to get the Tangyur and Kangyur.[4] They are in Rampur now, but the Raja has no children. They die.

October 19, 1933. Work on plate not yet finished but the end is definitely in sight. The old man, Tob, and his 16 year old son Namgyel were paid off and the bottom rim given to the others who are putting on padme design. The old man can't abide Puntsole and his workers. Puntsole speaks very kindly or nothing of him. Wang Gyel came back from Kanam. He got 100 rupees of silver from Rimpoche. The lama said he had changed 5000 rupees for a sahib (Tucci probably) and couldn't give more than 50. Wang Gyel said 100 or nothing. When the lama heard he would have to go to Chim for change he gave in and said I was a friend of his. The snuff bottle isn't done. The rings are but Wang Gyel doesn't approve of them. He didn't leave Belu and his son any work as a consequence. Rain set in in the forenoon and continued all into the night. Snow on the peaks. An old lady came this morning and scolded the horsemen for feeding a certain weed on her fields in the hay we bought here. The negi sent two charpie[5] but they wouldn't go into our tent and we had to return them. The dog is menstruating and has to be kept tied. Another friend of Rup Chand's back from Chimurti brought two cabbages and a little civet cat for sale. The cat we got in Kulu once. It was killed at Rushklam in the winter. Runtsok's (55 years) son is Bitapur, 30 years; other worker Dorje. Another assistant is Norbu. Yesterday and this morning the goats and sheep were very playful battering one another head to head. Rup Chand said a storm was approaching though the sky was perfectly clear yesterday.

Plate 51. Snuff bottle with silver work added by Tseppel's son at Labrang and the lid added in Colombo, Sri Lanka (UMMA 17226).

---

4. Sacred texts of Tibetan Buddhism. Two major divisions of the Tibetan collections of Buddhist scripture (Joe Leach, personal communications). *Kangyur* is "the translation of the word" and refers to works spoken by the Buddha himself; *Tengyur*, translations of treatises, are commentaries and treatises.

5. *Charpoys* (?)—traditional bed with wooden frame and strung webbing.

October 20, 1933. There are three castes here: zamindar, lohar, chumar and weavers. The zamindar will not eat with the others; the lohar not with the weavers and cobblers. Rain continued most of the day, heavily all night. Snow came halfway to the valley. Rained into the night. The work at the smiths finished. Gave the Nago solchok[6] to Puntsok to be mended. The people here have to buy much grain from other places. Paid for principally with neoza, gudmas, or wool. A horse got down on a nasty place on the river cliff and couldn't get up. A little movement would have sent him to his death. People assembled but couldn't attach a rope because the horse was unfriendly. Sent for a little girl of his family and then hauled him up. Several years ago a man went down the cliff along with the horse. In Lahul each house in each village has a name, no matter how humble the structure.

October 21, 1933. Negi Tundrup Gialtsin sent more rice and chuli, believing we must be short of food having stayed so long here. Puntsok repaired and supplemented the old Nago table. The other two made six padmas[7] for the plate. Heavy snow through the night and till near noon. Clear at evening. Snow melted promptly but the brown snowbirds came to the hills.

October 22, 1933. The day dawned bright and at 6 and we left Sungnam. The Negi's daughter brought a basket of sweet cakes fried in oil and a basket of buckwheat cakes. The latter were exceptionally fine but our men quoted a Tibetan proverb: *jocho tokna, draupe gom* (A gentleman when hungry will gobble down bitter buckwheat flour). A dog pursued our bitch nearly to Chiapan and then another left his father and donkeys and came five miles to Kanam. We arrived at Kanam at dusk and camped in a field. The people wouldn't give wood or grass, just as before. Some Tibetans told Rinchen Gyaltsen on his way up Kanam was the worst place on the march. Their houses are full of wood and the nulla near is blocked with fallen trees. In front of Chiasan a huge mound.

October 23, 1933. The Belu and son silver smiths made some monstrosities of rings. Two are so clumsy they can't be worn without bodily harm, having sharp prongs and points. They said they had never made such before, just for me. I went to Labrang then and found the snuff bottle not finished that was left ten days ago and promised for the 16th. We couldn't get any sort of food for ourselves or animals at Kanam so we moved to Labrang in the evening. Went to two Labrang houses and saw several interesting things: a tanka like the coarse Tetha one; one good Tibetan, one old style (poktal) and one well executed native (?) among some 20 native-made; a stone (marble) slab with Buddha and saints;[8] an old carved chest and a gold-marked copper stand for an image set with turquoise and white sapphires. Found an old digra said to be Tseppel's work.

---

6. Textile.
7. Lotus blossom.
8. UMMA 17012?

Excellent drawing. Tseppel died four years ago. Met the Lipe Galender Nabe Sonam Tobgyet. Yesterday there was a spectacular landslide of stones from a peak. The rocks struck the earth slope as they fell and clouds of dust marked the spot. They ended in the river. One struck the river in its middle and plowed out again and reached the opposite shore. We got some nice fine Tibetan wool from a burgi above the Chiassu Bridge. The Gyeshi Rimpoche is having a farewell party. There is another rimpoche at Labrang who isn't a Tibetan but looks exactly like a Hindu fakir. He isn't at the celebration. Tseppel made a silver chor for the Rampur Raja and his son made another using 1000 tollas (80 tollas = 1 seer) of silver to cover the original wood.

Plate 52. Small carving from Labrang, Kinnauar (UMMA 17012).

October 24, 1933. The black Rimpoche here came a couple of months ago to go to Mansorawar—a rather out of the way path. He is said to speak English and come from Sikkim. There are about 500–1000 maunds of neoza sold from this district, says a first officer. It seems about right. The neoza District is from Ropar to Chini. A man, Kargut Tanba, from here fled with another man's wife to Kulu and his brother, Sundrup Gya Tso, wants him back. They are of a good family and had a nicely carved stone plaque which we bought. There are red bears here. They sometimes come to the house to eat apricots. The village of Nessung has 35 houses and has a language of its own. They say that a man near here once gave a feast to the travelers going to Rampur. Several Nessung men came and after eating asked for food for some imaginary companions in camp. The donors said they wouldn't give food thus; the men should come if they wanted to eat. Whereupon the Nessungians shouted all sort of obscene names of their companions to come. This in front of the women of the assembly. They got the food for the camp. Our bottle was finished by our sitting all day and watching the worker. He has learned a few of his father's designs but has no artistic sense. The Kuluese say "*gana gushi e kurar*"—"grind a hammer and it becomes an axe." There was a big celebration at Kanam. They blew a couple trumpets and beat drums for a good half hour at dark. The big lama gives supplicants no advice. "Go ask your local god," he says. There are three kinds of buckwheat here and in Lahul: ours and two bitter, one grey, one black. The houses are built against earthquakes and more substantial than in Kulu. They sometimes slip without earth tremor. Went into two houses to look at things. The monasteries are little box-like rooms, circa six by six feet. Our men I hear burned some of the Kwangkam at Kanam, having got no wood from the natives. Chortens are scarce and mane walls are not frequent.

October 25, 1933. Left for Jangi. The forest becomes heavier below the Lipe nulla. Deodar[9] common now. Present to Ropar. Old fields in the forest overgrown with trees. There is an old fort at Labrang, partly ruined. They say there are two rooms locked and no one knows what's in them. The crows tonight circled high for some five minutes before going to roost. Villagers near here were returning from Kanam today with a red rag around their necks. A lama's old clothes put in the fire and the smoke inhaled is said to be a cure for some ailments. The Sutlej is crossable at Jangi to Morango. There is an old fort on a sharp hill just above the river there. The Tibetans call the river Langchen Kampa. The apricot trees are turning red. Some sheep here have no visible ears. Bought a green pumpkin here, but couldn't get neoza, walnuts, buckwheat flour. Since Sungnam haven't been able to get anything except two men gave us apples and walnuts at Labrang (Kiap, their language). Bought a piece of cloisonné so black that no design was visible and only a speck of color. A mate had been made locally sans enamel, also black with dirt of ages.

---

9. Himalayan cedar.

Diary: October 1933

October 26, 1933. We left for Pang, 15 miles. We should have gone across the river and followed that road but we wanted to get down as soon as we can. At Rarang we found the monastery we once photographed and couldn't identify. A lad from Lipe whom we knew before was going home from Pang. He couldn't control his surprise when he saw our six horses and asked if all those goods were ours: when we came to his village we had no tent, no servant, and only two horses. An old friend from Asrang knew all Rup Chand's family and thereafter wanted to leave his horse and come on foot with us. He asked Rup Chand how it was they could stay with me, with such a difference of creed, view, etc. Rup Chand replied differences between men as existed were manmade. The lama would find nothing about them in his sacred books. "True" he said. But why did Rup Chand kill birds. He was an example to the common people. Rup Chand said it was written that if a man did wrong he was sure eventually to repent. A bird I killed the lama asked for and taking it muttered a prayer of one minute, asking that it be given a good place in the next life. No word of protest to me. He saw in a dream an Englishman many years ago who said "mūm, prehm, gūon" and has never been able to find its meaning. He had a patent medicine pill with the letters JBR on it. I translated them to satisfaction and he memorized them. Arrived at Pang and the resthouse people had wood ready and three apples from Chini and walnuts in our room. The resthouse is by far the pleasantest and cleanest I have ever seen. It was overcast and storm in all nullas, so we put in. The neoza logs make a grand fire. The tree is most attractive too. It sheds its bark and looks like a sycamore trunk. Altitudes: Kanam 9740, Lipe 8915, Jangi 9242, Pang 8687, Chini 9468, Rogi 9508, Urm 7835, Wanpur 5120, Nacha 7148, Torand 7160, Seralan 6914, Guarua 6382, Rampur 3280, Ninth 2938. The old priest said I had once been a Buddhist and therefore was drawn back to my country. He had a pill that someone had given them with JBR on it. He asked what it was and I read the letters ཇ་བ་ར་ [*ja ba ra*; JBR] which he repeated several times.

October 27, 1933. Left Pang for Chini. The view from the resthouse is magnificent. One can see the hills beyond Chini and to Jangi in the other direction. Snow has settled to a little above the high villages. The Sutlej valley is only a gorge. Lateral erosion has been little so that the valley walls are rugged and steep. The peaks above the forest line are jagged and soilless. The river Betuta appears at Pang and [indecipherable]. Kabsha and Quakta live [?] in the now dense forest of neoza and deodar. At Chini there were some very nice apples and large solid cabbages. The climate is colder here than above, as they warned us. A native came to call who had spent ten years in Lhasa studying medicine. Met a leper in the road, the only one all summer. The grain is green in the fields here. Not seen elsewhere. Above Rarang the people have brought water in troughs almost [perpendicular] down the hills.

October 28, 1933. Rinchen Gyaltsen posted letters to Jaipur, Hajee, Lee, and mother and sent six rolls of film to Bombay. By the time he got back it was noon. We left then for Roghi. The lombardar who gave the nice head of cabbage couldn't be located, so he got no tip. Arrived at Roghi. We were deluged with apples, a few neoza, and little white peaches. The apples are of five kinds and first class in every respect. No worms but some core rot and some subcutaneous rot spots. We bought a bottle of the grape wine. It is perfectly colorless and as strong as whisky. The grapes grow along the river. The chaukidar here set out over 100 apple trees. What he will do with the fruit, I can't imagine. The wine is two rupees a bottle. First class wine is five rupees but the crop failed this year. That is made of special grapes. The wine is very pleasant but too strong for agreeable consumption. I didn't dare give our men more than a cupful or they would sleep the sleep of the just. Sonam Tsering finished the bottle and had to be helped home: he told our men they were donkeys, hee haw, for not appreciating me, that he would give them a sheep in the morning, etc. He knew we couldn't get grass in Kanam and had gathered up a supply for us. We got one and a quarter seers of neoza at Rangdum per rupee and two and a half at Chini. It is not so cold at Roghi, though the elevation is the same. The people here haven't finished cutting the buckwheat.

October 29, 1933. Sonam Tsering came again this morning with a plate of neoza. His little boy kept bringing hands full of neoza and once an anna. We put some mercurochrome in his eyes but Rinchen Gyaltsen had forgot to put in water after adding the tablet and almost burned his eyes out. At Labrang the tehlsidar after three days of snow blindness put off his goggles and wore a pair of reading glasses someone recommended and loaned him. A Tibetan complained today that T[ucci?] sat on his books and tankas. They told me to buy the sacred things but treat them with respect. A Tibetan helped load and asked what we were plaguing our horse to carry the apples for. We said we wanted them to eat. "What matter if you don't eat the sorry things," he said. He said he ate them if people gave them, but not willingly.

October 30, 1933. This morning a man came and asked when we were going to start. He said his horses were afraid of other horses. There are a few scattered neoza trees to below Urni. The scenery which yesterday was so splendid as we walked along on the crest of the hills today came to nothing. We followed the river after descending from Urni. A yellow composite shrub is beautifully in bloom on the cliffs. Many Tibetans have reached the river valley. One camp invited us to tea. A dipper diving in the river could stay under 10–12 seconds. Our Lipton tea had a drugstore taste so we put in some prianku[10] that we gath-

---

10. *Nebeta longibracteata*: a plant with woolly leaves and lavender-blue flowers that grows on peaks; used in flavoring soups, relishes, etc. and as medicinal (Koelz 1979:39).

ered in Pin Valley. Then we bought a cake of Tibetan tea that didn't have any taste. Our herb remedied that too. We got 16 [indecipherable] eggs at Uru last night for an anna a piece. The men got some green pumpkins for one and a half annas and some free. At Wangtu there was nothing else to be had. A few mosquitoes came at dusk, bred probably in the foothills area above the present river level. The water in summer is apparently six to ten feet higher than now. The snow that above here has descended below the upper tree line is out of sight at Wangtu. It is much warmer in the evening than anywhere above. A man came this morning with very sore eyes. Sonam Tsering and his companion were similarly afflicted. We slept as best we could in the hollows between the rocks on a ledge 500 feet above the river. Rup Chand used to be a somnambulist and for fear he might get out of bed they tied his leg to a bush. There are one or two caves along the river a little above Wangtu that have been converted into dwellings. Who lives in them?

October 31, 1933. We came to Nachar. There is a post office. In the forest officer's garden are some 20 apple trees, pears, and cherries. There were some roses in bloom. The gardener was asked for apples, but said he couldn't give them. "Send a basket and earn a rupee or two" said the shopkeeper. "They are sahibs but Kala[11] sahibs. I just now saw them with a gun." Several large flocks of goats and sheep came down from above to spend the winter. The goats were scabby and stunted and all the animals thin. Monkeys have been seen from Kanam down. The deodars hare are tall and have well shaped straight trunks. Ram Sham's pied horse was tied with a short rope to the other resting. Suddenly, it got in its head to crawl up on a large boulder dragging the other by the neck. Resistance was in vain and finally it got on top of the rock only to fall on its back on the other side, smashing the wood saddle. The Roghi Chaukidar was a very polite person and in every sentence added "Khudarnanad" once or twice. There is a tower-like house at Nachar with some nice conceits carved in several panels on the balustrade. The god's house has also some nice carving, but not so old. We bought four kinds of squash, one said to come from England. From Labrang down it is custom for people to say "the price is —. If you want it all right, otherwise, beat it." Heard an owl in the deodar forest like the little "lucktychuck" of Badhwar. Heard another something at Nangtu. Above Labrang and probably Wangtu dry farming is impossible. At Labrang there was great interest in my watch. One and all put it to their foreheads. A boy at Lipe asked the address of someone who could give flashlight bulbs. I said he could get them in Kanpur. He replied those were Japanese bulbs and he wanted American ones.

---

11. Black.

November 1933

November 1, 1933. We came to Tarand and camped on a spur that runs across the valley. There is a god house on it and nice fresh grass, like a lawn. Some huge deodars rise skyward and bound the view up to the snow peaks in front of Chini. Below Rampur one sees the river. The moon rose nearly full and the snow peaks glistened as in the sun. Straight behind here is a peak that seems to be the best hunting ground so far seen. The vegetation in the forest is now abundant. We met a flock of Spitians returning from Rawalsar. Most of these were from Lahung. One had seen us at Guling. They said be sure to come next summer too. Though we spoke no Tibetan and they talked about us amid themselves, not an impolite word. I gave them some squash seeds to plant at Tabo and Po. A strong breeze came from down the valley but died down at evening. The people here profess never to have seen a carved chest. There is a tall house with carved panels in the balcony.

November 2, 1933. A strong breeze rose during the night and blew down the valley and continued into the morning. We got Churezeg ready to go to Lahul. Wang Gyel will accompany him to Kulu and return via Anu. He took all the birds and dry plants. Poor Churezeg cried when they parted from us. He got rupees 120 for his two horses from Chanda Tol. We came to a field a mile or so below the Chaura resthouse, 6000 feet. Several flocks came so far too. Our Tibetans all stayed behind and didn't come to Tarand. Plants are abundant on the shady slopes, especially mosses. Barass begins at Tarand. Met three mules laden with cider for Chini. The government has the liquor monopoly. Hindus don't spill salt, otherwise in the next life they must pick it up with their eyebrows or lash. About 11 am, a thin cirrus feather floated across the sun and became what the

Tibetans call a Jetsun.[1] The broad area about the sun was colored pink purple and in the openings of their cloud sulfur green light and blue sky. Either some great man has died or been born.

November 3, 1933. We stayed in camp all day. There are quantities of monal, kabrha, and charman here and various species of small birds feed in the trees. Hunting is difficult because of the steep slope. The flock camps of last night left at daybreak and their places were taken a few hours later. Flocks are sometimes laden, sometimes empty. The sun rises on us very late and the shade is very uncomfortable. Had a little stomach disorder and opened the cherry brandy. The dog is about over her heat begun before Sungnam and she is fearfully afraid of the forest. At night she won't venture outside the firelight.

November 4, 1933. Came six miles and camped on a spur, some have sun until it sets. Before at this place there were many words. A Kampa "general" went by today, with huge horses, like Yakandis. He says they are Spitian. Heavy traffic again today. A group of men are carrying pine boards to Sagahand and one had a basket of cabbage from Chini. Can't abide sweet or greasy food so had to go to bed sans anything but apples. All we have to eat is squash and flour. Ropak language to here. A girl came for medicine. Her eyes have been sore for a year, pain with water running. Some Kanons who recognized us told her we had medicine. Heard a bird after dark giving two short whistles, majesty and stupidity are slow [indecipherable].

November 5, 1933. Stayed on the ridge camp. Rinchen Gyaltsen went to Senhar but couldn't buy anything but dal. No eggs or vegetables or lemons or raisins. A large flock of swifts flew over camp off and on all day. Saw three of the little owls. Got the R.B. nuthatch where it was the year before. A new Phylloscopus[2] with buff wing bars and a white tail. Grass very abundant for our animals. At night a beautiful line of fire on the opposite peak where the old grass is being burnt. Monkeys had a fearful row on the cliff back of camp. The silversmiths, to get a relief design first incise the design on silver sheet embedded in wax. Then reverse and beat between the lines giving relief. Then reverse and finish.

November 6, 1933. Hunted on both sides of the spur. Travelers continued to head down. Our new Spiti horse continued to be lame on his front foot and no cause is obvious. Grass is splendid for them. No poison grass as at Nachar in the forest. A big monkey gathering something in the top of a dead tree came down crashing and banging into a rosebush that had climbed to the top.

November 7, 1933. After moonlight herds and travelers continued along the road all night. One boy driving sheep sang very agreeably. Almost all the horses

---
1. An honorific term, meaning lord, master, exalted one (Joe Leach, personal communications).
2. Warbler.

for riding that go down have sleigh bells. An owl sang by camp just before dawn but though I got under the tree I couldn't see him (the "too too" kind). The night noise that I previously described as two whistles sounds from a distance something like two anvil strokes, an owl probably, swifts again over camp. Our neighbor called "he shot his sister one night for a porcupine but didn't kill her. She was hiding in a field from her husband who had come to fetch her." Rup Chand says over half his income goes to entertaining visitors who call. Everyone must be given beer. If too poor for beer, give boiled potatoes or something. Thus everywhere in Lahul. The Roksar people are models of hospitality. There are some eight houses and in spring and fall all Lahul stops there on the Kulu trip. The people go out to sleep with the cattle and give wood free. They have to carry it on their backs five miles. When one eats fat and is distressed the Lahulis give warm sugar cane syrup. There is also a severe stomach cramp for which they apply a cold iron to the abdomen (langtap).[3] When old folks or children sneeze, Lahulis say "Tsering."[4]

November 8, 1933. We left our pleasant camp with the view up and down the river and the snowpeaks on all sides after noon and came to Majoli. We camped here when we first came up. I went hunting in the morning and got one of the beautiful jays from a flock of three or four. Saw a fox in the forest. Travelers are still coming down. We can see 25 or 30 fires on the slope beside our camp and there were camps all the way from Serahan to here. They are building a new road to Serahan up the hill from the nulla, between Serahan and Majoli. There are already two roads, but the P.W.D.[5] must spend money. Baltis are making it. Rup Chand's brother was reprimanded by his superior for making a sound wall on a road embankment! All walls made crumble in a couple of years, else there would be no work and the money go to —. The Russians at Naggar had a beautiful retaining wall built in the spring and it was bulging in the fall. Eggs are plentiful here. One Mussalman has chickens but wants five paisa an egg. Nice big white squash but they wouldn't sell one. Birds are scarce on the road where on our first trip they were abundant. Gursals[6] here but not above. Camp here very cold. Sun sets early on this slope. Our horse had a bad smelling spot on the fetlock side of his fetlock, left front foot. Must be some disease. Two of the other horses have been lame before. We came empty today. The horses had their manes full of five or six kinds of burs and stick-tights. A little below Majoli where a summer torrent comes almost perpendicular down there are a series of potholes in the granite schist. They are about three feet across and as deep and the stream bed is not much under. There is no water.

November 9, 1933. A man cutting leaves from a tree had to stop several times on account of the travelers' goats and sheep. When they hear the familiar chop-

---
3. *Lang thab*, meaning unknown (Joe Leach, personal communications).
4. "Long life" (Joe Leach, personal communications).
5. Public Works Department.
6. Myna birds.

ping noise they leave the road and flock to the sound. There was a cherry tree loaded with bursting buds and bloom by Serahan and Majoli. The people here had little wood to sell. They collect it as they need it from day to day. We left and came to a village below Gaura. They were getting an evergreen arch ready at the Gaura resthouse to welcome the D.C.[7] who they said was going to Dankbar. We had great fun imagining the splendid climate he will find in the heights. The children at Bushorda were pleasant little creatures. One boy brought a log for us and when he got a paisa for it, he and a companion each brought a load from the forest. Then he brought the few daffodils they could find. The fragrance is the most delightful and refreshing I know. I gathered a lot of seeds of a pink-red fruited cotoneaster at Saura. The leaves are black-green, the fruit abundant, the bush well-formed and three feet high. It grows on the earth slope outside the forest and should be a first class addition to our garden. A man wanted to sell bear fat, black red bear fat is wanted by Mool Chand Gheal [Gheral?] merchants in Lahore. Ibex fat and red bear fat solidify only at relatively low temperatures. Interesting that both are cold region animals. Goat fat hardens readily and fat goats die in extreme cold. The Tibetans see six stars in the Pleiades—*min druk*.[8] Urdu—seven *sahelion ka jumpka* (seven sisters society). There were many people camped between Gaura and our camp. Space little.

November 10, 1933. Came to Rampur and camped in our old place on the sand beside the river in Kulu. A Ropak herd is in the place where one was before, no one else. There was a little lad watching the camp. We gave him a cartridge below Nacha. He was carrying a basin of water to camp and opened his lips to receive it. Met a young Englishman of no very sympathetic countenance going up on foot in shorts and white shirt. Said to be the D.C.; Mr. Gill has just left also. The D.C. is said to be going to the Rampur-Tibet border at Sangzam. Bought two beautiful earrings, a teapot, the nicest so far seen in India, and a Tibetan ladies belt from a Tibetan trader. I have never got so much first class material in a day. There are many people in the fair, but merchandise moves slowly today. The road was full of newcomers ahead of us. The dog rushed at a goat that ventured near the tent. The goat looked behind to see what was the danger, and seeing nothing, asked the dog what on earth was the matter, not frightened in the least. The wife of the man who sold the earrings objected to the sale. She said the money would trickle away.

November 11, 1933. Spent most of the day in the bazaar. Found several square སྲུང་མ་ [*srung ma*; protectors, guardian deities] (protector) and some goa[9] skins. One woman who sold a silver ornament put it to her forehead and then licked it before giving it (*yanglen*). Trade is dull. They give four seers kaccha wool per rupee. Found one of the big brown blue blankets such as I got before. They are made in the district from Gaura to Dakali and are much in demand so that they

---

7. District Commissioner.
8. *Smin drug* is the Tibetan name for the Pleiades.
9. Tibetan antelope.

Diary: November 1933

Plate 53.
Silver ornamented copper tea kettle, purchased November 10, 1933, from a Tibetan trader in Rampur (UMMA 17249).

Plate 54.
Women's dress ornament, purchased November 10, 1933, from a Tibetan trader in Rampur (UMMA 17231).

are expensive. I paid 11 rupees this time as before. If they were of Tibetan wool they would be first class. The man who brought home our wounded horse came and brought two letters from Lompo. We have to send L. 15 rupees more. The horse is all right. Kushog Ngari Tshang left Kulu ten days ago and is presently at Rawalsar. An old man who plays the sitar, an old Hindu stringed instrument, and who has recommendation from the rajas of Jaipur and Kapuntala is in the bazaar and wants to give a performance. Cleaned the burs out of the horses' manes. The horse-bringer says apples are selling for six to seven per paisa in Kulu. The Chinese are subduing the Yarkand rebellion. The inhabitants are being slaughtered and their land given to Chinese.

November 12, 1933. Bought a few things in the bazaar. Met a German who is also buying some things. He is going to stay tomorrow in order to see my things. Some school boys who are now on holiday stayed with me most of the day. They were particularly sympathetic—one of 19 from Tangi and one of 16 from Kila. They say there are carved chests in their country. No one in all this valley has enough land to live on its produce. What they will do after schooling isn't clear. One already had malaria. They said Kuncharang above Tang speak the Mussalman language. There are 35 houses in each village. Then tur. villages of similar size. Chitkul and Raklam between Darangath and Sangla (above Lilba) have another distinct language. The Lipe language group is only slightly different from the Kanam group (dialect difference). From Nachar to Sarand is a dialect difference from that above. Saw some Nessang people. They have good-natured flat-nosed homely faces. Crowd is thinning. Tomorrow many more will leave. Spiti horses didn't come this year, because the Muth pass is closed. The Rotang had only 18" of snow and was closed for two or three days. A man said 3000 sheep loads of wool have been blocked beyond the Tibetan Pass. Counterfeit rupees describes as "yellow inside" can be bought in Calcutta 30 rupees for 100.

November 13, 1933. Spent from early morning till 2 pm in the Tibetan camps. Got a nice pair of bracelets, a nice turquoise set gao and a beautiful necklace. Bought all the stones—beautiful pink corals—of one man and the three zees of another. The people were most kind and were ready to give anything I wanted. The three zees came from different members of his family. A baby boy had one we wanted to examine but the child protested with tears and we gave up. The Spiti Khampan have a camp of their own. They were weighing up their *pateesh*—six and a half rupees for four pounds (batti). The chief was buying it. Each man seemed to have lots of two to three battis. The Li Kushog has arrived and will go to Rawalsar and Nepal and then return. Ngari Tshang Rinpoche left promptly for Lhasa. Prith Chand went as far as Kaling. I met the young German on the way back and took him to our camp. His name is H.R. Deuster,[10] Solan KSR, Simla Hills. He is working on some Tibetan subject for his doctorate degree. He is a very pleasant broad youth, who is interested in the Tibetans. Hindustan he doesn't like. We talked till after dark. He may want to spend a year or so learning a Tibetan dialect! He is a poor judge of quality in some things but he liked our plate and kammerkass. The last is probably rare. Sonam Tsering ཚེ་རིང་ [*tse ring*; personal name] says he never saw one. At night Wanggyel came. Everything got home all right. W. is with Trashi Dowa the wool-merchant. Wanggyel says a dog near Ani barked at him and when he went to investigate he found two newborn kids and ma in a bush. The dog had stayed behind to protect them and the flock gone on.

---

10. Rolf Heinrich Deuster, *Kanawar: Grundriss einer Volks- und Kulturkunde* (Leipzig: Jordan and Gramberg, 1939) (English translation *Kunawar = Kinnaur* [Shimla: H.P. Academy of Arts, Culture and Languages, 1996]).

November 14, 1933. Sonam Tsering ཚེ་རིང་ [*tse ring*; personal name] and brother came to camp early with the three zees we bought yesterday. Then the brother presented us with two nice old agate beads and a beautiful old piece of turquoise *trukar* ཁྲུ་དཀར་ [probably means *khru dkar*; probably refers to a kind of spotted or multicolored turquoise]. He says he tried to deceive us yesterday by calling glass ruby and therefore the gift. We gave him a string of lapis and brother a nice pearl handled knife. Then they said they could get us a beautiful gold written manuscript, weight one maund, with carved sandal covers from Spiti and a nice carved shrine from Kanam. I offered rupees 100 for the book and 50 for the shrine. Book's name is *Do Tarfa*. They said they do no trade in sacred things and it seemed so. They offered absolutely none to me. Our men got a big basket of bean leaves for the sheep's intestines

ཚོང་བ་མིན། དོར་བ་ཡིན།

[*tshong ba min. dor ba yin*; not for selling, for throwing away]. Beautiful things sold, the money is generally spent and it is as if the goods were lost. The corals, etc. were bought from the people who sold the earrings, Kammerkass, etc. The wife said she wouldn't sell anything of hers till her husband gave the money for the things sold before. We left till they had settled matters. Some sheep coming down had two new lambs each sewn into their load bags with the head sticking out and asleep. Wang Gyel inquired if a Sahib were here when he came last night. They said a "gora" there was, not a sahib who daily crossed the bridge to the Kulu side. Rinchen Gyaltsen knew us from the description "a somewhat old man and a black one who shoot small birds." A funeral came down from the hill above and buried the corpse beside us. Afterward the musicians came and played for us two dance pieces. There was a reedy sounding wood horn, a brass flute, and a drum. The tunes were pleasant enough but like all the hill music monotonous. Then at request they played a little funeral music. They say it is "reversed song" (*ulta rag*). A huge avalanche came down from the hill and narrowly missed a shepherd and his flock. The stones continued to come; the man was not a bit startled but stayed and drove his animals out of danger. One goat lost a horn. I saw a nice looking Tibetan woman with a nice necklace and gold gao in the bazaar and asked Sonam Tsering who the beautiful refined looking woman could be. He said it must be so and so's wife. Later we came to her camp and there was a coarse butcher-like woman to whom Sonam Tsering told what I had said. She treated me with the utmost attention as a consequence. We gave a sheep some cooked mutton from our plate and the animal begged for more. It walked sniffing around the tent five minutes. Wang Gyel says we eat so much garlic that our beds smell of garlic. There is a god (Bima Kali) at Serahan who requires a sheep and two bottles of spirit a day. The sheep coming down empty to graze are required to give one from each flock at half price. The rest the raja gives. On one day in summer the god takes 108 sheep. The daily slaughter of sheep the Brahmins eat. Bought a huge bunch of daffodils for an anna. We gave two annas and the woman wanted to give all she had. Bought two brass shotgun cartridges. Had to sign a register but show no license. Have only four AAAA and buckshot. They poisoned the thanadar[11] last summer. Two others recovered. Our horse's lameness has been diagnosed as a shoulder ailment. They will sear the shoulder. If you sear the spinal ridge the animal dies. A raja's subjects a few years ago set fire to his buildings. They showed me a yellow metal today called *lartál* (*babla*, Tib) [unknown]. It is soft, somewhat like a soft schist, tasteless and said to be poisonous. It is abundant in Zankskar near Lingshol. Neoza are selling at 25 rupees per maund. I was standing in the bazaar with a small digra[12] for which I had given five rupees. A Tibetan asked how much I had paid. "20." He gasped "20!" "Is it cheap?" I said. "Cheap, cheap," he replied. Then he wanted to sell his earrings. I had unfortunately bought a pair just like them yesterday or I should buy them I said. How much? 200 rupees. "200 rupees" he repeated. Our dog has been sleeping on top of a bare rock to watch the basket of meat. Otherwise she sleeps under the blankets with one of the men. In Kenore we saw beautiful chrysanthemums: white, yellow, dark red, brown. None of the common sorts we grow at home out of doors. Lahul languages except Tibetan have no words above 20 except that there is a word for 100. They exempted wool goods from the usual tax for the fair this year.

November 15, 1933. We left for Arsu. The Tangi boy who wants to come with us says he will come anytime we want him. He is a nice honest looking boy with good horses and could be of use in dealing with people. His name is Negi Parmanand. He married this year but apparently has no strong inclination for domestic life. Dorje for example. There are cherry trees at Biari and at Arsu are a mass of pink. The blooms are not fragrant but the trees are most attractive. We bought a lemon apiece and ate it beside the vendor's house to the great interest of the occupants. They had several swarms of bees that went into the house through a small hole each. Around the hole, chiefly above, was a circular area of two feet diameter. Splashed with yellow and less red paint in elongated dabs, two inches. They said it was for ornament. They get up to a maund of honey from a good swarm. They need not kill the bees to take the honey. The native brought flowers again this morning and we had a huge bunch of daffodils. At Arsu the schoolmaster at once explained to the gathered crowd I was an American government employee sent to find out what plants were of use. See, I wrote where each plant was found so they could come and get them when wanted. The man who could get the most new things was given great honor in our country. Thus he said we prospered by taking advantage of the others knowledge. We got some fair apples at Arsu. The little zee we bought at Lobrang was laid by a snake along with an egg said the owner. The purpu Rup Chand bought in Lahul is very beneficial in crossing streams. Roll it in the air and call loudly and [indecipherable] on the

---

11. Chief officer of a police or military outpost.
12. *Digra*: shawl clasp (used by women in western Tibet).

deity and then go in the water. A zee with an eye on each of four arms of a cross at Rampur fell from the sky. There were several here, not beautiful. The people at Arsu brought me a root of *niarni* (Hindi) which grows in the mountains here. They put in incense, scent, and tobacco. A lizard at Wangtu went after a larger grasshopper and jumped a foot in the air striking at him with his paws. The grasshopper turned back and alighted. The lizard got him as he landed.

November 16, 1933. There were other bee openings besplotched at Arsu as at Biari. Rup Chand saw a bear this morning. An old man told him where to wait for him and the bear was driven from a thicket down a narrow gulch straight at Rup Chand's head. Birds are abundant in Spiti so we halted two miles below Arsu. A WR Harrier was hunting in a rice field. House flies most annoying. Cherry trees to here laden with masses of flowers. I never saw them so prolific of bloom. The trees are solid pink. Casting pearls before swine—*anotrika har bandarka gala pas. Tru-la khab jorma, nyingla tsuk* [if (one) gives a monkey a needle, (he) will stab (his) heart]. Give a monkey a needle and he will stab his heart. The pear trees are laden with fruit, nearly all of it ripe. The black M. Duong, the Kulu Khampas Garmo Tsering, a widow, has many zees and Dekit has the nice zee and coral necklace I have admired before. The valley here is very open with level fields of an acre. The river bottom has also settlements of several acres. The surrounding hills have oak to near the top and no snow on top. The climate is warmer than Kulu. Titar, chikor, and kalisha are very abundant. Bears they say are also very numerous in Ahichet below here toward the river. Four or five may be startled at a time. The hills must have many [indecipherable; ends "immu"]. Many natives may be seen on the roads carrying a satchel made of one's hide. All bags for grain and flours are of skin in the hills.

November 17, 1933. Made four bird skins. Hawks not uncommon. Bought two tankas a Mussalman had for sale. He said he had brought them from Spiti. I doubted that because the drawing, etc. appeared Bhutanese and the Spitians are Gelukpa. There were bamboo dowels in the place of wood and the silk of one was old Indian. Colors like the Puhtal group. He says one village behind Lasa is Drukpa, which explains the observations I had made. Certainly I had seen no such tankas in Spiti in the Gelukpa villages. Found a large drakshik[13] on the head of a apodacus. Our men report them on two other small birds. The people here make bread of the mild pears. They dry them.

November 18, 1933. Clouds overcast the sky all day. If we go to Serahan and it snows hard we'll have to come back and go through Suket, so we waited to see what would happen. Birds very restless so with difficulty caught three or four titars[14] today. All but one of the nine killed here are female. Never had a full

---

13. Louse.
14. Black Francolin, *Francolinus francolinus*.

meal of black titar before. All the flesh including legs white. A zamindar spent the afternoon with us. He doesn't believe in the gods' oracle. We asked him if the god would know if it were going to rain. A man on horseback in a flowered pink silk duster stopped with us and eventually made camp with our men. He is an intelligent nice sort who says he is a fakir.[15] He traded two horses and 30 rupees with Abdulla and got a trashy Yarkand horse which he thinks is Spitian and first class. He begs grain for the beast, gives four seers a day and has no trouble getting it. Beggars on horseback are not totally unknown to Rup Chand.

November 19, 1933. The rain began about noon but was only a sprinkle where we were. The birds continued restless and I had great trouble with my quota. An Indian bank inspector sent a servant to ask Rup Chand's name. First Rup Chand asked who the man was, where he had been, where going, etc. and then gave the information. Yesterday a lad came down in the valley where I was and said "come up to the road. The Wazir Sahib wants to see you." Who the Wazir Sahib was I never learned. The bears came and ate a large squash from our neighbors. Someone comes daily to beg us to kill bears. In the morning the neighbor children come before we are out of bed to show us titars. The men here called bakshish kashbish. The Kuluese have many such reversed words for strange words. Wang Gyel says the children are out at daybreak looking for titars in the fields. They are very abundant, one to five together.

November 20, 1933. Night sans rain and clouds broke but remained. Titars that haven't sang before now now sing in every field. Only one of our four horses will eat apples and he likes them. Two will eat boiled rice. Went again for titars in the evening but saw only four of the swarms of the morning. No sound of one. Wang Gyel went to Rampur for cartridges. Parmanand came back with him. He wants badly to go with us. He said he had dreamed of us last night. There are swarms of the grey-headed parrots. They eat the seeds out of the green wild pears. There are a few jackals but none sing. The Brahmin who worships in the little temple above us told us not to let the dog come near his basin of water he was carrying to the god. The Hindus sometimes wash their faces and heads in cows' urine. Cow manure they put on the floors mixed with sand to prevent dust. This in Lahul also. To insure consistency of wall mud horse manure and pine needles are added.

November 21, 1933. Parmanand went back to pack up. I tried in every way to point out disadvantageous possibilities of his coming and for a while scared him out of the idea, but he left with the intent to return in the morning. He was much concerned about the folks at home who would weep. He has a mother and wife and two small brothers (the wife is new and won't cry). Father wool trader. Father said "go at your pleasure." Parmanand says Gyeshi Rinpoche sent a lama to me in Rampur. Our nice zamindar friend called again. He comes only

---

15. Sufi ascetic.

every other day and won't let his children come for fear of trying our patience. He speaks in a slow interesting fashion, wears nicely spun cloths of interesting color spun by himself. He says the lemonade we gave him did him good. I gave him some more which he drank sans hesitation. He had many curious ideas that he submits for verification: bears' gall is good for any ailment internal and external. Gursal[16] and monal[17] fat are very effective in curing pimples in young men. Goat meat and goral[18] don't make the belly warm. Sheep, etc., are not so good, a sahib shot a goral one and a half miles away and made the men hold him down to take aim, etc. Heavy cold clouds floated in the sky most of the day. A wildcat and a kitten came near the tent. The half-grown kitten played in the wheat. The Koli children have amassed a good collection of empty cartridges. The other children have just begun to come for them. I had none today but promised to save them till evening. One to be sure probably that I would, waited to give me a handful of roasted barley and a poppy pod. He offered me a poppy pod and walnut in the evening. Two came breathless ahead of the rest and got two cartridges each for their enterprise. The bears came last night and ate all but three of our neighbors' squash. He and his boy were sleeping there to watch the goods. They say at Bangar a leopard fished a dog out from under his master's bed and ate him. At Noggri two years ago, a leopard killed and dragged off three large cows from beside the sleeping men and they didn't know it till morning. We had two requests today for titar meat. The doctors prescribe it for ailments that these two people have. If we are hunting titar some one always appears with an ailment that requires their meat, etc. Several Brahmins begged bird meat without an excuse and praised Rup Chand's skill as a hunter and his gentle appearance to all whom they met. It requires diplomacy to meet such situations. One man got two gursals cut up fine. Our men won't eat them, though around Sirsa they are much esteemed as strength growing medicine for children. Our zamindar (Dili Ram of Dugal) wanted to know whether Pathans spoke Persian and where ostriches lived. He knew they could run faster than a horse. Parmanand wants much to eat and sleep with us. I said "but your father wouldn't eat with me." "I will." He doesn't believe in gods, devils, etc. There is an owl? here that says something like "gwich, qwich, qwich." We asked the zamindar to come with us. He said he would if it weren't for the children. If there weren't a father they wouldn't get their food on time. His wife would eat it at once he surmised. Our horses are very happy with the pasture here. The cows all look fine. Opium is three rupees per battu says our zamindar. No profit at that price. The Lahulis make sweet bread of sprouted grain. The grain is wet and put in a bag in the sun till a small sprout starts. Then dried and ground. Beer also made of it. Rats store in spring yields some such grain. Especially good for

---

16. Name refers to several varieties of myna birds: Common myna, *Acridotheres tristis*; Bank myna, *Acridotheres ginginianus*; Asian pied starling or pied myna, *Sturnus contra*. Koelz's team killed several common myna in November 1933, and this is likely what is referred to.
17. Himalayan monal, *Lophophorus impejanus*, bird of pheasant family.
18. Himalayan goral, *Naemorhedus goral*, small horned ruminant.

Plate 55. Rup Chand with owl (date and location of photograph unknown) (Walter Koelz Collection, Bentley Historical Library, University of Michigan).

their animals also. There is a large grub that lives in horse-cow and sometimes men's noses. It grows six inches long and has a head as big as a paisa.[19] It eats the tissue and horses become thin. The best remedy is to give the horse no water for a day or so, then the beast comes down to drink and can be caught. Is very shy.

November 22, 1933. Wang Gyel was gathering some seeds from a tree. An old woman asked why. He said to eat. She told him they weren't fit to eat. Rup Chand was once gathering some specimens from a tree at Baignath. An old woman also asked why. "Dawai." She thought he said "Gu" and told him cows wouldn't eat the stuff. He kept on gathering and she departed in a huff. Parmanand came today and deposited 12 rupees with me. He must have left in great excitement. He brought no bed but one blanket and nothing to drink out of.

---

19. Small coin, equaled one sixty-fourth of a rupee.

He said he was sad when he left his companions but forgot it when he reached us. His lamas teach that eating chikor is all right but not eggs or chickens. We left after his arrival and came to Saraham. We sent the new man to buy a chicken, cautioning him not to buy an old one, telling him that the feet of an old one would have spurs and be rough. He brought an old grandfather that couldn't be cooked in less than five hours and it is therefore reserved for breakfast. We missed the money bag with 20–30 rupees and Wang Gyel went back to Rampur to see if it were buried in the sand. It's the first money ever lost. I left the nice wild pears with regret. There are some here but all green.

November 23, 1933. We left the broad plain at Saraham after Wang Gyel returned from Rampur. He came last night to Arsu. The money, 61 rupees, was where I buried it in the sand, only the bag was plainly visible. The tekhedar was much surprised that Wang Gyel produced the coin. Rinchen Gyaltsen left for Lahul to bring out our art collection before the pass closes. We shot a nice langur from a flock of 50 near camp. The fur is a lovely silver grey. The dog was extremely fond of the meat. The tekhedar thought Sunkyil a village dog and tried to drive her way. His marvelous attempts and Sunkyil's reaction kept us much amused. A Roopa who passed us two days before went on today with an ancient white horse that he bought for 20 rupees at the fair. The poor beast fell in front of the tent but got to just below the crest of Trishko Pass and fell hopelessly. Our men say chikor become snowblind in fresh snow. One year migrating geese were killed in numbers in Lahul on that account. Tried to teach the two men English. They learn as a child learns. Even a simple sentence doesn't appeal. We camped on the bear-ploughed meadow just beyond the crest.

November 24, 1933. I went hunting. Saw a dozen monal and three I. gurana but got nothing. Our new man bids fair to turn out well. He helped us skin the langur though none of his people will touch one. He and two brothers will someday own enough land to support them at Tangs. He says none there live above the timber line. The Rampuris subsist chiefly now on the stewed apricots and sattu. The former are to our taste hideously sour. Sometimes the god man hands out a few mustard seeds. If added good luck. Wang Gyel made bread with a filling of ground seasoned poppy seeds. The dog has for the last two nights taken to sleeping part of the night with us. She probably doesn't like that the men haven't made their tent. Parmanand says his mama, an old Tran told him by all means to come along with us. His people can't sew their shoes in public because of the caste rule. His people say a dog can hear a goat hair drop and a leopard can see one at night. He has buttons on his coat that are made from Dutch East Indies one-tenth gram coins. Bought in bazaar. Every zamindar in Rampur has to give 15 days for labor to government at any season or seven rupees. This year the tax of two annas on any putti or blanket offered for sale was excused. A tax of four annas is laid on every sheep load of wool that comes down. Not a sales tax but on things *offered* for sale. Doru are made for own use, not for sale.

Hence only a few show up at the fair and these have almost invariably been used. This is the long brown one. Made also at Tangi. Virtually all the Kenore boys who come to school in Rampur get malaria. Those from Arsu that go to Nirmand all get it too. There is no native dye in the hills that I am aware of except that at Tangi that they made a red of species of Galium (*bachot*) and a yellow dye; in Lahul a yellow from borage (*khelsbir*) and a Galium red (*shorr*); a black from geranium in Lahul and in Spit; in Lahul a yellow from a bear berry (*kerfa*), a black from a sulfurous mineral called (Naktsur); a brown in Rampur from walnut bark, in Lahul a yellow from a rhubarb (*artzo*).[20] A plant for dying butter for ceremonial uses and hair tonic from a borage (*muksi*).[21] Donn leaves (balsam touch me not)[22] and alum make a red stain for finger nails for women. No one will dig *ekkelbirr*[23] if there is more than one male in the family, lest rest die. There is a place in front of Nachar called Sutali (Sut = bed bug). Once a big bedbug ate most of the people and now only land there, no buildings.

November 25, 1933. Not so cold last night. Night before the pail of water froze solid. Lahulis don't like to hear an owl sing in the spring. Rampuris say an owl laughing on a house brings disaster. Wang Gyel says that normally some 15% of the sheep are lost from one cause and another on the Tibet journey. The Lahuli sheep men keep some sheep in Rupshu for replacement. Goats have a heavier loss because goats hide to escape the load. Cranes have small young when the sheep-men arrive. Usually a pair and young on each lake. Saw a quakta at 15 feet and then it walked off. In the evening, shot two. Crop of both filled exclusively with maidenhair fern. No lawyers allowed in Rampur, at least none have come. The Thanadar of Rampur died of suspected poison at Chiapon during the fair. The tehsildar was poisoned in the summer by the Thanadar. We came down two miles and camped in the forest. The little earless owl of Rampur was singing at Seahan, Kulu. We meet daily groups of three to ten men carrying kiltas of salt and a bunch of wool, which they have bought in Banjar. A Lahuli sells wool there. He doesn't give wool to single comers. They if not known would give a false name and village. Parmanand's board at school was thus arranged. The government gives a cook and his helpers. A master buys everything but flour and charges to the boys. The ata they furnish, each his ration per month. Two meals a day. Morning 8 o'clock red rice (*charru* from Roru) or sour bread (*butiru*) and dal. No tea. The dal has no butter and no spice but haldi.[24] Each boy puts his own butter and if he has any delicacies from home they will cook them especially for him four times a month. Evening 4 pm butiru always and dal. No tea. The boys get very hungry by night. The dal is a sort of soup with 40 beans or so. Rup Chand says boardinghouse dal is proverbial. Sometimes

---

20. Koelz 1979:16.
21. *Onosma echoides* (Koelz 1979:38).
22. *Impatiens* sp. (Balsaminaceae); including *I. edgeworthii*, *I. glandulifera*, and *I. brachycentra* (Koelz 1979:31).
23. *Datisca cannabina*; the roots make a yellow dye (Koelz 1979:31).
24. Tumeric.

the visitors give money for sweets but the boys never get it. He says he gives the money to the poor. Rup Chand says one master in Lahul stole almost all the money the government provides for wood and the children sat in the cold.

November 26, 1933. Rup Chand got a quakta this morning. Food maidenhair also. Saw a monticola.[25] Ram Shom went to Ani with a horse for oranges. Heard the little Rampuri owl here. A large troupe of Serajis came by at eve and camped above us. They shout and roar like a band of red monkeys. Shot a big langur and hung him in a tree. The flock was feeding in some deciduous trees. If the Lahulis want horse or cowhide they slit the corpse and the vultures eat out flesh and bones.

November 27, 1933. The night was clear till after sundown. Then the clouds came from the southwest and by morning we had three or four inches of snow. The een called all during the night. The Serajis went up through the snow. In the morning it broke and we came down to Bathad. No snow there. Droves of Serajis went home all day, chiefly with salt. One group had a local man who did their accounting. The wool seller gives him a tip and he helps cheat his fellows more thoroughly than the wool merchant would have done alone. A Kuluese who heard that one must beware of such people, thought he would steal a march on the man and slipped some stones into the scale when his wool was weighed. He put the stone on the wool side. We put up in the resthouse and got 20 eggs for a meal. Tangi keeps a man to shoo monkeys. He kills them also. There are bees at Tangi but not over four to eight pounds of honey can be extracted. At Riba opposite Rarang they get as much as four to five batti. One man has about 50 hives and has a house especially for them. He makes a wine from the honey (*digri*) that is intoxicating and sweet. Kuluese make it also. It may not be old or it is bitter. The best neoza grow at Purari opposite Chini. They are larger than others.

November 28, 1933. Came down five miles below Bathad yesterday. There was difficulty in keeping the loads on three horses but the minimum today. Serajis in droves again all day. One man obviously took me for a Tibetan trader and wanted to buy a maund of salt. Never point a finger at or turn an image (Buddhist) upside down. You may extend your hand toward the object if you wish to point it out. *Prangbo lamla mi chham; Khiki jongtong-la mi chbam*; "Fakirs quarrel over the faith, dogs over the empty dish." Mr. Asbo tore up Miss Bird's grand piano at Kyelang and made a table of it. We have been eating latterly a dish said to be a favorite with Sikhs, called *halua* or *prashàd*.[26] It has a base of wheat (*suji*). In hot ghee roast the "cream of wheat" and pepper korn, raisins, anise, zinal, almond. Finally, add sugar water to mass and it is ready to eat with roti. Kuluese and Rampuri women will not pass between two men, even if sitting on opposite sides of the road. It is equal to adultery in the women's opinion. Every year as many

---

25. Rock thrush.
26. Halva; prasad refers to the sacred leftovers of food offered to a god in Hindu ritual (often a sweet).

villages of Pangi as possible go to a peak above the village in Har[27] month. Those who have lost by death some member of the family within the year weep audibly. At least one consoler with the sort of comments common the world over on such occasions. After a proper period of crying the villagers give the mourners flowers and then all dance and sing. At night, go home and dance all night. At Tangi there is a ceremony called Pulaitch in Bhadun[28] month (common to Kenore). All the villagers take their own goods leaving a watcher in each house and for five days and nights dance and sing. Eat and wear best. Tibetans call this Yarkit. They had such a picnic at Rangum when we went the first year. Parmanand says Karrar is a good dish. Boil turnips, add barley, flour and salt and drink.

November 29, 1933. Came to Banjar. Ram Shom met us on way two miles up with our maund of oranges. Seth gave them for seven rupees. The Raja's men said 20 rupees was the price for them but came down to 12. The oranges were welcome as always. They are the best flavored outside of California. He stayed with a zamindar at Shoja. The children roared and wouldn't give their father his pipe or otherwise obey. "What awful family. My children are my enemies" complained the poor father. The shotgun today developed a new ailment. It won't stay closed now. A government employee came and said if we wanted land there was a nice piece near our last camping place. *Tatpa cherrar, kyiso la me bar.* If you have faith, fire will come from a dog tooth. *Kha(k)lo(k)* is a horse disease, usually fatal. May live three to four years. Pus comes from neck, ribs or legs. Also cough with blood, etc. Especially hard on mules. Cure difficult. There will be a motor highway begun next year from Simla to Sorabain.

November 30, 1933. Came to Manglor and camped at the sumdo. Tried to catch some of the numerous fish in the stream but they wouldn't eat meat, worms or grasshoppers. Some two pounds. Lahulis give children bird brains to make their brain steady on giddy places. Birds' hearts being small make the children cowards. There is a monastery above Tangi at Kuntsuming called Ranglik Srungma said to be founded by Lo Tsawa.[29] There are many old things there. On the Tangi Tibet road there is a large monastery, Lingo, said to be also Lo Tsawa's. Day clear and pleasant. The lemon trees are laden and there was nice fragrant red [indecipherable] in bloom along the way. *To-la gyangwa rangnor yur, gyapla Khurwa mi nor yun.* What you have eaten is yours, what you carry on your back is another's. *Sernachengi sakpa te, lama chikki khala song.* Stingily saved stuff comes to the lama's mouth. Parmanand's: *serna sakpa shulu lu* (stingily saved fades away). Parmanand, sex maturity at 13–14. *Makki chuz*—a fly fell in the butter and before throwing it away the butter was sucked off.

---

27. *Hor* is the term for "month" in the Tibetan calendar; it is not clear what Koelz is referring to here.
28. The seventh hor month *hor-zla bdun-pa.*
29. Most likely refers to Marpa Chökyi Lodrö, sometimes referred to as Lotsawa (translator), an eleventh-century teacher associated with the transmission for many Buddhist texts from India to Tibet.

December 1933

December 1, 1933. We came and camped with three tents of Khampas below Oob. *Parai leb lana apne rup guana* (Kulu). *Dusron ka slewar lagana apna shahel bigama. Beto-la kowa langto.* "To a cow a bull is fit." We met a wedding procession going up above Larji. The groom had a few strings of gold thread hanging on his face. He was sitting in a dandy carried by five men. A small boy had a necklace of eight pairs of leopard claws, fastened with silver in pairs. We wanted to buy it but the lad fled. His father said he had fed a sadhu[1] for three months and on his departure the guest had presented the claws. The silver he had put in himself. Such things were beyond price but he was ready to sell his gold earrings. They wear hear [sic] three small balls as big as a small hazelnut, covering with tiny gold grains. It is also fashion in Rampur. Went hunting in the river by Manglor: got three of the little dobes and of the dippers. Our Zankskar horse was tremendously interested in the half dozen motors that she saw across the river. The boys are learning Pheobe Cary's "Of all the beautiful pictures." They learn a verse a day. Some small beasts have taken to biting us since Banjar, such bites as we had in Kulu in June.

December 2, 1933. We came to Bajaura. Rup Chand and Parmanand wanted to go by motor but there was no room. Day warm and clear. The evening was also warm for the first time. The jackals sang in the evening across the river and down by Nagain. We wanted to buy some fish from a cast netter but he said the trout had all gone down stream or into the side streams. A wild pear

---

1. Holy man, renunciate.

had a few blossoms. Shot a white capped redstart eating a wild pear. There are three tents of Robajamanda[2] here. We met a dozen flocks of white clean goats and sheep that the Kuluese from above are taking down to Mandi for winter. In one camp several kids were trying to get milk from the dog and the dear beast was patiently letting them try. Parmanand says and Rup Chand confirms that the Kanors and Lahulis manipulate the children's heads to make them round. Also nose, fingers, etc. The Seraji women's hat is a small red oval bordered with black about twelve by six inches worn above the forehead. The Chamba Pangi hat is still smaller.

December 3, 1933. Came to Kulu. Rup Chand and Parmanand took the motor. The boy had seen one before but had never ridden in one. Not much tamasha.[3] Saw the first camels in Kulu. Day overcast. Phatti Chand is in the house where we stay, ill with dysentery. His *búa* (father's sister) widow of the old Kulu Raja, a lass of 35 or so has come to see him. She hides from me but sleeps in the same room with her nephew. A Brahmin called to give her religious instruction. She asked for constipation medicine but said she wouldn't take it till she got home tomorrow night. All my clothes are in rags and I can't go out of the building till Rinchen Gyaltsen comes from Lahul. The pass is still open to horses, we hear. Our things kept by Waslu Dawa were all in good condition.

December 4, 1933. Stayed home and fixed up the birds and put out our art collection. Sukh Das says our jewelry was purchased cheap. Bought a roll of *bu-re*, a sort of silk said to be very durable. Enough for four shirts at 18 rupees. The fine cloth I bought in Lahore is so rotten that a shirt lasts a week. Parmanand found two baskets of cauliflower destined for the Russians and bought them by giving three annas profits to the carriers. Our roll of waterproof paper delayed last spring is here. All good news from my letters that a Lahuli brought from Bandrole. Two letters written previously by Van Tyne and Miss Haglen lost. Van Tyne very enthusiastic over the skins that he says arrived in good condition. Mr. Wood went around muttering to himself and patting the hawks! Wrote that summer very hot and dry, President Roosevelt made economic dictator, prices rising, mother well, Pete wants me home, etc. At night, Rinchen Gyaltsen came with the textiles and my clothes from Lahul. Brought them on horses to Manali. Wang Gyel came also. He left this morning—early, to take Rup Chand's horses across the pass, but the pass is so easy that some Kyelang men will take them. Sunkyil was disconsolate when the horses went and came to our beds and whined. Ram Shom went home. Our family disbands after over five months. Parmanand had great fun looking at my clothes and adorned himself in my suit and outfit. He looks very well in our clothes.

---

2. Meaning unknown.
3. Word referring to commotion or fuss, from Urdu.

December 5, 1933. A Spitian brought one of the goa skins he was tanning as a sample, so distorted it won't lie flat. Rinchen Gyaltsen's brother sent garlic and a neighbor some bodunger.[4] The neighbor gave us tinder grass when we went to Ladakh first. Socks from Wang Gyel's house and the Raja's widow. Summers photos all good. Caretaker brought vegetables again, tomatoes, turnips, and radishes. We boiled the turnips, tomatoes, and apples and got a nice mess. Wangchuck came and brought some apples. The poor fellow has still a bad heart. Worked all day on the birds. Sent for 500 rupees and wrote Allah Buksh at Jaipur.

December 6, 1933. Wrote three letters to Japan and one to Tahiti. Finished unpacking the Spiti-Rampur birds. Bought some large seeded raisins called *mumáka*. Harichand tailor came from Naggar and is making a pair of trousers from the cloth that Rup Chand's mother sent me. He will come to Kangra with us and do our sewing. Ram Shom came for rest of his money and brought lemons and walnuts. Found some eggplant and cauliflower in the bazaar, so we are removed from starvation for a day or so. Wang Gyel went for honey samples from the bazaar, well diluted with water. Farmers often put in cheap molasses. The big trousers commonly worn by Mussulmans (called *shelwar*) are also worn by Hindus. Common Hindu trousers are so tight on the calves that it isn't simple to get out of them. P.N.'s account 205 maund. *Rongate*—body hairs (Urdu). *Mi tse la kyit sum duk sum; Chitke mye tungmas la drang sum tang drotsum*. Lahulis play cat's cradle. Known to the old folks (spring's long days).

December 7, 1933. Went to Bandrole. Mr. Lee very friendly but gave not much information. The Mandi aja and new wife and entourage went up when we were at Oot and came back when Rinchen Gyaltsen came from Manali. They had been on pilgrimage to Beas Kund and the sacred snakes at Rahla. Dream of rain and it won't rain. See a corpse, fish, sheep, and it will rain. Sowing fields in dream sign of quarrel. Dream of losing teeth bad for you. See someone naked or with ornament in dream bad for him. See red fruit being eaten bad for eater. Eating flesh in dream bad for you. Mud, bad water or bad things sign of something good to eat. . . . The dreams that come on the first quarter of night will not come into effect for a yr. Last quarter at once. Expressions: Ouch—*a tsa tsa*; Br-r-r-r = *Br-r-r-r*; *A chu chu*; stitch in back = yawn accompanied by *a-ro-ro*. Bad smell, puff-f-f; good smell with a sniff and "*ha*." After pleasant taste clack tongue; regret: *o hó, tsa tsa tsa*; *a kha kha*. Old women in sad recollections make noise like a cheel.

December 8, 1933. Packed birds for America. Tashi came and stayed night. Says the SDO tells people who salaam him "*Main mashlak hum*" for "*main mashkun kun*." I am beloved for grateful. His mother has been here and has

---

4. Meaning unknown.

taught him how to treat the servants so that they are all ready to leave. Tashi didn't get his full wages for Spiti but got government allowance. Now for the month after Spiti, he gets the extra promised him withheld on the account of this allowance. Rinchen Gyaltsen bought some dried mushrooms in the bazaar and cooked them with a chicken.

December 9, 1933. Arranged an exhibit for the Lees. Colonel Johnson also came. Continued bird packing. Parmanand's three uncles once were given some channa[5] by strangers. They became mad and one was robbed. Insanity temporary. The poisoners make an offering to the god of poison and name a day for giving the dose. It must then be given to someone, even a member of the family.

December 10, 1933. Mr. Gill came on invitation to see the Spiti birds. He says Whistler[6] was a very good collector but we have much more than he got, but then we had more leisure. He volunteered to promise again of 100 401 cartridges for the 50 borrowed one and a half years ago. No mention of the gun promised. Rup Chand wants the English wife of the forest officer to come. OK. An English missionary Honorable Miss Someone from Kangra came to the Lees. She said to Mrs. Lee she had heard the Kulu women were very bad, running from one man to the other. Yes, replied Mrs. Lee: "they have to work hard, are sometimes underfed and beaten and some do that." Then she went on sweetly and cited Mrs. Donald who with no such excuses, used to take her two young daughters and go camping in the public resthouses with another man, leaving the old gentleman at home.

December 12, 1933. Tibetans, gaddis, and some others churn by putting cream in a skin and shaking. Lahulis use their tea machine (*dash chum*) or the gudars churn with a water wheel dasher operated by cords. Kuluese shake in an earthen garu.[7] Today clear till three when clouds came down from Rotang and a little sprinkle. Previous two days with heavy nimbus clouds but no precipitation. Mr. Gill sent 100 410 cartridges for the 50 he borrowed a year ago last summer in Lahul. He came again to look at the birds. I showed him five or six fresh birds that are good records for Kulu and offered him the skins when made but he refused.

December 13, 1933. On the 11th home all day. Mr. Gill came in the afternoon for an hour or so. Today Trashi came to learn how to make bird skins. A man went to Naggar to get my shirts but they weren't ready on promised time. Letter from Leh says that many tankas and images brought by the Haji Yarkankdis and

---

5. Chickpeas.
6. Hugh Whistler, an officer in the Indian Police from 1909 to 1925, who was an avid bird collector, and who published some early works on Himalayan birds ("Obituary," *Ibis* 85, no. 4 [1943]:524–27).
7. Ceramic vessel.

bought by Ladakhis. Trashi says if he wouldn't get more than six months for it he would *murramat karma*[8] his master. The cows have discovered our copious fruit fields and come into the patio to eat the orange and apple seeds. Mr. Gill loaned me Whistler's papers on Kulu and Kangra birds. Later, *Ibis* 1926.[9]

December 14, 1933. Walked up to above Raisan hunting. Got a solitary snipe and a Siva. Day clear and warm. *Shikastadils sursan, gati.* The Indian teachers of English here and at Rampur teach the language with a vicious roll to the 'r' and peculiar choppy rhythm. They are well-aware that the English don't speak thus but an r is an r and must therefore be given its fullest value. I tried to tone it down but the boy said he wouldn't dare exhibit the correction before his master or he would say an English child has come to the class. Incredible, but my first-year babu who barely knew the rudiments of the language used frequently to correct my grammar with murderous incomprehensible slabs of his own diction. Mrs. Donald once met a number of illegal hunters in the woods and threatened them with punishment. Except for the aid of her servant she would have been tossed over the cliff.

December 15, 1933. Day clear and cold. Home all day. People complain we smell bad from eating so much garlic. Wang Gyel says our beds smell of garlic. Began cleaning pasham for a blanket.

December 16, 1933. Made up three species chickens. Man went to buy more. Lug Negi said we were ok to sell. Mr. Gill's Trashi came again to learn bird skin making. Mr. Gill wants me to go hunting with him tomorrow. Boys played game: Eagle and Sheep [drawing] two eagles at xx 20 sheep at oooo, no one may move into the center space. Parmanand at 16 went with a boy to Simla. Met a young Hindu on street who took him home and kept him 10 days and wanted to adopt him. Had a wife and two children already. Parmanand's pa said no. Lahuli physicians have a cure for syphilis which requires that no water or fluid be drunk for seven days. Sunkyil and a dog had a battle over a bird gut. The crows seeing that the said gut was a cause of strife removed it. Whereupon Sunkyil, having vanquished the dog, not finding the gut attacked a horse assuming he had eaten it.

December 17, 1933. Rup Chand and Rinchen Gyaltsen went to Naggar. The other boys brought down our three boxes specimens and my objects d'art from Bandrole. Couldn't go hunting with Mr. Gill but he and a young forester Benton came in the late afternoon. They wanted to know how much I spent, was I coming back in the spring, where spend winter, etc. Mr. Gill wants me to come over to hunt one afternoon before he leaves on the 20th. He is coming

---

8. Kill?
9. H. Whistler, "The Birds of the Kangra District," *Ibis* 68, no. 3 (1925):521–81.

again tomorrow. Some men came to look me over before selling—ok. Packed birds and plants. Day clear. Kulu people say if you don't believe in their gods they are powerless to harm you.

December 18, 1933. Mr. Gill and Mr. Benton came again. Got the packing well organized. Rup Chand came back at dark from Naggar. Things quiet with the Russians. Their own servants not so faithful these days. The darzi[10] who is making me shirts sent them with the dobi.[11] He won't tell, he said. The old patwari[12] Beli Ram has taken all our things to his house. Everything is all right. Mr. Gill got a card from home saying among alia if you get sick be sure to buy a piece of cloth for the god. The women from Lag are bringing wood six to eight miles for two annas a load. They get here about eight. All of our men have washed their head in cow urine only to make it clean. The god spokesmen in Rampur and Lahul drink it in cupfuls.

December 19, 1933. Worked on Rup Chand's plants all day. Gill and Burton came again in the morning. Won a rupee from Gill about my Kangra pitta. Phutook, our 22 year old neighbor, has a strong moustache and no beard. There are a few hairs on his chin that he pulls out. Rup Chand's relative, a bank inspector, came and told an interesting tale about the Roerichs. They wanted to buy his horse. Price around 225 rupees. They said they would pay later. When he went for the money they wanted to give him an old nag and some cash. He took back his horse and wrote a letter telling them his thoughts. They replied, asking pardon for the inconvenience, that they hadn't consulted the manager when making the deal. Manager now said no room for a horse. Later they sent manager to get back the letter. He gave it, the damned fool. The people at Gundla are very careful of their victuals and their serving. A cup is given supported on the palms. Elsewhere in my experience the thumb is put inside for safe carrying. They are also very careful about washing their hands before touching food. Botpa and family arrived and have been put in the room below. Popped some popcorn that I brought last year. Folks never saw such stuff and enjoyed it very much. They pop the ordinary corn by putting it in hot sand or ashes. Nirwal Chand says the bedbugs in the schoolchildren's boarding house are so abundant that you can beat them out of the charpoy with a stick. They almost ate him in Sukh Das's store when he stayed there last year. Parmanand says the fleas in Chini are beyond description.

December 20, 1933. Finished the plants and birds today. There are over 4000 plant specimens, not counting some from Spiti 1932 Bhagwan Singh and c. 4850 birds. Mammals not counted. Some nattily dressed man with the

---

10. Tailor (Urud).
11. Laundry person.
12. Village accountant: land record officer.

bazaar chaukidar called to meet me. I said I was busy. Man said to be a police. Rup Chand suspects my acquaintance is sought because Christmas is near. Mrs. Roerich asked Zotpa very sweetly about us, urged Rup Chand not to go to the US in summer. I would be surprised when I saw their new museum at Naggar. Ate caladium (*gandire*). Put in lemon and tomatoes or they would have been tasteless. Kuluese often half invert an unfamiliar word: *musala = sumal, madrasa = dumarsa; barackmaste = barmungastri; Kichar = chikar, banduk = dubuk, recrut = harkjut; hawa jahaz* is Urdu for rail (flying train). Our dog was named after a female acquaintance famed for her passion. Now the lady's relatives live below and we have tried to suppress the name but one or the other of the men bawls out the word at least once a day. My new *bu-re* shirts are very agreeable. I got four out of the 18 rupee sheet and paid three and a half rupees for making. They don't look like such elegant things but I am promised they will wear three years. Cotton Khaki such as is available here won't last three months. Rup Chand cites three pairs identical twins in Lahul. Triplets in Naggar. *Ka tsu berang ma cha cha Rar ha chiss ta cha do*. "If the lamb does not dance, how can it dance as a grown sheep?" Chunrezig's horse Norbu used to have to have a shawl thrown over its head so it wouldn't be afraid of the load at loading. The men used to call it a *dopata*.

December 21, 1933.   Almost finished the packing. The Lahuli lohars[13] and Rampur chumars,[14] also some other classes but not lohars, always scratch the head when you ask them a question. They used to cover the mouth when speaking to superiors, lest I suppose the fumes of their breath should offend.

December 22, 1933.   Went to say good-bye to Mr. and Mrs. Lee. Parted with them with regret. They have done everything possible to make my stay pleasant. Mr. Lee said in addition to Colonel Mahon others came and expressed surprise that he intended to give back the things I left in his charge. Must have been the B.'s, who are noted thieves. Trashi's master gave him a pair of shoes but before giving them ordered the hob tails to be removed for his own shoes. They cost at most two annas. Desisted when the other servant said the shoes would be ruined. Parmanand and Nurmal Chand are very fond of gum. They chew it all day. Day cloudy.

December 23, 1933.   Day cloudy. Closed and strapped the ten boxes: seven large, $3 \times 2½ \times 2$; three smaller. Sunkyil had six puppies, all giddar[15] color that got them thrown in the river at night. Sunkyil made an awful fuss. Before the event she got ready on Rup Chand's bedding and except for the whining she made she might have had her way. She has been uneasy the last week. Last

---

13. Smiths or metal workers.
14. Tanners or hide workers.
15. *Gidar*: jackal.

night she was restless all night. Last night there were many lights on the hills, Diwali. Some of them were processions of torches. At Holi (in March) they burn two huge piles of thorns in front of the raja's house. On one of them there is a pole with a flag. As the pole falls the people make great efforts to get it, even getting scorched by the fire. The pole is kept then beside the red figure of Hanuman beside the river.

December 24, 1933. Waited for the man to come back from Dhanwal but he didn't show up. Went hunting and met the Kulu post master. He was coming to see me he said and give me a letter he had detained on its way to Bandrole. He said he had an insured letter for me in his office, to call this evening. I went an hour or so later and found only a clerk who said positively there was nothing for me and I couldn't have it today if there were. The postmaster is out of the Christmas bakshish, I suspect. He couldn't look me in the face from his handling of my mail. Rain fell last night and for a while in the morning. Afternoon clear and pleasant. Sunkyil gave over her grief around noon. Wang Gyel made *gye-thuk*—coarse noodles served with meat. The girls cleaning the guard hairs out of our pesham brought back the last basket. They say it is too difficult to clean, the hairs are so fine.

December 25, 1933. Went hunting up the river. In my absence the motor manager and four drivers came to pay the Christmas respects, also our former jamadar.[16] Dr. B.S. Mohan, Medical Superintendent, STG Marwari Sanatorium, Garkhal (Simla Hills) called to find out what I was doing. He almost forcibly insisted on seeing some of my tankas, plants, etc. Then he wanted to know about the Russians. He left a note saying he had the privilege of meeting Prof Roerich and would write him about my work. The man came up to Naggar two years ago. *Len chik kuna kurnma; len chick dzunma dzunma.* "Once steal, thief. Once lie, liar." Parmanand says our tea is nauseating. Colonel Johnson's complexion he finds disgusting. Our men can't say "well." They make it a very heavy sound.

December 26, 1933. Went hunting across and below river. Got the canus woodpecker,[17] first one of the collection. A woman asked not to shoot into a certain tree, it was a deota's[18] tree. Nurmal Chand has lessons to do for vacation. Parmanand had lessons to do for the summer vacation also. If not done, got a couple of ruler slaps. Began to teach our people "Oft in the stilly night."[19] The Hindus don't say "he died" any more than we do. Some of the expressions are "*who ahanphani se rukhsat hogaya*"; "*who rurak bash hogaya*"; "*deant hogay*"; "*who alim ukbako sudhari*"; "*who guzargaya*"; is commonest. Lahulis say in

---

16. Military officer.
17. *Picus canus*: grey-headed woodpecker.
18. Local deity.
19. Popular song, with lyrics by Irish poet Tom Moore in collaboration with John Stephenson, and set to traditional melody in 1806–1807.

slang: he's gone for salt, apples, or pasham. Tangr: "He's gone to buy cloth." I asked Parmanand if he had given his people my address so they could notify him in case of trouble, etc. He replied he didn't want to hear of trouble. He would imagine them well as he left them until he was forced to know otherwise. The beehive's opening is △ about big enough for a bumblebee to enter. Saw three yesterday in a house.

December 27, 1933. Got a Tibetan pigeon alone in a field. Bought a nice old gao from a Kibor Spitian. Some children from Losar and Kiomo met me in the bazaar and were old friends. Pasham cleaners get four and a half rupees per battu. Our men clean about two chittangs a day (32 chittangs per battu).

December 28, 1933. Our motor came from below and we got the ten boxes put aboard. Sunkyil was very uneasy to see all the things go and went to the pipal tree until the loading was finished and whined. Two fakirs have been sitting beside the tree for several days. When the last box was put on the edge of the stone platform around the tree one rose and threatened a fight because the last one was shutting out the wind. He said he wanted it cold. Our men told him to put out his fire and take off his blanket. Sent a box of things to Lahul. A Lahuli boy wanted to stay with us for his board. Saw a plumbeous fish eagle on the Beas. *Jinne khuda wakhe kaum marsaki. Adhi chhor sanko dhai, aisa dube tana jai* (Hindi). Leaving half to take all.

December 29, 1933. Rinchen Gyaltsen had us fed at five and then we shivered around the fire till six. The motor said we should be ready at 6:30, so we went down there and shivered till 7:30. They finally got away at eight. Puntook and Wang Gyel slept on our load in the bazaar last night. Rup Chand went with it. The manager of the Imperial Motor Company told us they had monopoly of goods and passenger traffic in Mandi state and our driver would have to say our goods belonged to him. Rup Chand had to get out and walk past the police at Oot, Mandi, Gum, and the tax place, or the motor owner would have been fined for carrying passengers. The truck was completely full of our ten boxes and our winter baggage was on the roof. The motor was a brand new Ford, beautiful in appearance and a brilliant hum. All the people admired it. The rate to Pathankote was 75 rupees. Parmanand didn't enjoy the ride much on account of seasickness. At Paprola we all stopped and camped on a meadow beside the station. Poor Sunkyil didn't like her ride very much wedged between the boxes and tried to chew her way out. At Paprola they heard or discovered otherwise that I was a raja and asked Rup Chand my name. He didn't know of course. Parmanand met some people from a neighboring village and they were more surprised than anything to see him in Mandi. At Oot there were three shot gorals for sale. The motor manager denied having ever had any pheasants for us last year from this shopkeeper.

December 30, 1933. The motor driver told us to be ready at four so we were again torn from Morpheus' arms. This time we got off at six in the total dark. Till Palampur we were alone. Then a native gentleman got aboard. He knew Abichand and somehow took me for the Lahuli Padre. He said my work was very good and that I had been many years in Lahul. A little later a lad of 14 got in headed for Gurdaspur. He got out at Nagrota for lunch and told another stranger to keep an eye on his bundle. He didn't trust us. I gave him a tangerine and he gave me a copper. I couldn't return it so I told him I had no change and one got three tangerines per paisa. This was satisfactory, though he had a little doubt at first. Another passenger couldn't decide what I was, whether native or sahib, but the man who knew I was a padre enlightened him. Arrived at Pathaukote 11:45. Found Rup Chand at the goods depot with everything. Our men went back with a load of rock salt. They will get to Kulu tomorrow night. Rup Chand had no difficulty this time, had to walk only once at Nurpur. He got in an hour ahead of us. The police said if he were going to ship these things from Pathaukote, he would have to pay a tax of five as per maund. But Rup Chand didn't know where he was going to ship them from. A Mussulman wanted to buy Sunkyil. At Pathaukote she had a bad time watching because the train people were bustling about late and early. All were scared of her but one youth who came for fire. She rushed at him and bumped him but he never flinched and only smiled. One man there told us not to sleep in the open. At least under a tree or we should be met with frost. Very good advice but the night was partly cloudy and our bedding was only damp. The goods people said our things could be insured if they were paintings, jewelry, shawls and such like but they would have to be sealed and examined and we should have to get permission from the Div. Superintendent in Lahore by letter if the value was over 200 rupees beyond their line or 1000 rupees to Karachi. They advised passenger rate. We found the passenger rate to Karachi 7–13 per md vs. 6–9 by goods, and decided on that method. The babu said we couldn't insure the things but that there was no likelihood of loss. I made every sort of request for the safest method and finally had to take a receipt marked "insecurely packed," which meant, he said only that the boxes weren't locked. I gave him three rupees for a tip which he said was too little but quickly took when he saw no more was forthcoming. Our rings (Rup Chand's chlorastrolite[20] and my zee) attracted great attention from all the onlookers at the station. The goods 10 boxes weighed 25 maunds and cost 200 rupees to Karachi Cantonment. Returned to the dak bungalow I found a young Irishman McMann and his wife and the wife of Mr. Robson whom I met here in the spring. I showed him the receipt I had and he in indignation at once went to the station. We found our things in the van. The station master said as per original receipt we would get

---

20. Michigan greenstone; found in Isle Royale or Keweenaw Peninsula.

nothing if things were lost. He changed the receipt after inspecting boxes to Railway Risk and the train pulled out. Sunkyil found some puppies in Mandi and looked them over very carefully to see if they weren't hers.

December 31, 1933. My booking clerk came to mark off my shipping receipt "Form A Held" but when I wouldn't let him he brought the station master. Then another came to return the Form A to show me that the agent had sent word to Amritsar, Lahore, and Karachi to treat the baggage carefully because it was valuable. Tried to have the cap for the snuff bottle made and left Wang Gyel at the goldsmith all day to see that my orders were executed. At night when I came I found a monstrosity and had to leave it. The McMahons went hunting and got nil. Went with them to see the Robsons. Mr. Robson was called to the dak bungalow by the police and had a bad time with them about some money. Then their dog bit the M's baby in the leg, a nasty gash. A Mr. Kenyon there had a nice garden. He says our summer flowers grow well in the winter. In the summer nothing. Pansies beginning and roses. Last Tib. cuttings three feet high. The male erndkarbuza tree cut down a female springs up. Not yet ripe. The native pears and apples here are no good. Two crops a year in the Punjab—the winter one of wheat and chilies and the rains one of pulses, corn, cotton. Mr. Kenyon says pigs are very cautious and don't come out on moonlight nights where hunted. There are nilgai,[21] barasingh,[22] and pigs near Pathankot. Had supper with the M's. The Nepalese nurse put red pepper fried in butter on the child's wound. The Dalai Lama's death announced this week. The successor won't be known for several years, until the reborn child is old enough to talk. Hence the reborn people are so often capable. Posted letters to Zirbe, [characters], BFPI, Japanese, Tahiti, Haigh. Men like pomegranates eight annas per seer; amrud[23] five annas; cauliflower one anna; sanktra[24] three for two annas.

---

21. *Boselaphus tragocamelus*: large antelope, also known as "blue bull."
22. Swamp deer, *Rucervus duvaucelii*.
23. Guavas.
24. Oranges.

January 1934

January 1, 1934. *Mufoka sharab Kazi juti men dalkar. Pran leta hai* (Kasi—Mohammadan judge). *Male mufi dile berahm.* Left this morning on the motor for Amritsar. There was a violent fight at the start when our driver took two passengers from another that had had to stop for repairs. The struggle was carried on with shoes and ended only by the bystanders overpowering the warriors. The little motor driver fought valiantly for his passengers, but lost. Then at Gurdarpur for another such robbery, more war. The two had each lost front teeth, one one, one two, probably from shoes in previous battles. The roses in fields along the way are beginning to bloom and fragrant; cotton gathered. Arrived at Amritsar the driver said Rinchen Gyaltsen had withheld one rupee. Large crowd gathered, police thought it was a question of a bad rupee and picked out a 1918 rupee with a dot on one side and said that was false. I saw no chance of getting anywhere and gave the rupee. The hotel has changed from Cambridge to Imperial at eight rupees. Servants [indecipherable]. One bet I had been here in Room 9 three years ago and won. The dealer who once brought rubbish stones to Lahore when Miss L . and I were there came again. Had a long visit. Nurpur Raja has old textiles. German sterilization of 400,000 unfit; best sign of progress one could hope for. Went to Rada Kishan Bharanj and Gopal Das. Selected 20 phulkaris[1] from latter to be held till we looked over the other peoples' stock if our offer wasn't accepted.

---

1. Embroidered textile, made in western India by women as part of dowries.

January 2, 1934. Sent word to the cook to prepare for us only Kari. The sahibi doesn't want to eat kichar michar.[2] He prepares the kichar-michar as well as it can be prepared. The hotel management is new and everything is clean and pleasant. The servants are cleaner and nicer than Nedsu's. A gorka keeps watch at night. Rates eight rupees. Went to Radha Kisha Bharanj and bought seven old Bokhara suzni[3] and a couple of pieces of old papier mache, etc. He has a nice assortment of pictures. I am seeing more and more beauty in them. Many are uninteresting but the old ones are often lovely. The old man has been visited on every previous trip to Amritsar but never has sold me a thing before. Saw a magnificent illustrated Persian manuscript with 34 beautiful pictures. Said he paid rupees 25,000 and will take a little profit. Has a nice old silk print. Have looked at 40 old shawls, none worth buying. Many of the Chamba wedding scarfs. Alike in gay color, crudely drawn figures, finished on both sides, cotton cloth. Bought one for 40 rupees.

January 3, 1934. Gopal Das refused offer for phulkaris, so went to Radha Kisha Bharanj and selected. Bought three or four old Afghanis or Waziristan phulkaris (the red block) from Gopal Das. Others haven't any, or few. He had five fair or good ones. Brought me 14 suzanis and bed quilts, all old and very rotten. Asked 35 rupees for the lot. Refused 35 for one and 80 for four. Says that bears gall gives relief at once to shortage of breath. Parmanand thrilled with the Golden Temple. Met three Lahulis and two Rampuris. One Rampuri lad of 14 is a servant here. Everyone likes our dog. The hotel wants it. Men say trade is dead. See no one in the shops of any sort but people look clean and well-fed and the streets are full. Little rain night before last. Say that pneumonia will not be common now. Many cases in summer due to fans. Saw some old corals, very cheap. Large ones at six and a half rupees and some nice medium sized ones, odd shape for three and a half. No turquoises. Taj Dun is here. Has nothing but rubbish. Much glass jewelry. Parmanand has eaten his first mangos and pistachios today, also pork. Sikhs are fond of it. They say the fat-tailed sheep cost 16 rupees. Tail 12 annas a seer. There are a few mosquitoes here and we fastened the netting to a chair and bed post. The chair fell on Rup Chand's nose in the night and left a wound and swelling.

January 4, 1934. Our 14 suznis came back in different hands in the morning. We played we had never seen them before. There was a mass of rubbish, among though, a wooden plate I didn't know. They said in the bazaar it was from Dera Ismail Khan. Still being made. Got a nice iron inlay box. Our Lucknow man hasn't come yet. Went back to our old merchant and looked at more pictures. Have seen at least 1000 in various shops and only six or seven worth buying. Almost all are Kangra. The few Moghul ones are portraits. Bought one picture of cows and Krishna for 30 rupees. Bought also 12 phulkaris for 100 rupees and the only good Delhi embroidery seen worth buying since two years ago. A man

---

2. Gibberish.
3. *Suzani*: embroidered textile made in Central Asia; often made by women as part of dowries.

Ghulam Hussain brought a nice black Persian shawl in Lahore. Asked 500 rupees and got 75. He says also a German has been buying for the Berlin Museum. Met a man in the street who said he was Professor of French in Amritsar College. Speaks French. Name Tara Singh, Hari Singh, Mugal Gate, Amritsar. Came to hotel unannounced in evening but didn't stay long. Seems very decent. Has short hair and is a Sikh. Our old merchant is so elated that we deal only with him. He wants a separate check for each purchase though he pays four annas collection on each. Four little beggars stay with us most of the day nowadays. One after a tip just stays within sight, ashamed probably to beg again. The men's cinema last night was Indian made and to Rup Chand and Rinchen Gyaltsen intolerable. The others hadn't seen one before and were interested.

Plate 56. Inlaid box of Bidri ware, purchased in Amritsar (UMMA 17250).

Plate 57. Phulkari textile, hand-embroidered golden silk thread on red background, purchased in Amritsar, January 4, 1933 (UMMA 17358).

room and thrust a book in my hand and two necklaces of marigolds on my neck. They said they were from the Golden Temple and shoved a book in which was written by some police inspector that said person, former Golden Temple mahant, was a legal worker for the government and a useful informer and he wished him well. I should forthright subscribe 10–20 rupees. I declined with thanks saying I was not accustomed to such methods and I didn't know who they were. Flour here two rupees per maund (pka), Rampur four, Tangi eight rupees, Lahul five and three-quarter, Kulu three. Was roused at seven by the Lucknow merchant. Went then to see his shawls. He had four or five choice old pieces. I bought two more for design, color, etc., ca. half his stock. Then went to see the old man again and looked at pictures. The old man knows me now and gives a low price to start with. There is one picture I must buy because it has the colors of my old tankas. He took me to one very stylish place where there were two shawls. The proprietor asked 200 rupees and was promptly taken aside by my friend. The revised price was 30. The calendar maker lama of Lipe met Parmanand yesterday. He was shocked that he ate with us. Ladakhis use dog hair (pasham) for cloth. Zanskarians donkey pasham, Rampuris and Lahulis don't skin horses, except that the latter make sieves for separating grain and straw etc. Taj Dun says that in 1925 he bought many shawls in Paris. Old man says paper is relatively new. Old books on bamboo, leather, birch bark, etc. Bought a fine old shawl, much painted, apparently at the time of manufacturing [long line of characters]. ଓ ଚ ଧ୍ୟ ଅଧି ଓ ଚ –ଓ ଚ ଡ ଓ ଚ ଡ ଧି ଧି ଚ Diwan Sant Ram, Loghad Gate, Kucha Jas want, Katra Duld, Amritsar, old science books medicine, astrology, mostly Persian and Sanscrit [sic]. Also has a book on making gold. Two books manuscripts with above characters. Kuluese when death occurs say often *Kiska hota usika marta, Jo hona tha hogaya*. Bought several more pictures from my old friend. He stuck out for 80 rupees for one. I was ready to give 70 but no. Have passed up not more than four other ones on account of cost, and certainly have seen 1000. It is seldom that one sees a good one at other stores than his. The old book seller says he will show how to tell age of books from their paper. The coral merchant raised the price of his corals on our return and when I walked off he begged me to buy at the former price. The Gopal Das were ready to sell my phulkaris at seven rupees for 20 but I didn't need them. The French teacher came in the evening. He wanted me to try on his turban and wanted my photo and when was I going home.

January 6, 1934. Left via motor for Lahore and put up at Cecil's. The motor driver was new and was very happy to have beat the 10 o'clock train into Lahore. The usual nar with the tanga walas and coolies at Amritsar. At Cecil's there was a Bayaura man who was so delighted with us hill people that he couldn't leave us. Went to the bazaar. Found many sitar, sand grouse pintails, GW teals, and bater for sale. Several ruddys and two queer large ducks. One dealer had a lot of imitation Chinese porcelains. One had a nice Kirman nownazalik, one a nice royal Bokhara and rose design. Persian leopards 15–20 rupees. Snow leopards,

45 or less, wolves 10, stone martens 20 or less. Bought some boiled ham and had lunch on the street. All the men liked the ham. Spent half the afternoon trying to find the films that Mosel Chand kept to no avail. Bought the baby suit for Tonrup Gialtsin of Sungnam. A man yesterday tested my watch streak on his test-stone and said there were three parts gold to nine of alloy. Parmanand's school used to sing before the raja: "*ai shahi ali, tu duch huama, lakhon bara juta sale.*" The master sings the first line, then the children sing the rest, clapping their hands now in front, now behind (twice): "*maharanien mashnik sadha, maghrit soha, sikha tesa chalta rahe*," with the "maharanien" the children bend their knees forward and thrust out their hands with fingers thrown open; the neat is accompanied by a snap of the fingers and a bowling motion of the hand. Each stanza is finished with a salute.

January 7, 1934. This evening the sunset was remarkable. In a cloudless sky rose against a light mackerel sky lit with a pale yellow pink, long frail brushstrokes of madder purple, running from the horizon to near the zenith the background of the ether was baby blue. The clouds increased and all light faded. Then the mackerel-ing took on a strong salmon pink, or blue paler than the firmament, or these two tones bled together and the feathering became dark purple. The crows that fly in long lines to the forest nearby finished at 5:45. There were two lines visible from our house. The grey titar sings night and morning nearby. Went to the Museum and saw 1000 pictures. There are many of interest to the student, many portraits, all chiefly Indian schools. Two or three were very pleasant and as desirable as those we have. There is a very nice room of Peshawar Buddhist sculpture, black marble, some six good shawls, a nice Kangra brass image, and a few good rugs. An old wood carved doorway was most attractive. My yesterday's duck have been sold. There were common ducks at one rupee, peacocks at five rupees, guineas at one, some scrubby turkeys, white and brown. Hajee came in the morning and brought some rubbish. There was a silk doti, the like never seen, and a small bag made of pieces of the pink silk such as I have in the tablespread. Also a new rug of good dye and most pleasant color and pattern. Design like Kazak except for the medallion. Relative of Lee's rug. Hajee said our Lucknow shawls were all good, some very choice. Aman brought a lacquer picture, probably Persian but not very old apparently. Parmanand says our crackers are no good. Swiss cheese is good. The dealer who last year wanted to buy my zee asked for it again. Met a flock of Tibetans, probably Chang Tang. Parmanand recognized a Rampuri with them. Men went to see Jahangir's tomb. Weather yesterday and today cold. Lahore City, Deli Gate, Chauk Wazir Khan, Kuch Gondewala, Haji Karim Baksh Carpet Merchants. Gave Rinchen Gyaltsen some cheese in the bazaar. Arrived home with rest. Rup Chand always gives servants something of everything and when he did now, Rinchen Gyaltsen said he had had some. The Cecil Hotel where we stay was once a sumptuous place. Outside there are wood scrolls to delight the most artistic minded. Our room has nine sides, 10 × 14 × 14 ft high. Paneled throughout except for a strip of plaster at ceiling three feet wide. A fancy portico of wood scrolls is built in front of the outside door (inside). Over each window and other door is some complicated wooden device. The one over the door connecting with the house's interior has in its center a motto "Be Good." The ceiling is the chef d'oeuver. Frame after frame lead up the central lozenge 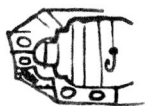 from which hangs a huge iron hook that once supported the lighting. [sketch] then there are 12 disks of wood and four corners of scrollwork, and outside this 29 framed pieces of oil cloth. The floor is parquetry with three pieces of wood per block of one square foot. Over the fireplace is a mirror, flanked by two small ones on each side and the whole ornamented with railings and scrolls. Red green and white hexagonal tiles border the fireplace opening.

January 8, 1934. Went to the bank and sold rest of my dollars at 252, the cheapest the dollar has been since I have been in India. Buy rate is 267. Saw my accountant. They keep a record of the payee. He has to pay 4–6 annas per 100 rupees for collection if [indecipherable] Lahore. Didn't see the manager and took my envelope out of custody. Went to Museum again. Counted 700 pictures and sketches, five percent very nice. Also two nice Kangra images, one of Buddha, very old. Saw the director and asked him if exchanges were possible. He says the museum has many Peshawar sculpt not on exhibit, and about as many more pictures. Wants to see mine. Says pictures will be gone from market in 40 years. If I had 50000 rupees, I could buy all good pictures available in India. Met the young Indian from the DM office. He said there was considerable trouble about me after I left here a year ago. He was ordered to write the Kangra DM about me, but didn't. Bought a small Indian rug of good design but for color and a few other pieces of Hajee. There is a nice Kurman prayer at one shop. Otherwise nothing of interest in the city. An orange-seller said he sold about 100 a day and made eight rupees. Our men say if their hair is two or three inches long it often hurts to put their hands even lightly on it. Rup Chand dreamt of flying the other night. I have had such dreams. I remember in the dream of flying, so easily that I determined to try it after I woke up. Ears from tip to tip: Rup Chand 73; Wang Gyel 66; Pramanand 61; Rinchen Gyaltsen 63; mine 70. Bank employee says counterfeiters are making genuine coins of silver that are indistinguishable from the government's. A rupee costs 10 annas. Says counterfeiters can't imitate the watermark in bills.

January 9, 1934. *Banahi ka sab yar (dost), bigari koi natrin.* "That which is made has everyone for friend, spoiled or incomplete, no one." *Bamyaki ramai mukan ya biya (shahdi) ne khaiye* (eaten). Mohamed Makbul Kashmir House, New Steet, South Gate, Bagdad, Irak. Five days from Karachi. Son of Abdul Samad, Suraf Kadal, Mahala is Haji Mohamad, Srinagar, Kashmir. Haje came in the morning bringing back all his goods of yesterday. He whined till I bought another chamba cloth. He said his daughter was getting married and needed 2500 rupees, and I being an old friend should buy enough to supply the money. I told him I wanted to hire him to sleuth out nice things with me and he is ready to go. Ghulaim Mohammad, the white bicycle Kashmiri came again with his exquisite royal Bokhara and I bought it for 85 rupees. Parmanand went to Amritsar to get the two pictures we left behind. Radha Kisha Bharanj was so pleased to see him he embraced him and told all the people within earshot that I had found his goods so superior I had sent for them from Lahore. He explained to Parmanand for the severalth time that his brother had thrown him out of a tanga and wounded his knee. The DC[4] at Lahore sent a chuprassi[5] once to summon him with his pictures. His brother got hold of the servant and bribed him to say nil of the pictures. Meanwhile, brother took pictures to the DC. Went to the Museum and showed our pictures to Director. Said all were choice. I asked to see his that weren't on exhibit but he made an excuse that the good and bad were all together and couldn't be seen short of much labor. Our gold inlay dagger and gold velvet print came from Jaipur. The [indecipherable] didn't want to let me open it at first sans paying 165 rupees but I was firm and it was opened at the manager's presence. Said individual was very gracious. Rup Chand was along. Everyone was most civil today because I wore my best clothes. People here like Sunkyil too. Haji wanted her and asked how much meat he'd have to feed her. The Baksh Elahee manager game me a letter to his branch in Karachi. Says I can buy a new gun by depositing my old one. Pohtums[6] available. Much rain in Amritsar last night. Nil here. Mosquitos plentiful in our room but not outside, nor did they trouble us at Shalimar last year. Shalimar gardens originally had seven parts, now two left. One man in the market asked if I didn't come from Kulu. I asked how did he know. He said Colonel Pennick's son used to wear a beret like mine. I saw white eggplants in the bazaar. The Abdul Samad said he had three stone figures three feet high buried at Srinagar. Nearly perfect. He says at Baghdad many old things. His son is there.

January 10, 1934.   Left on the 9:30 AM for Karachi. The hotel that last year charged one rupee a day and agreed to give meals for two at 1–8 now billed me six rupees a day. Arrived at station with my voluminous baggage was seized by the Br. Station inspector and had all weighed. Then had to bribe him 12 rupees to let me off of half of it. I went second [class] and the rest all third. An Indian agricultural agent in the compartment was going to Bawabshah to select a piece of newly opened land for an experimental farm. He can take up to 20,000 acres. It will be watered from the Sind Canals. An English company has taken up that amount. He wants me to come there and hunt, will see. A young Englishman is in the compartment too. He gave me Philip Gibbs': *Middle of the Road* and I read it through. Well written and not too tedious with its preaching. Boy nice.

---

4. District Commissioner.
5. Office messenger.
6. Meaning unknown.

Rest of compartment Indians and unflavored except one assistant state minister at Karachi who is very pious and prays abundantly. The journey all day has been through such country as one sees till Delhi. The sunset was a glorious red through clouds. Lived on bananas and four boiled eggs all day. Sunkyil got restless with me and went back with the men. One pays double for taking her with me, rupees 8–12. In the baggage car there is no safe place. One first class passenger had four dogs with him. We attracted our usual attention. Some Haji Yarkandis tried to talk to us in their language. One as usual said "of course you don't mind the cold" and "it's as cold in the hills physiologically."

January 11, 1934. Last night the men arranged inter selves to give me a lower berth. Marvelous consideration. I gave it to the Englishman instead and spent the night comfortably though sans sleep in the upper. Our station master got up at four to pray and prays 15 minutes at a stretch. He is fasting but couldn't get any food in the night. I got off at the city, intending to camp but the city is so open that one would have to go far. A scout for the Claremont Hotel got hold of us at the Cantonment station and when we agreed to go with him at four rupees a day he took us to the city station and we had to walk back. Two men pulled our luggage in a cart. We arrived in front of a huge house in a spacious well-kept walled in yard. Many such houses and buildings in the city. The household came to take a peek at us and then withdrew leaving us on the front steps. The scout was however resolute that we stay and finally after 10 minutes ushered us into a huge bare room, like Hill Auditorium. I was thoroughly amused by this time and demanded to see the manager before I let the coolies go back. After another 15 min she came, a young Indian woman who spoke very good English. She thought everything would be all right that I wanted, including rupees 5–8 a day for extra food for the servants to be served to us in our banquet hall. A bed was brought for me, linen and towels were supplied, plus a chest of drawers and a dresser. We cleaned up and then Rup Chand and I went to the city. First to American Consul. He met me on the stairs and went on down. When I got into the office, he came back hastily. He said he had seen Rup Chand ("A Tibetan") down on the street and knew it must have been I, so he came back (I didn't say to him I was coming to Karachi, so how he recognized me by Rup Chand's presence is clear). He said he was ready to help me take Rup Chand to America. I didn't say I wanted to, but said I was glad he would help. Oh yes, he said he was ready to help me in any way he could, but today he was going hunting and would be back Saturday morning. Would I come in then, meantime go to Cox and Kings and see them about my boxes. I went there and they were most agreeable, ready to do anything but come back Saturday morning. Then we went to see the gun shop to have the gun mended but their gunsmith was away, but would surely be back right away. The city has broad streets and nice stone stuccoed buildings with none of the slums that make the Indian city elsewhere. It has probably all been built as a result of the expansion of the seaport. The population has Arabian and Negro inter-mixture and there are Persian names to the shops. Went to the two fish markets. The one was the busiest place I have seen in India. It swarmed with women selecting a few fish here and there. A bunch of dania, a few onions, or bananas made up the provision store. There are the usual vegetables and fruits, but many ernd karbuza, papota, and a small potato looking fruit with two or three large black seeds that is eaten only dead ripe and tastes like a superior sort of medlar.[7] There are trolleys that run toward the harbor called Kiaman, a bus line that runs night and morn from airdrome to city. Otherwise transport is in a one horse Victoria that rides very comfortably.

January 12, 1934. I took a list of contents of the boxes to Cox and Kings, then spent the rest of the day trying to get the gun fixed, and buying a few things. There are no antique shops here. Yesterday I went to two rug dealers but they had nil. In one was a Persian old man, so cross that he couldn't speak without making some complaint to his son. He had bushier eyebrows than I ever saw in a human being. The hair must have been one and a half inches long, equal to his mustache. He had a few rubbishy pictures. I saw them again today in the hands of another dealer who has tried every year to sell me things either in Amritsar or Lahore. This year he has a nice prayer Caucasus and a queer sort of Bokhara. He had a gold ring with a beautiful turquoise set. We bought some crabs that had been kept for us in the market. The stock is exhausted at 8 am if you don't reserve them. There are no wild ducks in the market today but many titar, grey and black, also gurrals. There are silver smiths that make vases and things that the English buy. They are most interesting. The shops have poor stock. I had to visit four film places to get my films and could find only 350 .22 shot. The gunsmith agrees to mend the gun for 12 rupees versus 51 at first estimate. Bought some green coconuts, two pani wala, 1 matai wala. The water is no different or copious than in a ripe one. The malar has a little rind of flesh inside. Boiled the crabs. Only Rinchen Gyaltsen and I ate them. Wang Gyel and Parmanand ate only one, Rup Chand none. The guavas are becoming more pleasant. In Lahul a weasel going to house is good. Reverse bad. An owl calling in spring is carrying away corpse. In fall load of grain. A man in lifting a heavy load makes an hoo sound, hence —. Lahulis formerly had great respect for oath. Indians little. In Tangi before gathering zerra[8] all must take oath that they haven't gathered any previous to the opening day. If you throw a stone at an owl, he will pick it up and take it as evidence to Yamrap, the celestial book keeper. Our landlady is native Xian and wealthy and pays 200 rupees a month rent, the cook told Rup Chand. She takes in roomers. Keeps a school for Indians mixed creed and sex. There are three sisters. All speak good English, all with names like English: Mrs. Moore, Mrs. Green, etc. The father is Dr. Nazareth.

---

7. *Karbuza* is a muskmelon; the small fruit resembling a potato is likely a *sapota* or *chikoo*, medlar (*Mespilus germanica*).
8. *Bunium persicum*, zirra. Koelz (1979:35) describes this as an esteemed condiment in Lahul and India, and also used to treat neuritis in the arm (*tsakar*).

He has most distinguished consultation hours: 10:30–12:30, 5:30–6:30. Transport of goods in the city is effected largely by carts drawn by a single camel or donkey. The camels are so lofty that the shafts have to descend at an angle of 60°. The donkeys are small but the swiftest things we have seen. All our crowd admires them. Many of the camels are seen with loads of grass on their backs so that the camels' legs only are visible.

January 13, 1934. Finished formalities for shipping 12 boxes of things to America. Freight to here, rupees 200. Packed up the Amritsar purchases and the shawls brought down by hand from Kulu. The boxes have all arrived in apparently good condition. Nine of them are $3 \times 2 \times 2\frac{1}{2}$, the other three a little smaller. Three have locked cases inside, the lock wrapped in cloth and sealed. All boxes lined with waterproof paper. Insured marine $7000. Gave itemized list and got consular invoices. Sailing from Bombay, *President Polk* on January 26. Arrive New York, February 27. Charges forward. One boat left today for Bombay, but there will be two more arriving there in time for the *President Polk*. Saw the American Consul Mr. Riggs again. Told him how I have been treated by government here. Had a letter from Dr. V. Ghani urging me to come to Nawabshah. There is all sorts of game and he has a friend with a large estate. There are no American cartridges here, the cartridge dealer told me. When I threatened to telegraph to Lahore, he sent to the other shops to see. Went to the Kashmir House in Elphinstone Street. A nice Kashmiri who gave proper prices right off and didn't try to show rubbish. Has five or six old shirazis, much worn. An English woman and two men came and bought a rug made of some ten jackal skins with the tails sticking out at one end for 23 rupees. The dealer was much amused afterwards. He said formerly the sahibs were discriminating and bought good things but nowadays such stuff is in demand. Can find no Cutch work. Cox and Kings most obliging. The Kashmiri says if we go to Shikerpur, Upper Sind, we can get many old things. Tatta was once the capital of the Sind but there is nothing there. There is a dealer in Hyderabad, Sind. Abdulally Moosathoz says good shooting at Shahbandar.

January 14, 1934. They say that goats male produce a harsh wool sheep offspring x ewe in Lahul. Went down after gun. Ready as promised, 12 rupees. Went to another antique shop. Three shawls, rubbish. Virtually nil in city except two or three common old Shiraz. The men went to see Albertina Rasch last night in a movie and liked it, dancing, music, acting, and all. The audience mostly soldiers who smoked. Indians in the expensive seats. I stayed home all afternoon and the men went to Kiamari to see the ocean and ships. Thrilled. Got five species of sandpiper, a gull, and two herons. Saw *Literary Digest*, *Saturday Evening Post*, and *National Geographic* in a bookshop. Post one rupee. Nice stone martin skins tanned at 16 rupees. A peacock wanted to go along with a fakir and the man had a bad time getting away from him. The peacock always ran after him and wouldn't desist even when threatened with a stick.

January 15, 1934. Spent the day finishing up affairs in the city. Deposited trunk and suitcase with Cox and Kings sealed and insured against fire for 1000 rupees. The Sind laws prohibit the killing of about 20 common birds including the gursal which they sell for food in the market. So after being sent from pillar to post, Commissioner being out of town till the 21st, I phoned the American Consul and he phoned someone else who said shoot anything you like if you have a British India license. Went to a tannery where they said the foxes and langurs would be tanned for two rupees each. Women were turning over skins that were drying in the sun. Women dressed in red print clothes and good natured. All the inhabitants are accommodating and often go a long way out of their way to call someone or show a shop. A dealer from whom I once bought two rugs in Lahore came and showed me his stock. Has two or three semi-antiques of no first rate quality. He seems, however, to know rugs. He remembered the one striped shiraz I bought from him. Has three bad shawls, no pictures. Can't get American 12 bore cartridges here but got 1000 rounds 22 shot. The silver work that was to have been ready last night wasn't ready today and the shop being closed for some sort of holiday we had some trouble rousting the proprietor. He cheerfully gave back the things and expressed regret at inconvenience caused.

January 16, 1934. Went to Lloyds Bank and opened an account with 2000 rupees transferred from Lahore. Left a suitcase and trunk both sealed with Cox and Kings yesterday. They said they couldn't insure for anything but fire and they would insure them for 1000 rupees fire. I got nil to show for the transaction. My two last boxes were nicely strapped with iron bands. Went back to cartridge shop and bought 100 12 bore English # 8. A life insurance agent Mr. Mirza said there was fine shooting at Udaipur. A huge artificial lake there. Gave me name of a cartridge dealer there. A man from Jhimpir was at the cartridge seller. He said there was a large lake at Jhimpir and there were only a few ducks, all of which sounded more hopeful than the tales of unlimited game we had heard before. We resolved then to go to Jhimpir and then go on to Manchar Lake. Dr. Nazareth said that was fine and Lakana also. A friend at Lakana had a vast estate on which were many lakes. The daughters gave us a duck [indecipherable] and we went to the cantonment station to go to Jhimpir. We were sitting there when a station officer came up and began conversation. He inquired if we were gurkas and the men agreed. A little later a little whippersnapper and an old man came to me and asked for my passport. I demanded credentials and they showed CID cards. I gave passport but they couldn't make anything of it and another officer came. They wrote down my name and asked Ringchen Gyaltsen his name and went. No questions as to where I was going or about guns. Then went to book luggage. The agent made overtures for bribe but it wasn't understood and he refused to book on grounds name and address wasn't on the luggage. Then I said very well I'd take it along. Foiled there he charged 1–4 for the dog, the rate for third class being less. Went intermediate with Rup Chand. Two other Indians in it. Comfortable enough. A child had been struck in the hand by a flying stone just before reaching Cantonment Station. Arrived Jhimpir at 10:30 and camped outside the town.

January 17, 1934. Today is Eed. The shikari[9] who came along with us got us a cart of two splendid oxen and we came to the lake. The guide said today being a holiday we'd have to pay three rupees. The man himself said he was getting two. When found out, our guide said that was another fellow, etc. The oxen came as fast as a horsemen and the driver was a good sort. There is a small rushy lake that one comes to first. There were a few sirkabs, mallards, teal, etc. at one end and some terns and herons and snipe. They promptly left at the first fire. We camped beside a village of Mussalmans at the lower end of the small lake near the stream that goes into the large lake below. This one is about a mile long and bordered all around with rushes. There are fish. The other lake can be seen some miles away. The villagers are all 5'10"–six feet, friendly, and apparently poor. Their houses are huts, made of brush and grass. They have some nice cattle, sheep, donkey and oxen, camels. Fields we haven't seen. They are all suffering from colds, due probably to the cold wave and their splashing in the water after fish. They lie on a round copper vessel and splashing with their feet, propel the thing about the lake. Their net is of spun cotton supported on a frame of two crossed sticks [drawing]. This hive-like affair is dropped into the water and the netting entangles the fish. They are out in the ice water at daybreak and some spend all day and go again at dusk. They get a mullet ca. one foot long. Catches from two to ten. Some walk about up to the navel and some have boats. The boats are about 15 feet long with an upturned blunt prow.

January 18, 1934. I stayed home and made skins. Rup Chand went down to the other lake and brought back a Jacana, new sand grouse, new courser, and a new lark. Our Piru told us to look out for thieves. The people were a bad lot. The people said he was the thief in the lot. An old woman with a painful swollen knee came for medicine. Another with scars all over her arms and on her tongue. A child with sores as big as a dime and larger all over the body. I have never seen that disease before. The people are tall and graceful and for the most part nice-looking. They wear enormous pantaloons, the men of blue or white; the women of red or some deeper color and turbans and shawls of mostly bright red cotton print with small figures. They have a fearful spitting habit and we have to ask them 50 times a day not to spit in front of our tent, a space that is occupied by one to five people most of the day. They live in huts, they say, because in summer the lake rises and then they go down by the other lake where their fields are. The crows and kites don't come to our tent at all, though we throw out the bird bodies. Large flocks of ducks come from the big lake to feed at the upper end of the small one at about dusk. There are also a few

---

9. Hereditary hunter.

Kunj and a flock or two of ten each of spoonbills. An osprey or two fish nearby. The sheep are different here from other places, pale brownish. The natives don't wear wool and we haven't yet found out what they do with it. Some few speak Urdu, but most of them don't understand a word. Day calm and warm. Yesterday a stiff north breeze. A group of 30 people came down from a nearby village carrying the corpse of an old woman on a charpie.[10] Some six or eight walking ahead of the corpse were singing. Their voices are clear and harmonious but the song hadn't a hint of melancholy. Three men on horseback came behind. They were some sort of officers, we were told. The clergy man who officiates at interments lives near here.

January 19, 1934. Stayed home and made skins, over 50 today. Natives began bringing specimens today. The children catch the crested lark in nets. When the total got excessive we stopped purchasing them but they brought them with the crest pulled out. An old man ca. 80 came for eye medicine. Breeze from the north. All nights are calm here, but at the Jhimpir station it blew all night. There are no trees near here so there are no opportunities to shoot hawks except in the air. The white headed harrier *aeruginous* is common, hunting over the reeds. A few others fly over during the day, of several species. The wheat flower is ground very fine and makes the best bread we have ever eaten. The rice in this area is very ill-smelling cooked. The water makes bad tea so we bought a jugful from a well a couple miles away. That's a little better. Lahulis and Hindustanis say elephants afraid of ants. Parmanand heard an elephant would die if he ate an ant. We haven't small change to pay for our bird purchases, so I issued paper money. This to be redeemed when one rupee worth was present. This they took readily. They called it a "pass."

January 20, 1934. I stayed home all day. Parmanand and Rup Chand went over to the Indus and saw crocodiles. They got lost and didn't get home till 10. Meantime we walked about with lanterns to show them where camp was. Fires burn in every direction. Finally we heard a gun toward Jhimpir and answered with one of the native's guns. Rup Chand replied but the sound came from toward the big lake and he was over by Jhimpir, just in the opposite direction. A lad with a fearful scream got an answer from Rup Chand, while Wang Gyel with the lantern went on past him. Rup Chand thought the gun was some duck hunter and the lantern also. Wang Gyel said not a word but let the gunman do the talking in Sindi. A water boy brings water from another well and it is drinkable. We have tea copiously for the first time since our arrival, the accumulated thirst of days. Piru brought in 11 sandgrouse (*Peterocles*) and six hawks. Made 30 skins. The six coots were very fat and the work went slow. Rup Chand says that the Jhimpir end of the lake was literally filled with ducks last night.

---

10. *Charpoy*: rope bed.

January 21, 1934. Made 60 skins. All home all day and all the village here every second. Brought some jewelry which we bought at very attractive prices. It was all old, and they congratulated each other on disposing of it. No use trying to drive them away, but we have tamed them so that they don't spit so copiously in front of the tent. They asked us not to urinate near a "padre's" grave nearby because four sahibs that did so got sick, etc. They say they can't fish when the wind blows. Yesterday was a good day. They brought in several mock catfish that weighed two to three pounds. There are four species of fish in the lake they say. They splashed about yesterday in the reeds. They have gill nets with pieces of straw for float. They drive the fish by beating the water with poles. Piru killed 13 pintail ducks at two shots this morning. The crows began coming to the meat yesterday. No chus eat there, though they sometimes fly over. Dogs abundant but none come to camp. We visit all day with the natives. Rinchen Gyaltsen often speaks in Tibetan and they reply in Sindi and each derives a meaning. The Sindi is usually delivered in a high voice to increase its penetrability. Bought a

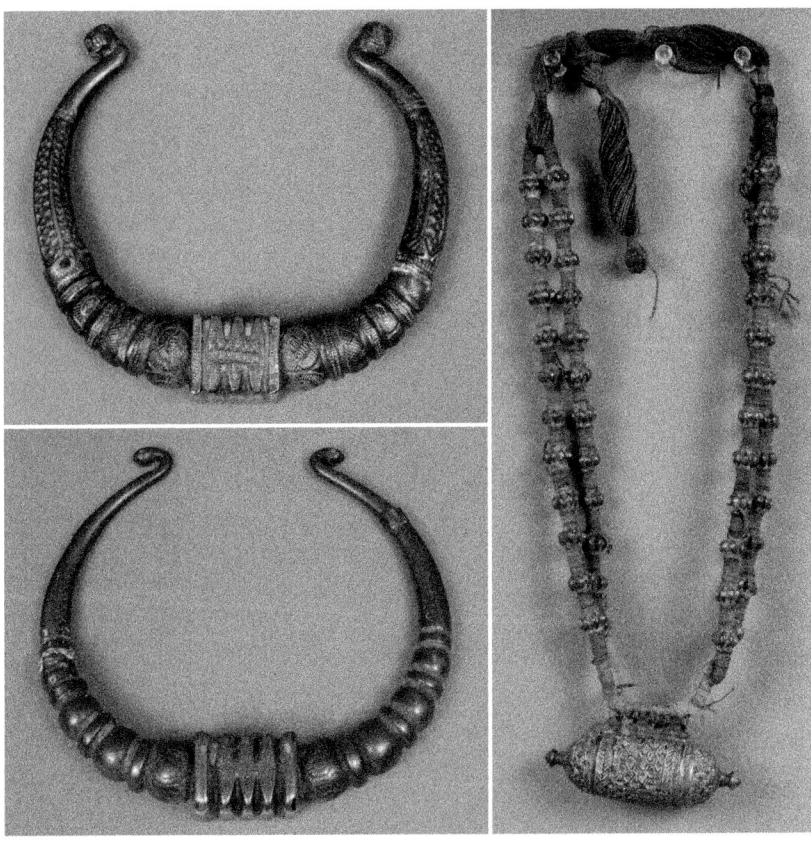

Plate 58. Jewelry purchased near Khaufer Lake, Pakistan (*clockwise from top left:* UMMA 17125, 17129, 17126).

telor at night. They said they could get four rupees for it in Karachi and I said they could take it over. They wanted the gizzard back. Our water boy wanted to take over the wood contract for three annas a day, but he was prevented by the local municipality. He has no right to interfere with perquisites that belong to them. All the silver things were priced and then as the price agreed put in one man's lap. Price given in a lump and the commission agent had to have a tip. Sand grouse excellent eating. Coots good too. The SE owls eating shrews and mice. The big striped eagle had a small heron in its mouth. The Dortse and Jispa Lahulis are famous for their accounting. They go en masse to Kulu and Leh and back and then cast accounts of expenditures on the way. Bookkeeping is mental and the accounting lasts a week or so with plentiful liquor. Fights are common and sometimes the magistrate gets into the business.

January 22, 1934.   Went around the little lake. Four kunj but not much else. The people came early and I bought 60 rupees of old silver ornaments. The very old things are very nicely designed and ornamented. They came back when I came back and a huge crowd visited and chatted as if they hadn't seen one another for years. Two children went along today to collect the empty cartridges. They say the gators here eat cattle and often children. Made 40 skins; 28 left for morning.

January 23, 1934.   The sheep here are large whitish to brown with small ears, short tails, big bellies, and very large udders. One today had a bag that nearly touched the ground. Bags as big as a four quart pail common. Were left alone for about half the day. If one comes and settles the rest promptly gather to see what the attraction is. We seem to have bought a sample of all the jewelry. Nothing new has come. Today a lot of pillows came for inspection. They are about two one half by one foot piece of cloth or embroidery. A hunter brought some spoonbills, godwits, and curlews. Piru is over by the river skinning two crocodiles. The skin here is called *moldi* and the upper lakes *Kanjiri*. We had a bad time getting the information and it may be wrong. The hunter was shocked that we would eat rabbits. He doesn't eat herons. Rup Chand had had no bowel movement for four days. Gave six cathartic pills last night and no purgative effect. Gave three more this noon. Ate four titars, two curlew.

January 24, 1934.   Made skins. None came to sell birds today but two new shikaris from Jhimpir, Ali Mohammad and Mohammad Ali, came on a camel and agreed to bring something tomorrow. The rival shikaris of this place are for taking us off into the rushes on the other side of the lake out of harm's way. They say crocodiles aren't dangers till summer. A good sheep gives two to three seers of milk a day; a good cow 16 to 20 seers a day. The hill cows don't give as much on average as a sheep here. A very good cow might give two seers a day. The horses and donkeys here eat bulrushes. When the water recedes they collect the rush corns and roast them for the cows. The cows get fat and give good milk. Churru = male yak and cow give two to four seers milk a day.

Tolmo = female churro x bull give about same. Garmo = female churro x yak gives about a seer a day, less than a cow. Drimo and a good goat give one to one and a half seers a day. The latter's milk must be half butter. They say the people wear their silver wares every day. The fishermen were splashing around in the lake today up to their navels in water almost all day. The water was bitter cold. Rup Chand's purgative began to work this evening. Wang Gyel's taken at the same time worked last night. Rinchen Gyaltsen had a belly ache today and took two pills. Sunkyil eating all the bird bones including duck. Those that are not eaten at mealtime she hunts up before the next supper. A boy ran off with one of our skinning knives. It was promptly missed and the lad gave it up when Wang Gyel went to the village to get it. The boy said it had flown into his pocket. One of Parmanand's worn-out tennis shoes is missing too. The people know that if you cut the shotgun shell around the wad all the shot will go in a cluster with ball-effect. A sahib told them once. Peter told the Rahulis. The natives here don't know "Hazur" but say "tum." One always say "*gharib log*" and "*bhuka backha*" and points out the great exertion for me in hunting, just like the Sirsa hunters. The latter always referred to me as *hazurka bacha* and called me "*mabab*." That hasn't happened here. The wild duck flocks go over to the little lake at 7:15, spoonbills go about an hour earlier.

January 25, 1934. Stayed home. Only a coot left from yesterday's collecting. Usually 20–30 birds to start on. Made 40 skins. New shikar brought five new sand grouse and a lovely male Kunji. Bought two large white herons and a sand plover. Our water boy has learned some Urdu and speaks it thus, aenga leanga—kareolenga. The store keeper had two meals of mungh dal so we bought that for a change; else we have only bread and honey, onions, and birds. He says no one buys it or good flour. The natives buy rice, oil, tobacco, kerosene, etc. The fishermen left on foot into the lake at 10 and came out 1:30–2. Then some went at three and came out at 6. Others came out at 8:30 pm. The water is decidedly unpleasant to my legs at any time of the day. In spring and fall when children whistle the Lahulis hush them. They are calling the wind and there's wind aplenty. At threshing when winnowing the grain all the workers whistle.

January 26, 1934. Home. Rup Chand went over to the other side of the lake to see if moving would be good. A man brought a lugger falcon and said if it were alive it would be worth 200 rupees. Rinchen Gyaltsen said there were horses like that too but when dead —. Fishermen went out fishing in the dark this morning at least two hours before dawn. The water at three pm was icy to bathe in. Those people are special fishermen. The other zamindars live in mud houses and buy fish from them. The men shot a dog that attacked them and I promptly had word of it. The dogs here are very savage. They invariably rush at one when passing a hut and can't easily be driven off.

January 27, 1934. *Kaguonko mo ma suar a mik; panditonko kata ma shernik.* "Don't draw an arrow in front of a crow; don't tell a story (fable) in front of a pandit." Tang Abdulla, Nanidi Village Khaki color, 5–6 years old, three and a half feet high, shot inside mouth, killed 10 paces from house on opposite of lake, has no paper for his land, house of sticks, lives there 12 months. Value 30 rupees for dog, refused offer of belgarri[11] for animal. The above man came this morning and claimed damage for the dog Rup Chand was forced to shoot yesterday. They threatened to go to court and brought all their relatives to the tent but I urged them to go at once as I had no intention of paying anything for the dog. The man admitted that he had held the dog and let him loose against Rup Chand's request. I was willing to give a rupee or two but Rup Chand warned me that all manner of cases would be concocted for us and those who hunted after us. They could set the dogs on us and force us to shoot, wound their cattle and say it was our shot, etc. The joke is Piru had come in the morning and said the dog was his. Now he said the man was his nephew, who claimed ownership. Piru's brother came from Karachi with 12 papoti. He said they cost 1–14. Actually they cost 1. I went hunting toward evening and a fierce dog rushed at me from his master's hut, 100 paces away. No attempt was made by the owners to maintain the beast and I fired a 22 shot at him. Very effective. Rup Chand shot his with a 32. shot cartridge and all the charge went inside his mouth. A white breasted chocolate harrier flew over tent this morning. A hunter eager for the bounty fired before we could and the animal went. Seven flamingos (*lahh*) landed on the lake near us toward evening. The water is to a man's navel. We tried all the day to shoot the crows that come to the corpses, but anyone laying hand on the gun say nothing of pointing it drove them off. No wind today and very warm. Evening cool. Winds till now always northerly. The fishermen make their nets out of milkweed fiber. It is white and very snug. They call it *"ak,"* the plant. The natives all say that ordinarily there are many ducks and kunj[12] here. This year nil. They all say kunj are first class eating.

January 28, 1934. We decided to leave for the other side of the lake. The natives did their best to keep us with them. They said that there were great thieves where we were going, no wood at all, no game etc., and then tried to dissuade the oxcart man from taking us. He was going too cheap. The oxen thought so too. One when it was time to go beat it and raced off a mile. Then he wouldn't go into the harness and both balked at pulling the cart out of a canal we had to cross, but thereafter all went well. They didn't like the looks of me and I had to stay behind. The oxen here are splendid beasts that walk as fast as a good horse will trot. The Hissar animals were very slow but larger perhaps. I took a photograph of the fish catchers. They were all eager to be photographed. One, rather

---

11. Meaning unknown.
12. *Koonj: Anthropodies virgo*, demoiselle crane (R.K. Gaur, *Indian Birds* [New Delhi: Brijbasi Printers, 1994]).

stylish, knew two English words "yes" and "good morning." When we left he said good morning instead of salaam. The fish catcher men shave a square about three inches square on the top of the skull. Yakut Muhammid Shikar seemed the best of the lot. He told them they were rogues for trying to get the oxman to break his bargain and said we did well to kill the dog. There was one here to be killed. Parmanand says that some of his people clip the hair close on the crown. It is an old fashion. We camped beside some reeds inside which there is a shallow lake. There are many fields here growing a stunted grain and some fragrant scattered sarson. The houses are huts as before. Dogs fierce but we found them escorted by the villagers. It seems the dead dog now belonged to yet another man. There are many birds here: lots of harriers, larks, cormorants, chul, tiliars, and hunting ought to be good. We brought along the water carrier to get wood and water. He got two pails of muddy stuff from a hill nearby but was told no more could be spared as the supply was low, etc. It was quite potable after the silt settled out. He was very eager that we shouldn't think him guilty of sedition re oxcart. We recalled the night we sat up for a leopard in Rampur and killed a fox. Paljur examined the dead beast and expressed unbounded astonishment that such a small beast had been able to kill and carry off two cows. An old cream colored stallion used to come every day to lie in the mud beside our tent and spend the day eating around it. The other cattle never tarried there. *Kha-ni mane, sem-ne yane; mun men rahm baghl men churri.*

January 29, 1934. Everyone stayed home today. Had some little cormorants, herons, ducks, sand grouse to make and made 30 large skins. Our neighbors don't come near us as yet. *Changbala cho mordanan* is the formula of greeting among the men here. Saw Red-headed vulture (2). The big black-brown gold headed eagles are not uncommon. A cold wind from the north all day.

January 30, 1934. It froze ice in our pail last night. I went over in the little acacia forest nearby and got a new woodpecker and found a kite (ws) dead in a tree. Mahmud, Asa and Ali and a couple other little zamindar boys went along. They were a well behaved lot and didn't scramble for empty cartridges and when they got them entrusted them to one boy's keeping. Huge flocks of *melanocorphe* roost beside the tent in the stubble. A group of men who are carrying charcoal from two miles below to Jhimpir stopped for a visit. They wanted to know whether the Sahib wasn't come and we told them tomorrow. They carry 20 sacks of charcoal a load and get 1-9-0. They made one trip, four carts in the moonlight and disturbed Sunkyil. Waru brought a pink feathered goose and begged a #2 cartridge, just one. How he knew I had such cartridges, and why he didn't buy them [symbol]. ♂ I recalled how Mr. Phailbus' officer brought me a pintail at Naggar and refused a rupee but wanted cartridges. A police there also asked Rup Chand for cartridges. In Rampur and Lahul when a man gets crazed with drink they assume some witch has entered him and take steps to remove her. In Rampur they beat with a shoe. In Lahul they smoke with pepper smoke, tie horsehairs on the fingers, give soot water to drink, blacken face, beat with thorns and shoes. Finally, the witch owns up that she is so and so from perhaps a nearby village. She entered his body in revenge for some slight or injury. There are a few such in Lahul and everyone is afraid of them. Rup Chand's ma avoids the one in his village but if encounter is unavoidable treats her with great respect.

January 31, 1934. A man brought a flamingo last night. Our people were enormously interested in the queer thing. We closed the hunting buying today and tomorrow because alum is low. Ice again last night. Moonlight brilliant and night calm. We have issued 20 or 30 rupees in paper money here and it is uniformly accepted. A compliment to the whites in these parts. Our water and wood gatherer goes home every night. He came yesterday with his pants ripped. A dog attacked him near our tent, and he had his ax so encumbered with lunch and birds that he couldn't use it in defense in time. Went down into the old rice fields near the tent, now well flooded and got a memsahib from a flock of 10–12. Saw a dozen curlews and many jacksnipe. A white-tailed kite was too shy to shoot, also a large white-tailed. w.k. eagle like a kurral with white in the wings and no black band on the tail. Rinchen Gyaltsen left on a passing camel for Jhimpir to do the shopping. It's his first camel ride. Rup Chand says several Lahulis can tell when eclipses are due and the time of duration. During an eclipse the Lahulis and Kunawaras make all sorts of noises, blow the conch shell trumpet, fire guns, sing, play instruments. The kamzat log beg. This is to chase away the rakshas[13] that are attacking the planet. No food is eaten during the eclipse. After the eclipse is finished, the Hindus clean pots and houses as after mourning. Tibetans on that day kill no sheep or other animals or commit other sin, because such sin is chalked up as you red[14] on that day. The Chang Tang Tibetans know more about the stars than anyone in hill India. The Lahulis who buy wool from them can tell time from the stars. The Tibetans say there will be early snow, bad harvest, etc., from a study of the heavens. Some Rampuris and Lahulis have butter stored up since 50 years. In Rup Chand's mother's house are a dozen old leather jars (fashioned like a *kuza*) filled with the ancient butter. It may be used for rheumatism ointment, burned in monastery, but primarily is a sign that the family was at least once rich. They cut the dogs' ears here so that thieves can't hold on to them. I assumed they had all lost them in battle.

---

13. Demons.
14. Meaning unknown.

February 1934

February 1, 1934. Shot a *Butastur teesa*[1] today with its wing feathers and tail truncated. Saw another similarly mutilated. A ban's[2] carcass near the river attracted some 50 WR vultures, 10 common little brown ones, two large brown striped ones. No redheads. The dogs had a bad time keeping the big birds away from the feast. While they were chasing off one lot the next rushed in. Sometimes a greedy or slow bird got caught by the wing feathers to no harm. Went over by the river to a village of huts. There were large acacias and parrots, woodpeckers, and the tree birds there. Two dogs attacked me and came within three feet. Sunkyil is afraid of them but when they got so close she rushed out from behind me and stayed their advance. There are numerous wild *bhir* trees around the village. The people try to keep the parrots from eating them. Acacia become larger and abundant along the river. They are burning charcoal in one place. There is a large dike running along the river to shut out the flood waters and canals often thru it. The man from Soneri who hasn't been here for two days came on horse last night with a who-who owl and nothing else. His horse had a film over one eye and he had sewn a thread in the opposite ear. Rup Chand told him to buy a large conch and pulverize and blow the lime into the eye. He said there weren't any large ones this side of the ocean. I told them to burn the

---

1. White-eyed buzzard.
2. Unknown.

Plate 59. Walter Koelz holding vulture (date and location of photograph unknown) (Walter Koelz Collection, Bentley Historical Library, University of Michigan).

little *campelomas*[3] that are everywhere. Palden lamo[4] keeps a bag on her back and when people get dirty (she hates dirt) and bad she turns out the diseases that are on it. When she isn't offended, she is very nice. Kyelang language for owl bubu. White nape-spot on dog bad omen also white cow, a hawk and owl sitting together, white back goat. This refers to Lahul.

February 2, 1934. We expected Rinchen Gyaltsen back early from Karachi. He was to have reached Jhimpir last night and spent the night with Mohammad Ali. When he didn't show up we speculated on all the possibilities and of course thought of the most awful ones. He came with Mohammad Ali on the camel

---
3. Freshwater snails.
4. Tibetan Buddhist protector deity.

after noon. The American Consul got the money for him and sent a note. His name is Lloyd Riggs. Rinchen Gyaltsen said he has mindagyella.[5] He brought two dozen limes, one dozen guavas, two pounds of fruit drops, same of peppermint, cotton thread, and alum and a cup for Parmanand. The consul said it was unusually cold this year and everyone says no game. We redeemed most of our paper money outstanding. Allah Buksh sent photos of three beautiful knives and a nice gold thread and silk cloth. The prices of course were absurd. A man brought a pet fawn that had met with an accident and died. We made a skin. There are deer near Jhimpir. A shikar was taking game to sell to Karachi. They pay one rupee four annas for a telor. He also had rabbits and sand grouse.

February 3, 1934. I went down in the rice marsh and got some snipes. There were clouds of the fantails. Found where something had eaten a little eared owl. The netters brought a dozen ducks this morning. The early night is dark now and the nets catch the birds when they come into feed at night. Large flocks go toward the river every night. They go back to the big lake in the morning. Bought five species of ducks. A man brought a Pernis.[6] Made 47 skins. A group of urchins followed us every inch through the mud hunting snipes. I sleep behind the tent now that the night is fairly dark lest someone robs us from behind. I have a nice bulrush mattress. The giddors come to eat the birds every night and worry Sunkyil considerably. We get our water from the Indus now. It has the combined waters of Kulu, Lahul, and Kunuwas, being composed now of the original Sind, Chenab, Ran (Jhelum), Beas, Sutlej. A man came from Jhimpir and brought three halaled rabbits. Bought one for a skeleton.

February 4, 1934. Stayed home. A sandstorm wind blew all day but didn't make any difficulty in maintaining the tent. *Ancaetus gallay* and the male pernis came in. Both pernis' had stomachs full of green mess in which insect heads and legs (bes?) and masses of wax. The male had sticky neck feathers. Whitetailed Harrier eating fish. 35 skins. The Emberiza otol. found here last seen in Spiti. Wind blew away all the bulrushes of my bed. Rinchen Gyaltsen went to river to wash the accumulated laundry. Ate the male pernis along with a curlew and two ducks. Netters only got five ducks today. A man brought two bater and two black titar intended for market. All primans pulled and blood dripping from the bater. In Lahore markets all bater and most titar thus mutilated so they can't fly off. Mrs. Mahon might do some of her reform against cruelty there instead of distressing herself about terrible slaughter of scientific specimens.

February 5, 1934. Breeze continued but moderated in the afternoon. Wasn't unpleasant. Rup Chand brought in a pair of Luggars and three eggs. The female would have had three more. Saw a flock of 15 ± of the big white storks with

---

5. Unknown.
6. Buzzard; does not appear on the University of Michigan Museum of Zoology species list.

black primans. A man brought four white-hipped cormorants. Got four rupees for them and went away violently encouraged. Yakut called and wanted to forward a rupee. Very necessary today he said. No ducks in the nets. The men say the netted birds have the wings and often legs much inflamed. Filled with lymphs due to the long strain and bruising of the net. The old man who brings the best birds is always buying a cow for 25 rupees with the money we give. Have used one of the Japanese needles now for three days. Otherwise, two or three broke daily in sewing. 155 species of birds. 35 new to collection (species, not races). 709 skins since leaving Karachi.

February 6, 1934. Wind light and out of wind very warm. At night some large hymenoptera[7] one and half inches long came to the lantern and made the men very uneasy. They built a fire and the creatures flew into it. Wang Gyel is saving the hearts and eyes of crows he killed for a lama. Rinchen Gyaltsen, a rabbit heart for "dawai." Rinchen Gyaltsen wanted to eat four nice cormorants, but Wang Gyel not. The old hunter who specializes in hawks sent his boy today as he always does when the game is no good. Had a black eagle of some sort, prodigious fat, eating fish. I can't tell one from the other. A nesting eagle had five large rats in crop and stomach. Shot three of the crows and the rest promptly came back. Got a wryneck. Men went over to Elahaiga and beyond into the dense forest. 160 species.

February 7, 1934. Wind very little all day and so warm that I shed my coat. Cool in the evening. Sunkyil got tired and went to sleep somewhere and came home after the men, who went to Soneri. A group of men came from Elahiga brought some live ducks, titar and bater all so mutilated that they were of no use. Men scared up two goatsuckers and got both. The three kinds of white herons are in a flock together in the bulrush patch nearby. Yakut brought a goose, which was lucky or heaven knows what we would have eaten. We ordered a fish but they brought two suckers that had been in the sun all day and there was nothing else. A high pitched cricket sings at dusk by the tent. He is a fine songster but Rup Chand can't hear him. 163 species. The men yesterday saw a gador pursued by a dozen dogs. The gador wasn't at all afraid but trotted ahead with the dogs close behind. Every now and then the company halted five minutes in a bulrush patch and then resumed the march. Finally all but two dogs left and these encouraged by two boys with sticks went on out of sight. The animals all seemed tired. None of the dogs attacked the prey nor did the beast seem to fear the pursuers.

February 8, 1934. Calm and warm, relieved by cirrus clouds in the afternoon. Went hunting in the rice fields at evening. Got two Deob. sandpipers from a flock

---

7. Insect order, including bees and wasps.

of 20. Saw 20 curlews in a flock. Six or seven nets put up for ducks. They know here how to twist the ducks' legs up onto their back so they can't walk, though the practice is not common. At Gundarpur they always treated the ducks so and locked the wings together at the elbow. Here they more often tie the secondaries together on the breast. Waru says we should go crocodile shooting. They give one rupee eight annas per foot for the body part of the alligator. A flock of chiel of at least 100 circled over camp some so high that they were mere specks. Two sailboats came with the river. Their sails were visible from camp. The Lahulis and Kunawans make a tonic pill for old folks composed of musk, butter, honey, sugar, many herbs, including *aru*, *baru*, *kyuru*, and an animal (*chakdhul*) from Tibet. This last is rare and expensive. The whole is *manmar*. It is boiled long and made into pills. Can be made only in the cool weather or the pills melt. There is of course literature as to how to make the stuff. It is said to be a good tonic to the aged and to improve vision. Taken internally. It turns out that the dog we shot and for which several people claimed damages didn't die and isn't likely to soon. Waru says they also spear crocodiles and fasten a line and a straw float to the spear. The locusts came to Tang one year and ate the crops and killed some trees. The sheep ate the insects. They also came to Koksar one year and abundant in Dortse. Did no damage. On Rotang walking was difficult for corpses. We bought an extra water jug so the men now don't have to go to the rice pond after supper for water. They used to have a bad time finding the water and then their way back. The other men had to stay up and watch the big gobar fire they built to light them home. I woke up last night to hear the men reading their Hindi "*tumhare pitne laike hain*." They read in a peculiar sad disconsolate voice. Rinchen Gyaltsen has learned in the time he has been with us. Wang Gyel bought a book in Lahore this winter. Both can write and write figures since Bhadwar. It is almost necessary that we wear native clothes of the districts we visit though we have never done it. First, the birds all know us as strangers and keep out of gun range. Here we approach the hawks by wearing a sheet and look therefore like natives. Then, the natives recognize us as strangers and attempt to cheat us in everything we buy and to extort money in every conceivable fashion to say nothing of hanging around and watching us by the hour. Here at this camp there are few visitors and they seldom stay long. All have the spitting habit though. The Tibetans train their fingers to be strong and can break ribs by a thrust of their fingers.

February 9, 1934. Saw 10 Arocets and a verg. rail in the rice marsh today. Also a flock of 12 godwits. Quiet, clear, and warm. A large assembly of small grasshoppers and tailed crickets came to the lantern tonight. Parmanand coming out of the mud calmly remarked his legs were black like the Lahore statue. I asked what statue and he said Queen Victoria's. Saw 30 curlews in the rice marshes. Four or five flew off with something in their bills when frightened by a hawk.

February 10, 1934. Last night very warm. Restless in eider robe. Heavy dew and fog in morning. That hasn't happened before. Sunkyil from not having gone hunting yesterday spent the night chasing giddars and mice. One or two went under the bird boxes and she had to dig them out and woke every one in the tent. Men went to the river. Saw 27 turtles three feet in diameter sunning themselves on an island in the river. Also some dolphins jumping in the stream. Shot five little eared owls in a small copse, two species. The two large long eared ones had mice, also one of the small ones. Other two had insect remains. Yakul brought an osprey and some Jacanas, three or four alive. I wanted him to help kill them, but he said he didn't kill things except at a distance. Our water carrier we find has been bringing the drinking water not from the river but from a pond beside it. The buffaloes bathe in the puddle and the people wash their clothes in it. The people of course also drink it. We accused him of fraud and he replied, "but you said it was good water." I had objected to washing in the rice field water which is now becoming very stagnant and he said he couldn't carry water farther. Before he went however, he said he'd carry stones or anything if we'd let him stay. A fearful visitation of small earwigs and small beetles around the lantern while we were finishing the birds until a fresh delightful breeze came out of the west. The men have changed to their wool clothing including coats and vests. Removed as many as possible forcibly. Wang Gyel has cotton shirts but wears a heavy wool one and a vest over it. The natives are partial to a black tight-fitting cotton shirt that must be hotter than hell. Sunkyil wanted to go with the big gun this morning but wasn't wanted. Then I with difficulty persuaded her to go with me. As soon as she heard their gun a mile off killing a W.T. kite she made for them. Our Sonen owl expert has a job as cook for the dike overseer. He was much interested in the little eared owls. Saw 30 geese P.F. (?) in the river. A flock of 200 to 300 W.F. cormorants came back from the river to the big lake around 10 am. 168 species, 838 skins. The men found a field with some danya and since we have a couple of limes left, we ate some of our favorite chutney: danya, ginger, onions, limes. The tamarisks were long dripping with dew this morning. There is a delicious queer fragrance along a pepper bush now by the forest.

February 11, 1934. The five rupees that Yahut got for birds yesterday fell into the lake and only one was recovered. The water carrier returned very chastened and brought nice river water. He says he can't eat here or the people will beat him. Our dishes may not go into their well even, because they often contain "dead" meat. Cats, says Rup Chand, curse people that they may become blind and thus not see the cats' thieveries. Also the cat works to reduce the number of humans in the house so that there'll be more room for him. It is of no value in the hereafter to give a cat food. He eats with his eyes shut and therefore no recognition. To give a dog food is beneficial. There is a hot iron bridge to be crossed in the next world and the dogs you have befriended lick the thing cool. Last night a breeze came from the west and wet everything with fog. The tent dripped water. The fog continued until eight in the morning. The earth became gumbo where the salts had absorbed the water. The vegetation seems not to mind the white salty encrustation. When dry the earth is velvety to the bare feet. After a rain walking must be impossible. Men went to river and brought back a baby otter with its eyes not yet open. They killed it before bringing it, to the great regret of us all and of the fishermen who say they search for the puppies and can't find them. The den was in a crack in the river cliff and only one puppy. A runal had a nest in a single acacia in a field, an old tree about 20 feet high. The nest was a good man's load of sticks and large enough for a man to sleep in. Two or three pairs of P. domesticus had nests in the sticks. Female shot. One egg. Hunters brought eight species of hawk today. Men report three blue pigeons by the river.

February 12, 1934. Both stayed home and made skins, 42. Had six black fish eagles, 14 rb sandpipers, avocet, etc. Lost my white pearl-handled knife that has served me so well ever since I have been in India and don't know where ever. Got six curlews at two shots this evening in the rice mars, all male. Saw 200 geese go to river and a flock of ca. 100 plovers, golden (?) on the dry field near camp.

February 13, 1934. The west breeze continues every night and in the morning everything is wet. This morning the sun didn't penetrate the fog bank till 9:30 am. It stayed pleasant all day. This evening the cool damp breeze sprang up as usual after dusk. Went into the little forest and got four wood shrikes. A fisherman brought two dozen limes from Karachi and we shall have a nice chutney for a few days. A goat sucker got caught in the duck nets and we bought it from a child.

February 14, 1934. Went toward the brushy jungle and Lahiga. Saw a small red and white Periculotus[8] flying. Shot one of two white ibis. Got a new skylark from a flock of ten. Testis O. Curlews 50. Godwits 50. Stilts 25. The little sand-flies have put in their appearance last night, also a few mosquitos. Ocean breeze and fog as usual. A yearling buffalo from a flock of six others of the same age saw me lying on the grass, came up, sniffed my hair and rubbed his head on my legs. The dear. Tomorrow I'll bring him a present. Ordinarily the buffaloes dislike our odor and threaten or flee. Yahut brought two [indecipherable] sand plover from the river. Saw another. Parmanand and I made tea and had a little siesta on our hunt. The people have dug a little well but use the green water of a drying fish pool instead. Some BC terns were catching out the fish.

February 15, 1934. Home all day. Rup Chand got yesterdays red birds three and saw another. Bought two flamingos. A pleasant breeze all day. At night a brisk wind w. Full new moon. Water carrier paid out half his wages for a donkey assistant which he rode to work today and took to the river. [indecipherable]. 189 species.

---

8. Or Penculotus?

February 16, 1934. A stiff breeze till morning. Then it calmed completely and a few mosquitos came. A large rat jumped on my head in the night and when I sat up he jumped on the ground. Our visitors this evening are some large gnats that filled the tent. Three strange looking men came this morning and wanted some bird meat. We hadn't any but said we would give two flamingos if they came at evening. About four men on two camels stopped and inquired about these men. Said they were wanted for thieving. One, apparently a police in plain clothes, was very decent. Rest villagers. Rup Chand's mother once was staying at her small Gundla house with the female servant. While they were sitting around the "ash fire" one night they heard a child's voice from a direction in which no village or house. It must be the devil thought they, and hastened to their room. This they locked and bolted and thump, thump on the roof. Then sweet violin strains in the outer room. This was clearly the devil intending to lure them out. Rup Chand's ma fainted and recovered alternately. Poor servant wanted to get the ax but daren't get out of bed to reach it, etc. Thus passed the night till morning brought an old man who always called for fire. The women hysterical fell on him and he told them they had no business being alone at that season. The devil was always about. Next night a group of men armed with axes, swords, guns, sat up for the devil. Plunk came something down the smoke outlet and put out their light. Manfully they rushed to guard the exits and then on searching found a cat. It is customary when the new year begins to ask the lamas if there perchance isn't a devil somewhere in the house. If so, he must be got out and the lama undertakes a five day shinker.[9] Then they attempt to find the said devils hiding place. He may be in a necklace or a teapot so everything has to be opened. By a sort of lottery they find out which devil it is and his hiding place. They put the names of the various devils in a bag and then bring several articles. The bag is emptied and a name is drawn. If the certain devil's name appears then he is in one of the several articles brought. He must be disposed of. A paper bearing the devil is now sealed in a dog or fox or weasel skull or yak horn. If the house owner has no children a weasel skull is needed. Nine seals are applied to the head; the owner must press each seal. Then a 9 × 18" long rope of dog and donkey hair is made. Both are securely put in a sack with much noise put in a hole in the ground and burned with due shunker.[10] Owner must be present at interment. Ate flamingos. Very good. Flamingo stomach had small quartz pebbles size of large pinhead and some small black seeds. Our slave left his donkey here for the night and gave it a bundle of green bulrushes to eat. Oxen won't eat bulrushes. Some children brought a dead eagle, black and spotted. No shot marks but very thin. Buckwheat flour or pulverized rotten cedar wood thrown onto a fire burns like gun powder. Children burn off the fly's wings by blowing the lighted powder at the swarms.

---

9. Meaning unknown.
10. Meaning unknown.

February 17, 1934. Rinchen Gyaltsen went to Jhimpur to get change for his 100 rupee note and to buy boxes for our birds. We were firm at 1–4 for an oxcart and all refused but in the morning two carts appeared at our camp for one rupee that wouldn't come yesterday. We finished the bird quotas (35 skins) by dark when Rinchen Gyaltsen arrived. Two cotton teals, and a short-eared owlet (food insect) and a fish-tailed tern (laying female) notable additions. Today ca. 100 white storks flew over, circling slowly as usually. Day calm and warm but west breeze at night.

February 18, 1934. Day pleasant with a light breeze most of day. Last night as usual but less dampness and this morning no fog. Least Cormorant, little Passer., redshank and a new Larius came in. The children here swim well with breast and Aust. Crawl stroke. Parmanand has been having small boils on his left cheek and now Wang Gyel is getting them on his left arm. Saw a thin snake like a ribbon snake, 18" long. Fled precipitately. Man brought four black titars he had caught with a horsehair snare. Also a live telor. A rostrutula[11] caught in the duck nets. Large gnats horribly annoying at dusk. Rinchen Gyaltsen said he worked for a contractor who built something for Mandi state and didn't get half his pay. The contractor said the state didn't pay. Rup Chand says this is a common practice of contractors. The Manali Barons are classical examples. Giddars sang beautifully from the rice marsh at evening. Voices rich and chorus varied; black titar say to Gundlawali: *Tra(k) zahi kitsir (jan Khaki dast)*. Once a titar, a cock, and a chikor undertook to plant a field together. The cock said the chikor would eat the seed to look out. He did. The titar sad. *Sat tseral rakhiel* (7 measures of grain thrown away). Cock said: *mui ta boli agi* (I told you before). Chikor said "*dokhre, dokhre, dokhre* (zanim, zanim, z.). Rampuri *boli*.

February 19, 1934. Yesterday's promised camel didn't show up this morning. This evening the man and a companion each on a two year old camel that didn't want to mind came to apologize. A relative had died. Rup Chand brought in six swallow coursers and five of the Rink sparrows. The last were feeding along the edge of a sarson field. Flocks of 20. I went in the grain fields nearby. The fantail streaked warblers are in there also. Many motacilla[12] and larks. Day pleasant with a stiff west breeze that continued into the night. The duck nets catch nil when it is windy.

February 20, 1934. Day cool with stiff west breeze. Night chilly. A pelican came in today. The natives wanted the fat. There was about four pounds of grease in the entrails. The skin has very little fat. The fat is said to have great penetration and is useful for articular pain. Natives eat flesh. Say pelican great enemy of monkeys and monkeys flee their sight. Dive 10 feet down in water

---

11. Painted snipe.
12. Wagtails.

and scare fish to surface to eat. Camel came and Rup Chand went to Sonde. Got two blue pigeons from a flock of eight and saw what I take to be the little Sirsa gull. Godwits and jacksnipe abundant. 207 species.

February 21, 1934. The stiff breeze continued all day and into the night. No insects come then and the day was delightful but such a mass of cinders in my eyes at camp. The zamindars have begun plowing. A boy of 14 is plowing briskly nearby. The cattle move splendidly and easily, a contrast to our hill beasts that have to be beaten and coaxed and led and can be manipulated only by a strong man. Here the earth is hard as stone besides. The natives generally are kind to their animals. I have seen no one beating any beast. The Punjabis beat the draught cattle without reason. Rinchen Gyaltsen went to Karachi to get cartridges and alum. The men ate the pelican and said it was good and not at all fishy. Saw a G. nebulan and Rup Chand saw a white headed and tailed gull, also a large white duck with dark primans.

February 22, 1934. Our camel came today and I went to Amirpir and along the lake there. There is a priest's house along the cliff edge. Nicely kept with several nice acacias. The bed strata of basalt? and conglomerate crop out here in low cliffs 50 feet high. Large clumps of a very prickly euphorbia grow on the banks. The bushes are ca. eight feet high and about as wide and totally impenetrable. If a bird falls into a bush, it can't be recovered. The lake must be around four miles long. There are hundreds of common coots all over it. Day pleasant with brisk breeze from west. Went swimming in lake. Pleasant temperatures. Saw a Virginia rail. Male camels urinate backward and the stream is no more than a man's. Urine colorless in the stream. Our camel has a nice stride. He slows and takes short steps on wet ground.

February 23, 1934. Rinchen Gyaltsen came home around midnight on a camel from Jhumpir for eight annas. He would have stayed in Jhumpir but a couple people asked if he had money and the station master told him not to sleep in the serai. I had a cable from Ann Arbor: "Doctor Ruthven and I believe your return in May necessary. Cannot now secure funds for further stay in India because depression and collections demand museum attention. Winter." Consul Riggs (Lloyd) sent a note saying he would come to spend a weekend if roads were motorable. Rup Chand went to Jhimpir to send Consul a letter asking him to get a passport for Rup Chand so he can go along to America. Poor devil wants to go so badly. Got 12 sand gr on the way (2 species). Piru sent a jacksnipe. He probably saw a flock and shot his last cartridge at them expecting to bag half a dozen, which at two annas each would net a little profit. Rinchen Gyaltsen brought four pounds of lemon drops and three dozen limes from Karachi, also five guavas, so we are in clover. The guavas are pumpkin flavored but wet. Evening with brisk cool breeze. Saw a three feet lizard lying on the bulrushes at Amirpir yesterday. He went in the water when I disturbed him, at three feet.

February 24, 1934.   A donkey went by trotting with two men on his back, at least 300 pounds. Saw one day one with three tins and a man, ca. the same weight. Went on the camel to the river. Beyond the dike are large acacias and much tamarisk. There were a dozen anocets feeding on the silt mud of two lakes. The mud was up to my hips when I waded after one I killed. Many woodpeckers, little passers, parrots, etc. in the jungle. Pleasant breeze all day, stronger at evening. 36 skins. 205 species. Yesterday no new ones.

February 25, 1934.   Sunkyil saw the camel coming this morning and ran 100 paces to meet him. She likes the donkey too but the donkey kicked her sideways in the head. Home all day. First time since the alum gave out that our birds dropped below 35. 29 today. Two hors came in and a flamingo. A lad stoned a new warbler and small owl. Rice marsh drying up. Last night heavy cool breeze. A little chilly in my one blanket. Ate flamingo. Went snipe shooting toward dark. Many in the marsh. Also curlews. Thought we might go by oxcart to Karachi but no one interviewed today had ever gone there. The people have begun to burn the bulrushes and dig the corns.

February 26, 1934.   Night cool. Wind died down toward morning but mosquitoes not troublesome. The day was completely hazy all day and very pleasant. The evening decidedly cool with wind. A child nearly blind with trachoma came for medicine. There were three other children along and all four rode off on a donkey. Two A. monachus[13] came in. I went over to Amirpir again and got another wring neck and a new Embiriza.[14] A man came to take us to Karachi. He says he will get us there in three days, two carts at 30 rupees. The men are all eager to go thus. Sunkyil is shedding hair and likes to have it plucked. The men's boils have disappeared. I opened all new ones promptly and thrust in pieces of dry mercurochrome. That was more than they could stand. The largest A. monachus has a wing span of nine feet. The A. nipalensis[15] have become common the last three or four days. Saw five today.

February 27, 1934.   Neither the camel nor the oxcart man showed up today. Rup Chand and Parmanand went to the river on foot. Beautiful pleasant day, quite till afternoon. Then chilly in the evening. I am writing in the moonlight though the moon won't be full for some days. Huge swarms of flies in both tents last night no mosquitoes at all. Four flamingos came from Soneri Lake at one shot with muzzle loader. Three bitterns, a new boggy, a new chul (shot by the tent), a lovely blue brown linestrus male harrier, medium gull, carpodacus [rose finch], large Passeriformes (5 species new). The flamingo had small seeds and fine white quartz pebbles in gizzard. Streaked fan tail warbler common in

---

13. *Aegypius monachus*: cinerous vulture.
14. Red-headed bunting.
15. *Aquila nipalensis*: steppe eagle.

grain fields. Fly high singing and then drop into grain. One gull had only one foot and another had his tongue stretching out through a hole in his chin. Our four boxes with hinged covers were first nailed on all six sides, then the top sawed off. Rup Chand says Lahulis make them that way too. Punjabi proverb: *lakh bhi chhori, Kak bhi chhori.* (A lakh is thievery, so is a trifle.) A Kushog once came to Lahul and stayed for a year with great honor. They later learned he was a lohar.[16] He had great welts where he had been horsewhipped for theft in Tibet. Even the pious in Lahul believe that the priests are going to hell, along with their wives. They must thus be punished for the abuse of their office. Physicians and wives go to the oven.

February 28, 1934. Men went to river again hunting. Brought back two of the little sand swallows. They have a hole in the bank four and a half feet long. The bank regularly falls and probably that's why the holes are so deep. The hunters brought in little today. Misri came this morning with a gugu and sent a boy after dark with an eared screech owl. Parmanand said the blind boy got four annas worth of benefit from the mercurochrome in his eyes already. The meaning is 25%, not an unusual way of expressing percentages in the hills. The titars had their usual numbers after dark; we assume the giddars disturb them. Two more flamingos. We are undecided whether to go to Datta and risk getting an oxcart from there or to go by train from Jhimpir. Some Tibetans won't eat meat the first day it is killed because it is still alive. Mussalmans won't eat dead meat. Rinchen Gyaltsen has a trembling little hum that is heard only in mornings. Lahulis commonly repeat the ཨོཾ་མ་ཎི་པདྨེ་ [17] when it is cold instead of our br-rrrr. The Kuluese call it the "*tanda rag.*" The wind has constantly been from the west now for days. Morning always damp and with a heavy fog. Nights always comfortable. Our food has [indecipherable] here. [indecipherable] mornings: bread and honey, Rampuri shorba, and rice or halva. Usually nothing unless a chocolate bar till night when dal and rice or game and rice, a chutney of onions, dania, and lime juice for the latter when limes could be got. We have made 35 to 60 skins a day. One man went hunting with a companion every day and since small birds decay quickly we often worked into the night to take care of them. Large birds stayed all night. Parmanand has never lived half so luxuriously. His people have halva sans raisins or almonds for a big treat five or six times a year.

---

16. Caste name: smiths; includes Hindus and Muslims.
17. Koelz wrote the first three Tibetan characters for the Sanskrit mantra that invokes the bodhisattva Avalokiteshvara. The full mantra has six syllables: *Om ma ni padme hum.*

March 1934

March 1, 1934. Day unusually mild. Night distinctly chilly, with fresh breeze north of west. We finished four hawks that were left over from last night, a little owl, and two galeris, and Rup Chand got a black Titar after breakfast. Then we finished labeling the birds and the men went for a swim. Parmanand can swim 30 paces. We decided to go to Tatha and take a chance on getting a transport to Karachi. We shan't come this way again and may as well see the country. The birds all go into eight boxes. There are over 1500 skins of some 230 species, 65 new to the collection. Toward dusk a stranger came with a Rosy Pelican that he had shot by the river. The pelicans have an immense quantity of fat in the intestines but little on the skin and that is easily removed. There is an abundance of air under the skin so that the bubbles can be felt by pressing through the feathers. The bones are also very light so that the skin is of very little weight. The Peregrine falcon group are the contrast. Our preservative was finished so we bought an anna's worth of salt from a neighbor and made up the skin in the early moonlight. Lovely full moon with a rare bluish light. The men with the two carts (8 rupees) will come before dawn, so we pulled up the tent and got everything ready before going to bed. [indecipherable], 236 rupees, ca. 50 [indecipherable] otherwise not represented.

March 2, 1934. The oxcart men wanted to go last night to Tatta were to come at daybreak. Rinchen Gyaltsen got up at 3:15 and had breakfast ready for us sometime in the night but we refused to get up since no oxcarts were visible. We got up as usual and then hunted up the men. They finally showed up around eight. They are the two most murderous looking individuals that could have been found in a long search and they whaled at the poor oxen all the way without reason or rime. They are building the road we came on and for the first six to eight miles we met several gangs of workers making a bed of stones and carrying dirt on to it in baskets. The first ten miles was through dry bushy country with copious low mesas of limestone face left by the ancient river. We crossed several and often saw to one side or the other small fields. Toward the river there are dense forests of acacia and tamarisk bushes. About eight miles from Tatta is a nice place with a little lake and the river near. The country becomes totally fertile and pleasant within eight miles of Tatta. Large acacias along the road filled with parrots, gugtis, and [indecipherable] and near the city gardens of fruit: mango, banana, pomegranate. The houses have all a sort of funnel of boards on the roof facing west. We assume to catch the west wind. Tatta has some high buildings and one street. A bazaar we found some coconuts at Sauktra and feasted heavily. The men suffered considerably from thirst on the way. The thermos soon was empty of tea then it was filled with water from a little lake. The ox drivers drank nothing but once at the start when they first watered their feet and then drank. One ate red bread made of rice and [indecipherable] for lunch. That should have made him thirsty, but he had saliva to spit along the road. We put up in some sort of building near the road with pleasant location and empty. They told us it was for travelers and that the dak bungalow was two miles on. We had to hunt up transport on our next stage of the journey so we couldn't go far. Got nine skins on the way, nothing new but several species rare in our collection. They called Sunkyil to eat in the middle of the night but she didn't think of such a thing. To their repeated calls she answered one "woof," looked to see if there was danger and went to sleep. As soon as the men got up she came to lie by our head. All day she made for all the water there was and went swimming. The Tatta houses have doors that open in pairs, each 18 inches wide, one having a carved board ca. three inches wide which comes in the center of the door when closed. The carving is in [indecipherable] conceits and effective [drawing]. The houses are often of two, three, or four stories in the city.

March 3, 1934. Sometime in the night two police came and almost got eaten by Sunkyil. They wanted to know if we were Pathans and the men said yes, go away and let us sleep. They went. We arranged for two carts to go to Gharo in the morning for ten rupees. They appeared before daybreak and a little after sunrise we were off. Outside the city a mile or two runs a ridge on which are a dozen or more elaborate temples and tombs. The rest house is there too. There is a small lake below the ridge in which several of the medium cormorants were fishing. On the ridge, the roads branch, one to the rail station Jungshahi and the other to Gharo. They are building this road too and workmen are met all the way. Road however, very good. For about six miles the country is chiefly euphorbia, et al. and then near Gujo (10 miles) there is some farm land. We halted at Gujo and fed the oxen and gave water. There is no more water till Gharo. The country between Gujo and Gharo is all dotted with hummocks, running east-west where the wind has thrown the loess drift against the pepper bushes. The bushes are against the wind and the earth has often been carried out from under them leaving the roots exposed and no mound. Mounds four to ten feet high. The day was very pleasant until near Gujo when the wind was unpleasantly laden with dust. Got a new wagtail near Gujo from three or four and a new titar from flock of rosy Ped. near Gharo. There are large flocks of the latter feeding on the pepperbush. One flock had a dozen or so of the green and purple ones. Poor Sunkyil suffered from the want of water and when we got to G made for the brackish river beside the town. She tried it in three or four places, but couldn't drink it. At Gujo there were some remarkable ant lions. They don't make the funnel depressions but wells, one 1.5 inch deep. Out of none of them can an ant escape. Found some two dozen guavas at Gharo; otherwise no fruit except the wild bher that are much in demand everywhere. We got two white titar at one shot (female had an egg ready to lay) or we shouldn't have got anything to eat. We put up at the serai. The ox driver said when we told him to take us to the dak bungalow that was the government's and we couldn't stay there. He thinks we are flunkies entrusted with delivering these boxes to Karachi. The chiefs of our party have gone on by train or carts. He hasn't the remotest notion that I am a Caucasian. 10 bird skins.

March 4, 1934. There are few oxcarts at Gharo and these go under contract to Dabej, the rail station, we were told. There are many camels and our baggage would go on them. Our men after long debate among themselves and consultations with us, decided to take us to Landhi for 13 rupees. We agreed and at two they got us out of bed to start. We met four camel caravans on the way. The drivers were often asleep and one camel got loose and was coming along alone. There were also pedestrians. At sunup we got to Dabej, seven miles and stopped to cook breakfast. The country so far was Euphorbia thicket type. Here now a great abundance of desert plants appeared and I collected some 50 species just beginning to make bloom. A little out of Dabej there is a track of the reofl. gran bush that grows here into a small tree. Then there is a tract of brown creeping shrub, now leafless and more Euphorbia. About eight miles from Landhi the road drops off a little plateau and we are back on the [indecipherable] plain where there are few plant species and scattered farm lands. This it was to Landhi with some slight elevation and more numerous plats of cultivation. There are many large lizards on the lower half of the journey. They dig holes in the earth and rapidly get into them. Many animals are 18 inches long. There were flocks of sand grouse near Dabej, one with black breasted birds (Imperial?). Flocks of the large galinda and the dry Anthus, plenty of O. Desert and the black and white species, and sunbirds and gugtis ([indecipherable]?). The oxen grew very tired and we got to Landhi at dusk and put up in the rest house. The distance from Dabej is 19 miles, so we made 26 miles which distance Rup Chand and I walked. The carts were ready to take us to Karachi providing we started out again in the night (carts here are often beautifully carved. One of ours was very prettily decorated on all available areas except the tongue). This we can't do.

The beasts were dead tired and so were we. We tried to keep them till sunrise to give the animals a rest, even promising a tip, but they made off. They had no license to appear on the roads and thus were subject to arrest, so it was as well they didn't go to Karachi. The rest house is of two rooms, rather dilapidated. The ceilings are about 14 feet high, lower than most of the hill rest houses and the walls are painted a dark green, black in lamp light. There is one window and a latticed veranda around the whole house. One pays one rupee a day and as many people as you like may stay with you. If the other room is vacant you can use that too at no extra cost. Men are very partial to a shirt and jacket tight fitting to the waist and with short tails. Color often black; buttons down the front. Huge trousers, usually white. Turban usually red with circles and squares printed on in a nightmare of designs.

March 5, 1934. We stayed in the rest house all day. I put the recent birds in the sun and diligently changed blotters on the deluge of plants we got yesterday. The men went to Maleri, three miles below to get provisions. There are fruit gardens here and papaya and coconuts can be bought. We ate the former gluttonously. Oxcarts want eight rupees for two to Karachi so camels were engaged at two rupees each. Then the driver agreed to take the load on two camels and then came to ask to see the load. He at once decided that three would be needed, and when we agreed to that he thought four. The day was cool. Last night was chilly with a rain sprinkle and thunder. I didn't hear a word of it. The rest house has three dogs, so starved that they ate papaya husks and fought with Sunkyil whenever she showed up. Sunkyil is dead tired and slept all day. Last night she wouldn't eat. Yesterday, three miles above here on the right we passed three tombs and some pillars covering an acre. Parmanand managed to get the axle grease from the cart on all his clothes fore and aft and today got it on his face. Wang Gyel met a man on the road who said he had stolen the fruit he was carrying and the man intended to take it away from him. Wang Gyel suggested he try and then the man moderated and said why didn't he buy the fruit from him. But Wang Gyel was not to be pacified and answered savagely.

March 6, 1934. The camel driver that came to inspect the load last night didn't show up till 10. Meanwhile another came and said he'd take us in for four rupees. We paid six, and two oxcarts wanted eight. We finally got off but minus a riding camel. Two camels were offered at one rupee each, but the load drivers scared them off and we had to walk. The men now were Baluchis. They took us by an old path till near Karachi, mostly through grassy acacia jungle. Saw a Black Buteo, a flock of W.C. Black Larks (*Pyrrhulanda gusea*?) and a pair of the new larks. At Karachi there are flocks of the little black-banded plovers. Saw in the city a female sparrow hawk with black primaries. We camped in front of the cemetery just outside the city limits and beside a dairy farm that is, an enclosure in which cows spend the night. Rup Chand and I went in on the camels and deposited the boxes with Cox & Kings. We bought three bottles beer and two of orange pop for Parmanand and Wang Gyel and a cake of ice and came home and had a rest. The day continued delightfully.

March 7, 1934. Soldiers came out to drill nearby and did some target shooting. Went into city and saw the American Consul who will go to see about getting Rup Chand a passport. He will come out tomorrow evening to the camp to report. Mr. Lee writes that Professor Roerich and George had gone to America, Col. Mahon to Delhi, Banon discharged the day I left Kulu and Ms. Lichtmann in Naggar. Dorje wrote that he had heard we were in Karachi. Deuster wrote about the Chitkul language and quoted from an article by T. Graham Bailey, "Linguistic studies from the Himalayas," London 1920 *Asiatic Society Monograph*, Vol. XVIII, p. 78–86. Found nothing in the Linguistic Survey of India. Went to Cox & Kings and packed up three boxes of skins. Took to the [indecipherable] for further drying some vultures and flamingos. Left gun license for renewal. Saw a large black vulture flying over camp. Wang Gyel says if dead sheep are placed with the legs in the air, vultures won't touch the carcass. Rup Chand says vultures drop bones from high to break them. One laminergerer killed at Shedupatar had a sheep leg bone eight inches long in its maw. The camels here carry up to 12 boxes 1½' × 1½' × 2½' strapped on each side. On the caravans the driver may sleep as the camels walk along. The going isn't smooth and the body rocks and bounces. One man's head had slipped from his pillow and was jerking back and forth like a bird's head that has come off its supporting stick. The man was fast asleep.

March 8, 1934. The men all went to see the whale in the zoo. They were much interested in a kylang, a garmo, and a yak herd. I went in to get some vegetables in the evening and to get the renewed hunting license. The American Consul said he was coming this evening but didn't. Got a can of sweet corn and evaporated milk. Our folks liked it much. A can of sauerkraut was spoiled.

March 9, 1934. Went in with Rup Chand and Parmanand and got the things deposited with Cox & Kings. Everything in good condition. The office man who has been so helpful refused a ten rupee tip. Mr. Riggs not in all day. Bought the men each two shirts and Parmanand a pair of shorts and pants. He says he won't dare wear the shorts in his village out of modesty. Sat for passport photos. Went to the Mahatma Gandhi garden—the local zoo. Very nice flowers, all our summer flowers now in bloom. Cages clean and animals well kept. All very pleasant. Most of the people were around the monkey and otter. The water birds were especially clean and attractive. Two species of pelicans, two of cranes, two of swans, several of geese. Night herons were loose. A bear, lions, tigers, zebras, llama, ibex, etc. Some soldiers shot a dog in the city. Say many are rabid. There are said to be men in India who eat for tamasha. They can consume unbelievable quantities of food. Called on the Xian Indians with whom we stayed before. They pay 200 rupees a month rent for the house. Our ox driver from Tatta wouldn't eat our fruit, hungry and thirsty as they were, because we kept it in the same basket with our dead birds.

March 10, 1934. We all went to the city. Mr. Riggs said we could have a passport and armed with a note from the Commissioner that said "A Tibetan of British origin wants a passport to accompany an American scientist to the United States. The Punjab government should be consulted by wire if necessary. But since the American Consul is prepared to vouch for bonafides of the trip please give every facility." We got the passport after bribing the police force to hurry. They didn't intend to give it till they got the money either. Fee six ruppes, valid five years. I had to give a statement that I would be responsible for his good behavior and financially. Bought a nice bidri[1] cup and two old Bokhari pillows. Mr. Riggs came out in the evening with two Afghan hounds, like Russian wolfhounds, but smaller and more intelligent. Had a nice visit. He wants to go back to Poland where he spent 10 years.

March 11, 1934. Planned the summer for the men. Rinchen Gyaltsen will go to Ladakh. Wang Gyel and Parmanand will collect plants in Rampur and Parmanand will learn Tibetan. Sunkyil goes back and will spend the summer in Rampur. All the men wanted her, and I thought they didn't like her. Went back to the three rug shops, but found nil worth buying, not even a decent fur piece. Gave out medicines for the summer and packed our clothes. We ate karela,[2] a bitter warty cucurbit that you fry after dehydrating a bit with salt. Also a tuber a foot long and as big around as a thumb with five canals running the length of the center. Fried. The last starchy and with a rather interesting flavor.

March 12, 1934. Sent a wire to Jaipur to PM Allah Baksh to meet me at Marwar Junction day after tomorrow at 9 am. Then went and finished packing the four boxes that leave tomorrow for Bombay and from there on March 24 on the *President Johnson*; arrive New York, April 24. Didn't have to pay Cox & Kings for the boxes or the storage of my things. They will charge on the other end. Then got Rup Chand's visa good for 12 months in the United States. Went to District Magistrate to get permission to leave my superfluous cartridges with the police. After sitting two hours got the permit. Then the consul said what nonsense. Could have left them in the consulate and Mr. Riggs said he'd take them home, so Mr. Riggs got them: 1900 32 shot, 800 22 shot, 200 410. Made a vegetable salad for supper. Everyone liked it but Parmanand couldn't eat more than half his. Too rich with mayonnaise. Gave the men their wages to November 1, Parmanand to Dec 1. Also fares back to Kulu and Rinchen Gyaltsen 300 rupees to buy tankas in Ladakh. Paid Sunkyil's board up to November 1 too and her passage home. Left two pounds of preservative with the men in the hope of getting

---

1. Silver inlaid black metal alloy.
2. Bitter gourd.

a golden eagle and some Tibetan owls. Swifts at Karachi. The kachenar trees are in bloom and the buds can be bought in the bazaar. Parmanand found some small round flattened tree seeds that look like drakshuk [insects] and Sunkyil was terribly puzzled with them. Sunkyil has taken to sleeping on my bed the last two nights. Nights cool and pleasant. Usually a breeze in the [indecipherable].

March 13, 1934. We got off on the 9:41 for Hyderabad. Our people will leave tonight and are staying in Bhadwar til April 1. Then all go to Lahul. Parmanand is afraid of the Lahul journey and begged Rup Chand to intercede for him. Sunkyil wanted to come to the station too when the tonga went with the baggage. Parmanand and Rinchen Gyaltsen saw us off. Parmanand tried hard not to cry. Until Hyderabad the journey is familiar. From Jhimpir to Hyderabad is new but of dry low hills like Jhimpir. We had to wait two hours at Hyderabad and then got a through carriage to Marwar Junction. The land from Hyderabad till dark is well cultivated and sometimes there are nice gardens. We came intermediate all the way. There are less people than in second class and none of the passengers ask you your business. After Hyderabad we had a whole coach divided in two by a bench and little square compartment three by three with a hole in the floor for excrement. Three benches along three walls and two perpendicular to the fourth. The seats are upholstered but can't be made into a bed. There were six to eight passengers from Hyderabad to the first stop: Mirpur Khas. Then four till dark and all night but three of us. Train travelled at good speed. Got good food, cold beer from the restaurant car. Last night a donkey at camp brayed 55 times in a row. Day donkeys haven't such good lungs.

March 14, 1934. Night very cold, much colder than at Karachi. This morning we stopped two hours at Luni Junction and man told us gran[3] were to be found within 3.5 miles Mir Khan [indecipherable] J. Railway, Luni Junction say he can kill them for us at any time. They are common around Udaipur. One man said also at Jaisalmer. The country on the way to Marwar is certainly favorable in appearance—much brown grass and bher bushes and small trees with a little cultivation. Saw ravens, peacocks, cranes, flocks of red deer, nilgai, black deer. Here and there are knolls of smooth rock, rounded rising 100 feet from the plains with sometimes a village against its foot. At Khairla a man says if you play the flute deer will come to the sound. The crops were just turning; some very green still. At Moldi they were harvesting when we left. At Marwar Junction a nephew of PM Allah Buksh of Jaipur was on the platform with a letter from his uncle in which it said to speak loud to his nephew as he "listens badly" and that my rare old shawl was sent with a price 40 rupees above what we had agreed to last year. He had the three knives, two unfortunately with new ivory handles. The long one was a beauty. The shikar curved one wasn't so promising as the

---

3. Himalayan weasel?

picture. The old "gold and silver thread" piece for 450 rupees three × two and half feet was not particularly lovely. The background was dirty yellow with the design in red and green. It is undoubtedly old, but as Radha Krishan Bharawj said "purane juthe bhi hai" [old is also defiled].[4] Had an old sari of green and red, no interesting pattern and disagreeable color. A Kutch embroidery, silk chain on quilted cotton (red). Design very nice, but price rupees 300. Offered 100. Says many pictures. Paid 15 rupees travel expenses and left. The hills become larger, more numerous, and a chain begins at Khairla on the left. Before Erinpura a chain begins on the other. The rocks at Erinpura come to the track and are a coarse granite. Hills with patches of grass on half, rest bare rock. The mountains become higher until at Abu Road there are good peaks 2000 feet and bushes and trees on slope. In the streams coming down from them is water. At Chitralsam there is a very nice place to camp for hunting. The next station is Palanpur and there the mountains end. From Palanpur till dark the country is rolling with trees and fields like home. Grain harvesting beyond Abu. At Abu Road were some soldiers and two police with "Watch and Ward" signs on belts. What is there to watch and ward, I wonder. Before we get to Abu two inspectors came aboard and seized our things and weighed them. They loosened hold of my bed bag only with protest and then wanted to know if there weren't something else in it too. I patiently let them weigh and fine a charge of 9 rupees, 10 annas from Hyderabad to Ahmedabad for 1 maund 30 seers baggage. They hoped I would tip them to go away but not a suggestion did I make. We have had to go third class from Marwar Junction since there is no intermediate. There were two or three Indians in each second class compartment and what happened but that one of these came into third class and rode in order to start a conversation. I went to sleep before he made me understand he's a gentleman, is it you are coming from America. Third class was full up all the way but not crowded and nothing to complain of but the cheap perfume that the men use. No other bad odors. One youth smelled so high that it was almost distressing, but he stayed only till the next station. Yesterday one police had dirty feet that achieved the same effect. The line from Marwar Junction is apparently an insignificant one. We can't get supper except a "side dish" which the man says is a "cutlet." We don't want to try it. Bananas are non-existent till Abu. There are of course native food vendors everywhere. One at Abu dished out some milk stew with his fingers into a plate made of leaves or if not too fluid on some waste office sheets written in ink of which he had supply on a hook. The orange leguminous tree with large round leaflets is often in bloom in groves. Saw two white cranes near Khairla. Blue ones have been more or less common all the way. Pleasant wind all day. Wore a light sweater all day. Mostly Hindus in the district. Each and all (men and women) carry the anus washing brass bowl for that use and for drinking on train. At Luni one had gone off with his and the train went off too. Women often have nice bracelets on arms and legs, four or five of each on each

---

4. Matthew Hull, personal communications.

limb. The Luni shikar says you shoot gran from an oxcart. They are afraid of a camel. There are monkeys colored like a langur from Abu on. There is a lake ca. one quarter mile long near Khairla. In two places in the dry sand bed of the summer torrents they had planted crops. In one, in trenches about ten inches deep were melons and cucumbers. Two hours out of Ahmedabad, the country becomes flat and there are some crops I don't recognize, with yellow and white flowers, three feet high. Mango trees are common now. Semi-desert trees have been common all day. Desert vegetation has faded since Luni. Saw few hawks all day, and these mostly in the dry country. The Mohammadans in the desert Sind to Hyderabad make often prayer places in the jungle. They clear away all stones from an area ca. six by ten and outline the space with small stones [drawing]. Hyderabad has so many wind catchers that it looks as the city were covered with banners. Arrived at Ahmedabad, we created a great sensation. Our money was smelt as usual and one man undertook to get us nice berths in second class. He went to a place marked "Reserved" and gave us two lowers in a six bed compartment. Later an old Parsee took the other. A well-educated native who took me for an Indian and started a conversation in Persian came in to talk and told us gran at Larkhana, abundantly. Our old fellow passenger was also very loquacious. He told us how to get to Madras by the one o'clock mail and where to check our luggage in Bombay and then wanted us to sit up and have a cup of tea with him at the next station. The train is bustling full, all classes. No intermediate and very little first class. Am so pleased over the old shawl. That's the last thing needed to crown my buying this year.

March 15, 1934. The night passed without anyone routing us at all. We had a very nice compartment. My old Parsee friend took me by the hand and led me to the information desk where I got all the information about Madras. The best train left at 1:45 pm. I got a shave then and took a carriage to see my old shops. Buksh had nothing first class. Two Persian pictures good but not striking, a needlework sari in queer lavender tones, and a couple other saris. Offered 25 rupees each but he refused. The nephew, a handsome young man, was in charge. The uncle went to the Industrial Fair in London. Not a leopard skin to be had. My last year's old cloth merchant doesn't open till 2. Met a woman who came with Mr. Patterson. Mrs. Patterson is very well. Bought Rup Chand the same sort of NYK[5] Round the World as I have for 1516 rupees. They allowed half second class railway fare to and from Colombo. Japan transit visa for me free, for Rup Chand nine annas. Temperature in city very pleasant. Outside in train hot dry wind. The hills began at once and the route is very scenic. The train after an hour out runs in a hogback through series of tunnels with broad watered valleys on either side 2000 feet below. The mountains beyond are covered with trees. In the ravines they are green. At Khandale the country is very pleasant and a good collecting site. At Poona there is water for irrigation and beauti-

---

5. NYK: Nippon Yusen Kaisha, Japanese shipping company, founded in 1870; still in business today.

ful fields and fruit gardens. There is a reservoir below Poona. Water is carried down in pipes. Train electric till Poona. Intermediate very comfortable and only two others in the coach after Poona. Dining car and drinks. Our tanka which I have been trying to buy for two years is not in stock, at least the Parsee's wife and children couldn't find it. There was a small piece of old Persian tapestry with all over crude human figures, but they wanted 5000 rupees. There are flocks of white herons, small and intermediate, wherever there is water and in the irrigated fields. The country smooths out beyond Poona and at Dhond you have the dry acacia [indecipherable] plain. The supper chicken curry and rice [indecipherable] special was very tasty and with a bottle of beer we felt royal. Last night was cool and tonight became cool at dusk. We have been charmed with the beautiful situations in the hills and gorges and valleys from Poona and Bombay. I had no idea it could be so scenic outside the big mountains—like a bit of Kangra. Some grapes bought, good flavor, from near Bombay. At night we were left alone with a young Catholic Indian priest, dressed in some sort of white frock with full beard and a little bald spot shave onto his crown. There was some trouble yesterday because a Hindu drank some Muslim tea from a station vendor. The bystanders said he should have made clear what sort of stuff he was selling and he said he had been crying "Cha Islami." At Poona, there were groves of palms with leaf stumps on the trunk to which a pot was tied near the top and a wound made. Horses near Poona not military.

March 16, 1934. Night pleasant. Slept sans blanket and interruption. At six we had come to a country with scattered hill cones (granite?) and red soil. Palms common. Crossed a nice river at Tungabahadra,[6] elevation ca. 1200. Around mile 325 the hills are picturesque jumbles of smooth granite (?) boulders and rise to 500 feet and run in chains. At Adoni (mile 307) the soil is black and nice *roche montonnér* among the numerous "rock pile" hills. At 295, chains going to right and left with a 20 miles plain between. There are standing crops here in all the fields, usually a stunted cotton a foot or 18 inches high and an equally poor crop of some kind of sorghum or kaffir. All land cultivated at this [indecipherable] or bushes. Villages few and small. Houses varied: near Bombay of brush and matting; sometimes whitewashed here or stone and mud, square, flat-topped, latter remind strongly of Zankskar. Picking cotton in many fields. In Rajputana they had it harvested and in one place and a mountain of it piled up unsacked in front of the village. The dogs that do duty at the stations are about all one sort, mostly yellow or black, short haired, and pointed nosed. Some have lost a leg or so but most are in good condition and sans scab. There is often a goat or two also. Those are fat and friendly. At one station on the J. railway there was a huge billy of high family. At Karachi one goat had ears over a foot long. Faces like those of the K. Sheep. Saw a picture of Ranjir Singh in Bombay, done on paper with about the same compositions as ours, but with the men going in the

---

6. Tungabhadra River.

opposite direction. At mile 270 the country becomes hilly with stony knolls, covered with euphorbia and often date palms growing on the flats. When such plains are broad there is irrigation of fields of grain. At mile 265 two nice lakes, one half mile wide and soil reddish again. Pigs common, also some red sheep. At Tadpatri, mile 227, elevation is 777 and a broad cultivated plain bordered by low earthen worn escarpment. The Mandi type of temple is seen here and at other places from Bombay to here, not often. Agaves often planted along railroad right of way. Bought a large yellow muskmelon with flesh like a honeydew. Fragrant but flavor weak. Watermelons in Bombay common, not tasty. Also got a green orange, with white flesh. Skin tough but juice better than the sankta (local). Two kinds bananas seen, both yellow and large, fair. At Yarraguntha, mile 185, slate quarry. At Kamalapuram there is a reservoir pond in which were flocks of snipes and herons of some sort. We bought there a little yellow melon, size of a grapefruit, called sharbaamar. They brought us lunch then at noon, chicken curry and rice and a bottle of Beck's beer, rupees 1–4 × 3. At mile 145 there is a nice temple with towers of half dozen stories, a door or window in each. Houses there of stone with four-sided thatched roofs. Country stony hills with grass and bushes. Date palms in the ravines. Cultivation very little. At mile 136 a broad valley of rice fields surrounded by smooth hill wall 500 feet (Mandalur) and two impounded old lakes. A dry sand riverbed one-eighth mile wide. Got some coconuts with delicious water, best ever. No flesh. Houses small square, using four sided thatch roof, mud walls often whitewashed with broad red verti-

Plate 60. Virupaksha temple at Vijayanagara (Hampi) along the Tungabhadra River (Walter Koelz Collection, Bentley Historical Library, University of Michigan).

cal streaks against the white. At mile 120 a nice pond with a few ducks, snipes, gulls, and herons. Heavy groves of mango, coconut, date. Surrounding hills smooth, brushy or woody, and a high range 2000 feet toward the east and west. Mangos about as big as a small plum; some in bloom. At Koduru the scenery is again picturesque. In the fields are small onions, peas, and other crops and a pink lantana blooms along the hedge rows. The groves have also an undergrowth of brush and the mountains, now near, are often solid granite. The landscape is as/more lush than Kangra. There is even a nice little stream. And at mile 105 a lake Kodura would be a good collecting site. The mountains are very near (1 mile) from 105. Settigunta station at 102. And large lake and streams and dense thicket of bamboo [indecipherable]. No clearings on hills. Many mango groves, many young. Temperature all day cool, not nearly so hot as yesterday after leaving Bombay. No dust to complain of on all our journey. Rengunta (mile 83) has few bushes and groves and much open cultivated country. Several ponds. Hills as before but perhaps rougher, heavily forested at basal half. The coconuts are brimful of water, sweet and if cool, delicious. At mile 75 there are many ponds, one large reservoir with 200 ducks at the foot of a grassy rocky hill, 1500 ft. A better camp site can't be imagined. To one side the mountains are covered with brush. The valley has fields and groves and three or four miles to the north there are ranges with sheer cliff faces of 500 feet above the heavily wooded talus slope. Station Toduku at mile 73. The water for the fields is dipped out of the reservoirs into the canals. At one place they were pouring it out by hand. At another was an arrangement like this [drawing]. A man at b transferred his weight toward a and raised the bucket full of water to the rim. A man below manipulated the bucket. At mile 50, Tiruttam, the mountains are ended and the country continues as before: much rice and mango and some palm. People black, some with beautiful physique but not face; graceful carriage usually. Before Arkonam Junction an Indian sans uniform or badge came aboard and took my tickets and wanted to weigh my luggage. I give the tickets and then bethought that I wanted his credentials and he could only show a railroad pass third class. I demanded back the tickets and told him I was going to hand him to the police at Arkonam Junction. When we got to Arkonam it turned out that he was a railroad officer, but not authorized to collect tickets or anything else. He did succeed in calling attention to my baggage and I had to pay excess at Madras of 13 rupees, this is spite of fact that a similar officer in Bombay had booked it for 5 rupees. We took a very dilapidated old tonga and drove around the city. They tried at first to load us into a two-wheeled "covered wagon" drawn by a cow or ox. They said that was the regulation thing. The tonga driver had an assistant who stood behind and kept a look out for obstructions and well it was or we should have certainly been assassinated! We tried to find some curio shops so they took us to the second-hand clothing stores. On a ride through the town we went one place but they had nil. Two corpses covered with flowers except for face were marched through the city to a place of disposition. No music. The town has some very large municipal buildings that date from the 1850's. There are also broad streets and clean electric lighted shops in the main part. Plenty of European goods. We were a great curiosity and never failed to draw a crowd. The people seem very kind but completely childish. They understand English better than Urdu. At nine we left for Colombo, second class. We won't have to pay baggage now. Were allowed a second each. One Indian youth in our eight berth compartment. Madras is not particularly scenic. I shan't want to see it again.

March 17, 1934. The train is the jiggliest I have ever been in even in India. I tried the top and the [indecipherable] and in all got such a shaking that my innards aches. The Indian got off at Tangore Junction and I got up then, daybreak. The country is flat with mountains visible for a while this morning to west. It seems to be better watered and all the fields and dry stubble have some green. Ponds of mango groves, palms scattered as before but often rather dense. Natives have beautiful deep well [indecipherable] chests and various coiffures. One has the head shaved to the occiput and the long hair done in a knot at back of head, said to be one of the Tamil casts. There are 26,000,000 Telugus (newspaper) most (17,000,000) of them in Madras Presidency, all speaking one language, third largest language group in India. Today is their New Year 1856, era Salivahana, name Bhara year, 1,955,885,035 since creation according to Hindus. The natives' carriage is not so graceful as that of Sindi men and women who swing along from the hips, almost invariably with a grace noted by our men even. Beggar children are very numerous, almost at every station a flock. Some have borrowed a baby or a hideously deformed child, etc. to soften susceptibilities. Fruit offered has only been sanktras[7] and bananas, not a coconut or melon obtainable. All they could be had at every station yesterday. Around noon three blacks got aboard and went on to the end. An old one blew snot all over the furnishings. One man whom I had seen in Madras came in sans uniform and started asking about my baggage, if one box weighed one maund, 30 seers, or all must weigh, etc. the mention of baggage was too much and I snapped him into silence. The doctor, a native, got on at three a little above Mandapam, and gave a certificate after looking at us only. He wanted to know if we were sailors on the Hakozaki Maru. The ocean is visible on both sides at Mandapam with small islands near on the right. (All the cattle since Bombay have been measly; small and generally in bad condition.) We cross then by a bridge a mile (?) long over stones over which the tide was running to our left to a large island and stopped at Pamban Junction. The country is sandy with palms and flat topped acacias, also dunes with totally new vegetation; got first coconuts at Pamban Junction, ¾ annas. Breeze now delightfully cool. Large flocks of snipe, gulls, flamingoes, for a couple miles above 381 and all the way to Dharmashkhodi. We arrived at Dharmashkhodi at 4:25 pm and left India after four and a half days of train ride. Not a thing disagreeable, if you don't mention the baggage vultures, but

---

7. *Santra*: orange.

there is much to be taken for granted in the arrangement—no dust, no heat and food, drink and fruit all the way. We departed at 4:55 on a steamer "Goschen of Glasgow" and got an export certificate for the shotgun. They say there is good duck and snipe shooting at Dharmashkhodi and hunters seldom come. Water is brought down from above. Elephants in Mysore forests. Michael Gomez, first officer on the *Goschen.* A Xian Goese took me to his cabin and showed me some old coins and told me about the linguistics of the peninsula. Tamil is spoken from Cape Cormorin to Madras, Telegu in Madras Pres. [sketch map of peninsula]. West of the mountains Malayalam. To the north Canarese. Madura, Ramnath, Trichinopoli are old culture centers. At Annamalaya, a University. Old Tamil was written with lines and dots [drawing]: modern in parentheses. Gomez says he was struck with the similarity of Japanese writing to the old Tamil. Water is 40 feet deep between Ceylon and India, says Gomez. Customs came aboard when we landed at Talaimamar. They looked in my trunk (I had declared only my two guns) and pulled out the dagger. They passed it without a word, and asked if I'd not revolver or pictures. Then they wanted me to open the tanka bundle. These I said were too sacred to be put into their hands and couldn't let them go. I showed them the University's letter and they said that helps a lot. Your appearance wasn't quite up to it, or something of the sort. The guns they took and I shall get them when I leave. All dutiable things they take from you thus and you pay storage. Immigration said my visa expired yesterday. He let me go after taking away my passport and ordering me to report to the police Monday morning. The train was waiting in the dock. These trains are English fashion with cloth upholstering, two decker compartments. We had to pay one rupee each for sleeping accommodation. In our compartment was a young native who in spite of the heat wanted to shut the windows but refrained at our protest. Dining car served excellent tea. A cold began to manifest itself and I didn't get much sleep.

March 18, 1934. It was cool last night. Ride agreeable. Arrived in Colombo about seven am. A swarm of hotel pirates attacked us. One "globe" with advertisements in French seemed most attractive except that they said they catered especially to "soldats et marins." I wanted to go to the best one, the Grand Oriental, it would be a recommendation if there were going to be any dealings with the police, but the boy wasn't keen on taking me. I needed a shave and clean clothes and anyhow am queer looking, so he was a bit uneasy about my reception at the hotel. Arrived there they looked me over dubiously, then hastily assented to my request and showed me fourth floor to a room overlooking the harbor, except that something had been built over the front of the balcony so you had to scramble up and look around the obstruction to see the ships. The room was small but had a bathtub and running water. Rate 13.50. We went out to get Rup Chand some clothes. The first order for white suit was 14 rupees. The next, nine. Later we learned that you could get them for seven or less. Bought him some shorts and then went home to sleep. Both dead tired. Snoozed on till tea time when a boy brought in tea and then till dinner. I descended to the dining room, a large, high room with many tables. A few people (six or seven) only were there with me. A wide bill of fare and well-prepared food. Waiters barefooted, wearing dhotis and some with tortoise shell combs in their hair or on bald heads. The men wear a better-looking dhoti[8] than our Hindus. It is long (to the heels) and fastens in front. Hindus show the back of the legs. The combs are semi-circular and open to the front. We got Rup Chand's clothes in the evening and fitted him out. He looked fine; shoes we couldn't find at all suitable. No word from our boat.

March 19, 1934. Went to see the American Consul this morning and he had my passport. Said I'd need a new one or I couldn't get off the boat in any port, so for my ten dollar bill he made me out one. He is from Rochester, named Buell, a pleasant optimistic young man. Went then to the Secretariat and got a visa to "all British territory" without any difficulty, but that cost 26 rupees more. Mr. Buell says the United States has offered to give up our visa fees for the British if they will reciprocate, but they won't. The visa officer was a native and agreeable. A mother superior was there filling out a passport for a four year old girl she was sending to Australia. She tried the man's patience a bit but both were calm. Looked for Kandy knives. Abdool Kaffor has a beauty but wants 200 rupees. I looked for a good star sapphire but he hadn't any first class. One lovely color but poor stone. The cat's eyes interest me much. They have a line of light down the center. One was a perfect beauty, in black-brown and honey. Looked in all the shops for knives (they even took me to the place where they make knives for opening coconuts) but found nil (saw plenty, 50 or 60) till one man said he'd show me and took me out a mile or so to a house in which there were some nice old things, mostly furniture and china of the Dutch occupation. He had two nice knives, but owner wasn't in, so the son would bring the things in the evening. Got some good pineapples at 25 cents each, large and nicely flavored but not so nice as Hawaiian, and mangos 50 cents a dozen in the bazaar. Two kinds, one green, one yellow. Former more watery but with a distinctive flavor. Neither equal to best Bombay mangos but Rup Chand says better than the Kangra mangos he ate in Lahul or even in Ladakh. The boy came at night with the knives. Father is an old collector who is old and has given up his shop—now sells to special customers. Name M.J. Perera, Number 1 Elie House Road, Mutwal, Colombo. Says he helped buy things for the Museum. Asked ten rupees for each, which I paid and he left promising to come at 6:45 tomorrow morning. Left our snuffbox coral to be fixed up at one shop. It can't go home like this and some sort of work they do fairly well here, the silversmiths.

---

8. *Sarong:* wrapped cloth.

Felt none too grand, a mild grip and a nasty pulse that I can't keep down with aspirin. Our boat due tomorrow at six am. Got Rup Chand's ticket at Carsons. A nice young Englishman very agreeable. Got £2 at the bank.

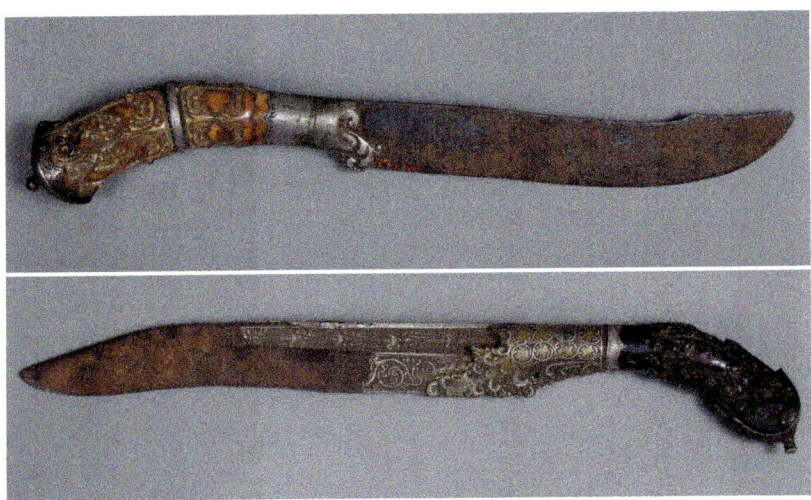

Plate 61. Knives from Colombo, Sri Lanka (UMMA 17260, 17261).

March 20, 1934. The boat came in at 10. Sailing time 5. Delayed by fog in London. Went aboard with my things at once and then went back ashore. In the American Express Company launch was a man called Arthur Lord Tourist Agency who said that on the east coast at Batticaloa there were "singing fish." Music like a pipe organ could be heard under the water at night if the moon isn't bright. At Mannai five miles outside Trincomalie are small hot springs five feet across, some too hot to put your finger in and have fish! Took a car and drove out to the museum and European quarter. Nice bungalows with large yards. At the museum are some local animals. On the lawn were some spoonbills, pelicans, herons, and storks loose. It was feeding time and they all scrambled for the feed. The museum has some biological collections on the second floor, cases of peculiarly Ceylon birds, butterflies, corals, etc. There is a room of Dutch furniture, of china and glass. Considerable exhibits of knives, ivory carving, brasswork, etc. Saw nothing very interesting except the ivory carvings and knives. The native art was soon spoiled probably. 1505 Portuguese came, 1658 Dutch, 1795 English. Island 275 miles long by 140 miles wide. Most cultivated, rice, nine lakh acres coconut palms, four and a half lakhs acres in tea, and three and three quarter lakh acres in rubber. In Ratnapur District, sapphire, topaz, aquamarine, zircon, amethyst, moonstone, tourmaline, garnet and alexandrite. Population

4,600,000. Singhalese Buddhists, Tamil Hindus, Moslems. See in Colombo Pathans who are money lenders. World's largest sapphire 466 karats bought by [J.P.] Morgan. Colombo 360,000 inhabitants. At Kandy in August, the Peraher of ten days. Buddha tooth ceremony. Peradeniyan Botanical Gardens near Kandy. Anuradhapura, chief of the ancient cities, 400 BC capital; there is the Bo tree, sacred to Buddhist, now 2259 years old. Sigiriya, a high rock, once a fort remains of beautiful paintings. Polonnaruwa has large Buddhist ruins, beautiful huge Buddha, Gal Vihara in the jungle. A number of beautiful Buddhas hewn from rock; Nuwara Eliya, 6240 feet elevation, has golf course, frost in winter, trout streams. Bandarawela has dry air and fine woods. Adam's Peak is pilgrimage place for Buddhists, Hindus, and Muslims. At summit is Buddha's foot print. Pearl fisheries between Mannar and Puttalam. Divers Arabs and Tamils. Average 60–70 seconds under water, sometimes two minutes. In 1870s no tea. Tea came in when a blight ruined the coffee. At Ankana a well-preserved Buddha carved of rock 46' 4" high. Ceylonese money all paper except 50¢, 20¢, 10¢, 5¢, 1¢. There are a few wild aborigines, rapidly disappearing. Mosquitos none, flies in Colombo few.

March 21, 1934. Continued miserable with grip. Swallowed aspirin copiously but brought pulse down with difficulty. The *Hakozaki Maru*[9] is a coal burner, not too elegantly provided. Passengers in second class few, mostly young English. Weather clear with pleasant breeze. Rup Chand got seasick. Stayed in bed all day. Read *Taming of the Shrew*. Petruchio delightful speech to Kate: "our purses shall be proud, our garments poor; For it is the mind that makes the body rich. And as the sun breaks through the drakes clouds, so honor peereth in the meanest habit. What is the jay more precious than the lark, because his feathers are more beautiful? Or is the adder better than the eel, because his painted skin contents the eye?" We have a table to ourselves except at night when a young second officer joins us. There were many small jellyfish pea to walnut size in the ocean in places. Abruptly ended, then reappeared in another colony. Saw a couple flying fish.

March 22, 1934. Influenza about finished. Days cloudy and pleasant. Read *Winter Tale*. Some very dainty speeches. *Two Gentlemen of Verona* and *Twelfth Night*: "I can no other answer make but thanks, and thanks, and ever thanks; for oft good turns are shuffled off with such uncurrent pay. But here my worth is my conscience firm, you should find better dealing." "In nature there's no blemish but the mind; none can be called deformed but the unkind, virtue is beauty, but the beauteous evil are empty trunks o'erflourished by the devil." "What relish is in this? How runs the steam" "Or I am mad, or else this is a dream: let fancy still my sense in Lethi sleep; if it be thus to dream, still let me sleep." "Who is Sylvia, what is she that all our swans commend her? Holy, fair and wise is she;

---

9. Steamship, of the Nippon Yusen Kaisha (NYK).

that heaven such grace did lend her that she might admired be | | Is she kind as she is fair? For beauty lives with kindness: love doth to her eyes repair to help him of his blindness; and being helped inhabits there. | | Then to Sylvia let us sing, that Sylvia is excelling; she excels each mortal thing upon the due earth dwelling; to her let us garlands bring."

March 23, 1934. Very pleasant breeze and nice banks of clouds, down to the horizon. Saw some very capable flying fishes, which go almost a furlong. Read *Merry Wives of Windsor* and *Measure for Measure*. The Falstaff and food page mess a scream. *Yasumuni Maru* passed at 6:30 with a Japanese prince aboard. Air temp 29°, water 28°.

March 24, 1934. Day cloudy and pleasant. Rain last night at dusk. Flying fish again today. Read that 120 kind of potatoes grown on Chiloe Island, Chile and that it is home of potato. "Think naught a trifle thou it small appear. Small sands make mountains, moments make the year and trifle life." Young say there is a spider in Australia so large that it can catch a chicken. Saw two lights of Sumatra after dark last night and today till nearly noon we ran within sight of the Sumatran shore. A couple of fish boats with a crew of two or three on our course. The S coast line is mountainous, mountains perhaps 3000 feet, much eroded, and bare? Jungle along shore. Thick clouds like snow hang in one of the valleys. Went to a movie in the evening. Rubbish, but the audience, especially the steerage Malays liked the Ben Turpin comedy. Temperature at noon like yesterday but warmer after M. Nice sunset. At night on the starboard three small rocky islands, densely covered with jungle, largest like city block. Saw a swift nearby.

March 25, 1934. This morning small islands covered with vegetation on the Port. Malays cook on deck over brasier. A Japanese aboard has a German wife. "Look thou character. Give thy thought no tongue, nor any unproportioned thought his act. Be thou familiar but by no means vulgar; the friends thou hast, and their adoption tried, grapple them to thy soul with hoops of steel, but do not dull thy palm with entertainment of each new hatched unfledged comrade. Beware of entrance to a quarrel, but being in, bear't that the opposed may beware of thee. Give every man thine ear but few thy voice; take each man's censure but reserve thy judgment. Costly thy habit as thy purse can buy, but not expressed in fancy; rich, not gaudy; for the apparel oft proclaims the man. . . . neither a borrower nor a lender be; for loan oft loses both itself and friend, and borrowing dulls the edge of husbandry. This above all: to thy ownself be true, And it must follow as the night the day thou canst not then be false to any man."[10] Before noon land on port, later on starboard. All hilly with heavy vegetation. Water changed from sapphire to dirty green. Small sail boats scattered. Read *Othello*.

---

10. *Hamlet*, Act I, Scene 3.

Iago: "Who steals my purse, steals trash; tis something, nothing; Twas mine, tis his and has been slave to thousands. But he that filches from me my good name robs me of that which not enriches him and makes me poor indeed." Iago: "Trifles light as air are to the jealous confirmations strong as proofs of holy writ." Othello: "He that is robbed not wanting what is stolen. Let him not know't and he's not robbed at all." Cha-no-yu — tea ceremony. Ancient observance which is developed by certain Buddhist priests of tea drinking art. Every movement precise, sans error. Said that delicate and accurate performance of each act teaches poise, etc. Situated in every large garden is a "C n Y" house constructed on prescribed lines, and in homes of many wealthy, a room. Doll Festival or *Hinamatusin* in honor of girls is held in March all over Japan. Tokyo has 92,000 cherry trees. Best in Ueno Park (2,400). At Edogawa, 1300 single-blossomed, make a tunnel two miles long along the Edogawa River. Asukayama, a hill, is electric-lighted at night. Arawkawa upper part of Sumida River, is famous for many colours, some even green. Best April 15–30: 1000 of 42 varieties, extend two miles. Kumagaya farther up has trees along river. Got into Singapore at 6:30. The entrance is narrow between islands and on both sides beforehand one sees collections of buildings, gasoline tanks, etc. on low green hills. Inside the harbor the scenery is the same with a long row of wharf-age. There is nothing tropical about the place except a few palms; otherwise the greenage might be Michigan in summer. There were some half dozen ships in the harbor; one an American, *President Monroe*. I went aboard to see what she's like. There were some nice looking men and women aboard, but I felt no great sympathy with them. The boat is full said an officer. Costs some $600 round the world. We lost our old Malay woman, the two fair maids with whom an impudent steward lodged them at table and the young couple with the two boisterous girl children. The old lady put on her most glorious clothes, a light blue creation with huge yellow butterflies. Three black plump companions met her, no less splendid in green and yellow dresses, with appropriate designs of checks and waves. The two girls were dressed beyond this earth, in a diaphanous figured silk, one in blue, one in wine, so gorgeous that you couldn't help but pity them that they had to put foot on earth. Rup Chand in bed all day with fever, grippe like mine, I supposed. Some 12 natives in little angonts[11] ten feet long came paddling up to the boat as we approached. They dove for coins, I suspect. The surplus water they splashed out with hand or foot and sat on the bottom. Physiques beautiful, especially chests. The Ceylonese are all deep chested, like chikor, says Rup Chand. *Love's Labor Lost*: "These earthly godfathers of heavens lights that give a name to every fixed star have no more profit of their shining nights than those that walk and wot not what they are." "Many can brook that weaker that love not the wind"; "That sport best pleases that doth least know how." In *Twelfth Night* mention of the bed of Ware, England. Saw the bed in Kensington Museum.

---

11. Small boats?

March 26, 1934. The ship's officer got my Japanese visa. Then I went out to look for a bank. I passed a European, ca. 60, and was told the bank was far, to take a trolley. I said I'd have to walk, I had no money. The man took out 20 cents, said he was walking in quest of work, but begged me again to take the money. Then I got into a rickshaw. The boy understood nothing but merrily trotted me through the town. When we had passed the banking center and got to the museum, I asked if he know where the Charter's Bank was. A European overheard and came over to direct the man back. He offered to come along to show me. Later on I met one of the men who had been aboard. He wanted me to get into his car. Most friendly people. The town is mostly Chinese, broad streeted, clean, winding streets and nice buildings. Shouldn't mind a bit living there. Went to the fruit market; bought limes 10 cents a dozen (42 cents a shilling), pomelos (a small grapefruit sans seed, very sweet), bananas and a red skinned sweet tandarind[12] sans seeds. Arrived back at the boat; 18 divers were alongside in their little craft. They never missed what they dived after. One old man smoked a cigar and dove with the lighted end in his mouth. Several were white haired. One youth had a magnificent torso and was most agile in the water. Left Singapore at noon. A dozen boy scouts (Japanese) came to boat with a small group and fastened streamers to the boat, sang till she was off. Turned cool after leaving Singapore. At dusk a conical island to the port. The Second Officer called in our cabin after supper, a very decent person (Osamu Sakamoto). One of the other officers got me a Japanese Visa. Read *Pericles*.

March 27, 1934. Read *Cymbalene* "Hark, hark, the lark at heaven's gate sings, and Phoebus 'gins arise, his steeds to water at those springs. On chalic'd flowers that lies, and winking Mary-buds begin to open their golden eyes; With everything that pretty is, my lady sweet arise: arise, arise!" The Malays that were on deck between first and second all got off at Singapore, and their places taken with Chinese. They cook also over a brasier with charcoal. Air and water at noon 27°. Radio announces President Roosevelt signs Philippine Independence Bill.

March 28, 1934. *Much Ado About Nothing*. Friar: "for it so falls out that whatever have we prize not to the worth while we enjoy it, but being lack'd and lost, why then we rack the value, then we find the virtue that possession would not show us while it was ours." *Man schatz nicht was man hat. Das ist schon langst enviesen. Ers wirm due wholtat fehlt. Dam wid sie nich gepuesen.* D. Pedro "Good morrow, masters. Put spent torches out. The wolves have prey'd and look, the gentle day, before the wheels of Phoebus, round about dapples the drowsy east with spots of grey." Leonatus: "for there was never yet philosopher that could endure the toothache patiently however they have write the style of gods and made a push at chance and sufferance." Saw a dolphin or the like and

---
12. Tamarind?

three shearwaters this morning. Water after Singapore sea green, now sapphire. Shakespeare makes out of his heroes some fearful villains, and then the heroines gracefully take them back and forgive them (*Two Gentlemen of Verona*—Proteus and Julia; *Measure for Measure*—Angelo and Mariana; *Much Ado*, Claudio and Hero. *All's Well that Ends Well*: Bertram and Helena; *The Winter's Tale*: Leontes and Hermione). Sea calm, air 28°, sea 27°. Read *Mid-Summer's Night Dream*.

March 29, 1934. Day cloudy and pleasant. Air and sea at noon 27°. Read *Antony and Cleopatra* and *King Lear* and *King John*. Dreamed last night some one gave me two beautiful blue muskrats, a beautiful purple blue. Rup Chand got up today from his influenza. Saw about 20 large shearwaters whitish below with black brown primaries. Flew in flocks of one to five alongside at dusk.
  Tatuo Musawa
  c/o Mizunu. Kanda Tokyo, Japan.

March 30, 1934. Began to get a bit rocky at bedtime and til the morning was a bit bouncy. Stayed in bed all day and got sick, though the sea was never bad. Only four passengers came to the dining room. Rup Chand was very ill and ate nothing all day. Got into Hong Kong at 10 pm. The sea calmed about 8:30. Some of the passengers went to shore but it was drizzling and I went to bed. Rained a bit during the day.

March 31, 1934. This morning the harbor was full of junks, four or five small warships, the *Chichibu Maru* and another NYK boat for London, *Corte Rosso*, a German boat with a swastika on the stock. The junks are grand. They are manned by men and women, often only women, two sculling from the stern and one on each side. They also have sails more or less patched. On one was a coop with two chickens and on all they were getting breakfast. They are apparently the owner's homes; the women range from white-haired to youths with babies strapped on their back. They brought baskets of vegetables, peas, eggplant, fish, and some long sweet potato like thing, eighteen inches long [drawing] covered with lint. Hong Kong is situated on the slope of a mountain ca. 1500 feet high, and has a fine harbor. Small islands form a bay. Somewhat like Naples. The city has some very nice buildings, banks, hotels and shops. Even the native shops are large and clean and stocked with foreign goods. The Chinese common people, male and female, often wear black smock and trousers. Saw few good looking, mostly the sort the Tibetans call རྟ་སོ་མ་ [*rta so ma*][13] sans chin. Went to the meat market. An abundance of fowl and beautifully dressed meat. Many duck heads, feet, and intestines for sale. Also pintail, [indecipherable], a number of teal, mandarin, and another duck I had never seen; also some kind of Titar and speckled dove. Fruit chiefly mango, pomelos, mandarins, bananas. The latter

---
13. The meaning of these characters is unclear in this context.

of saccharine sweetness, like those in Singapore. Got many new passengers. One old lady, painted like Polly and dressed like 16 came aboard with a huge bouquet and three lobsters done up in a newspaper. There were lovely flowers for sale, all our summer kinds: roses, dahlias, snapdragons, freesias, larkspur. We got about a dozen French Jews and a few others in our class. Had Sukiyaki for supper as an extra. The captain of the Japanese ski team that is on the way back from Wengen Switzerland came to the cabin and showed his photos. A nice little fellow named Tatso Misawa of Tokio. There are nine boys aboard. The Germans got first and Japanese second place. He bought all the pictures of the snow and peaks he could find, it seems. I gave him a tie and he picked out a nice red one with an overall check. We got as addition to our mess a Canadian girl, a nice little thing. She has been ill and went to Hong Kong to get the sun and it has rained the whole month she has been there. The Hindu got cranky one day and complained there was nothing he could eat. He said he couldn't eat meat (he has been eating it up to now) and couldn't find fish and vegetable curry sufficient. The steward said he could have eggs and that ended the matter. Air yesterday, 23°, sea 22°. Temperature after Hong Kong whence we departed at noon cool enough for overcoats.

April 1934

April 1, 1934. A stiff breeze all day with a choppy sea. Temperature air and sea 14°. The second officer said that if there wasn't a headwind in the Straits of Formosa at this season there was a fog. Saw flock of five little petres about as big as a Phalorops this morning. There was a fleet of junks at nine or ten this morning. Said to be trawling. Day cloudy but barometer 760. Most of the passengers sick. Rup Chand stayed on deck all day and got along famously. We still have some of the purple orchids bought at Singapore. From Colombo to south had carnations. At Hong Kong we got some summer flowers. There are three aeroph. orchids hanging on deck, now in bloom. There were some on the Chinese junks in Hong Kong harbor. The two new men at Hong Kong are American, six feet youths spending a year in travel. Rather nice sort. Officer said we were making nine and a half knots and water coming aboard [indecipherable]. At night a lighthouse on the port.

April 2, 1934. Calmed during night. Water in the morning murky green, probably from the Yang Tze. Cold and cloudy, air and sea 17°. Calm all day. *Troelus and Cressida*. Cressida: "women are angels, wooing, Things won are done; joy's soul lies in the doing. That she beloved knows naught that knows not this; Men prize the thing ungained more than it is; that she was never yet that ever knew love got so sweet as when desire did sue. Therefore this maxim out of love I teach; Achievement is commend; ungain'd, beseech; Then though my heart's content firm love doth bear; Nothing of that shall from mine eyes appear." Agamemnon: "Light boats sail swift, tho greater hulls draw deep." Ulysses: "Pride hath no other glass to show itself but pride, for supple knees find arrogance and on the poor man's fees." Ulysses: "one touch of nature makes the whole world kin, that all with one consent praise new born gawds; they are made and moulded of things past, and give to dist that is a little gilt more laud than gilt o'er dusted. The present eye praises the present object." Read *Coriolanus*. Passed during the day at intervals clusters of small fishing boats, 100 or so in an area, scattered about like flocks of ducks. The water since noon 250 miles below Shanghai more muddy than before. Islands on the port commoner. Flocks of gulls followed the boat in the afternoon. One of the men thought it would be interesting to collect match boxes. He knew someone who had a collection of them and of toilet paper. One of the American boys, Brinkerhof, Englewood, New Jersey, came to the cabin to see my knives. He bought some glass rubies, sapphires, and aquamarines at Jaipur, only $1.50 a piece. The other is Pete Fortune of Chicago. Two of the youths don't smoke; the big he-man sort too. Have read on the trip Edgar Wallace: *The Man at the Carlton*, pleasantly exciting and not at all crude. Went to Library. The first class has some 50 passengers, all old, including the lobster lady.

April 3, 1934. Calm all night. Arrived in Shanghai and docked at 7 am. After breakfast went uptown with the three Canadian men. Took a car for three Mexican dollars an hour and drove around the town an hour. The European section has some good shops, nice banks, and high buildings. There were several foreign concessions in our itinerary. The Japanese port still had much of the war damage visible. Exchanged money at $14.60 for one pound. It took 27 minutes to get the money from the Chartered Bank. I bet with a companion it would take 20: that's the rate of speed in India and Singapore. We went to a Chinese restaurant for lunch. For one dollar each we had five Chinese dishes, all of them good, with rice and ivory chopsticks. Rup Chand bought a Chinese cap for $1.80. Cheapest leopard skin $45 as against 18 rupees in India. Looked in a first class antique shop. Had no porcelain bowls older than 75 years. Said seldom obtained them and then bought at once by the Chinese even. Old cloisonné rare and expensive. Had some embroideries, a nice priest robe of ca. 150 years, and an imperial wedding coat, nice but not exciting. Also an old wong picture, a man in wong on a silk background. The apricots are just coming into bloom, also mangnolias and poplars. Urelond beginning to leaf. Palms grow here in the gardens. Asked two different nice looking young Chinese where a Chinese restaurant was, one said "I don't speak English" and the other "I don't know." Wherefore I gathered they didn't like foreigners. Rup Chand attracted much attention, but no one asked anything. Saw one of our zosterops[1] in a cage here (and at Hong Kong also). People large stature and better looking than at Hong Kong. Some very nice looking. Clean and no beggars. Day a nice spring day. Said to be three million inhabitants. Shoemakers came aboard and repaired the passengers' footgear. Plenty of Sikh police. In banks gunmen, armored cars

---

1. Genus of birds, "white-eyes."

in streets. They were eating in a hat shop when we went in and had 15 boxes of most tasty looking kinds of food. In food shops often a duck roasted, bill and all, and even a whole pig. Many ships in the harbor, including English, French, and American war vessels. The little sampan very common. They are little craft ca. ten feet with a semi-cylinder roof in middle and upturned stern. They unloaded from one ship many huge bales of dirty rags. Most of the Chinese wear kaftans of wool or silk, sans kammerkass. Skull caps of black silk or serge with a cloth top knot. At the dock there were three Chinese boys three to four high that did handsprings, single and double, and bent their head between the arch of their legs and collected coins from the passengers. With the help of a youth of 20 they did some very good stunts. One lay flat and the youth picked him up by the ankles with one hand and set him on his shoulder.

April 4, 1934. Finished Robert Chambers: *Beating Wings*, a tale of a girl who succeeded in everything and became a famous sculptress sans lessons at 22. Written in a pleasant, often effective, way that covers up the implausibility of the plot. Read Talbot Mundy: *Ramsden*, a fantastic tale of the Mahatmas in Tibet, neither plausible nor exciting. One of the passengers recommended it strongly since I knew the country! The day was partly clear and the sea without a ripple. Sea air 8°. During the day at several times saw two or three swallows, snipe, or other small birds all headed as we are. The four new Hindus were put at our table. One eats beef, the other three only vegetables. The poor things eat the water-logged stuff they get with the minimum of complaint, except once they said in Punjabi pigs wouldn't eat such stuff in India. The Japanese distributed forms requesting a detailed statement of what books we had and what tobacco, a form for each. Then another form asking about some 15 other classes of goods, including curios, wools, silks, jewelry, cameras, toys, new flowers at Shanghai, hyacinths, and sweet peas! The young German is a Sumatran tobacco employee going on holiday around the world. He got a seven day tour in Japan from a Dutch tourist bureau for 344 yen, which seems very high. He says Sumatra has a nice climate in the mountains. Highest ca. 7000 ft. Grow strawberries and potatoes up there and there is a beautiful crater lake. Doesn't like duranis, best season in September then December–January. Many other kinds of tropical fruit. Got a book *Ancient Ceylon*, by H. Parker, 1909. An account of the ancient inhabitants with minimum of reference to architecture or arts.

April 5, 1934. Promised to pay Gin Kichi Tsuruta, Second Class Bath Steward 5 yen at Yokohama. Day heavily overcast with rain after tea. Sea 12°, air 11°. Rough outside and many passengers sick. Calm inside the narrows which we reached at 12:30. The narrows area, Shimonoseki Straits, had many small craft and small steamers. There is the coal mining area and iron mills, also a brewery, cold storage for ships from the China and Japan Sea. On both sides low rolling hills or soft rounded peaks, usually conifer clad. Near shore, some green fields, probably wheat. A few peaches in bloom. The cherries showed buds among the pines. Inside the big sea it was calm but nothing to see on either side. Paid my bill which amounts to £74–9. Gave bar two Singapore dollars (he is also deck steward); the bath three Singapore dollars, and promised five yen more; the cabin and table steward $55 Shanghai between them. £ = $14.60 Shanghai = 16.40 yen = $8.40 Singapore. Hindus say eyes of rage are blind. The boat gave a farewell dinner, an elaborate menu with the room decorated in flags of nations. The company gave us tickets to go by rail to Yokohama via Rasto and Tokio instead of going up by boat and I arranged to have my surplus baggage sent over to the Chichibu Maru in Yokohama. I packed up the Gladstone for the travel overland.

April 6, 1934. Almost everyone got off the boat to go by rail. The Customs came aboard this morning to examine hand baggage on deck. He searched my bag minutely and then let me go. I could have filled it afterward with all manner of contraband from my cabin. The doctor only looked at us and let us go. We got to dock around 8:30 and went up to NYK office to arrange from Chichibu. Had to pay ¥234.32 for the two tickets. Got ¥16.90 for £1. The newspaper men came aboard to see our ski team and then made a great fuss over a plump little man in an otter coat and leopard skin cap. They photographed him with a huge gold medal, along with his wife in some sort of cheap wild cat skin coat. I discovered he was the ex-cook of the King of Abyssina and went over and got his photo too. He was very pleased to be snapped again. The young German Baron went off with his tour and he walked about the city till two pm. The one Englishman, McCartney, is most helpful and friendly and hails everyone who looks as though he might speak English. He often gets hold of a Swede or Dutch or someone else who doesn't understand a word of it. He and the other, Howard, are much interested in haberdashery and are always on the lookout for such in the windows. The town of Kobe is a queer mixture of large buildings and rummy ones. The workers' houses are frail wood structures with the roofs that would do for a cheap garage in the US. They are small but have good lighting as a rule. On the train we met two old women going on a Cook tour. McCartney discovered they were from Australia and that they were going to Nara. We intended to go to Kioto. So it was decided we should go to Nara too and we got off at Osaka and again somewhere else, and arrived in Nara at 4. On the way there is dense population and much cultivation. The hills are forested and the crops overgrown on the flats between. There were many old rice fields and the green crops were wheat and several sorts of other unrecognizable vegetables. Here and there small patches of yellow mustard. The crops are almost all, including grain, planted on raised ridges with furrows between. Not half the land is made use of thus and I don't see what the six to eight inch wide trench between rows is for. There were plenty of plums, nicely trimmed in arbor fashion, and some plats of small plum? Trees also heavily pruned. A few peaches in bloom. Otherwise a March landscape. The Australian women got off at Osaka and McCartney worried about "the girls" as he called them not coming aboard again. At Nara we went on foot into the town after checking our baggage. We found the deer park by following the rickshaws coming down the hill with Europeans, mostly old or women on the old maid side. In the park entrance, we came onto a dozen deer at once and bought rice cakes to feed them. The deer are utterly tame and very nice people. We went to the various temples and walked about the park till dark. There were many pilgrims, apparently peasants, mostly old women with the nice wholesome expression of the soil worker, each group with a special ribbon and paper flower and a standard bearer and guide. The park is said to be one of the nicest in Japan, 1250 acres with several hundred deer. Trees of conifer species and evergreen shrubs. One of these, an Erica,[2] ca. six feet with clusters of huckleberry blossoms, was in bloom. Birds rare all day. In the park, a young girl suddenly walked up to us and started talking in Japanese and then in a few English words, trying apparently to tell us that certain temples were worth seeing. Her mother and father looked on with pleased smile, apparently proud that their daughter was so accomplished. The people are all friendly and go out of their way to help. There are numerous photo and knick knack stalls in the park, the whole park not particularly impressive for that reason. The men didn't want to eat in the Japanese restaurant in the park so we went to the Nara Hotel, up on an eminence overlooking the city. The view is very nice, but a look at the hotel reinforced by consultation of a folder convinced we'd have to pay dear for food and they didn't want to stay. McCartney expressed a regret that by going we wouldn't see "the girls" again. Then we tried to get into a Japanese hotel where you leave your shoes outside, but they would have none of us and we finally ate at a little place on the way to the station. It was humble but clean and the meat was like leather, but it tasted good and all the family gathered to watch us eat, a set of pleasant friendly faces. I went and made a hasty survey of a dozen curio shops and found a fair piece of Satsuna for 3 yen and a piece of Japanese cloisonné for 10 yen. We took the 8:10 train for Kioto and put up at the station hotel. I and Rup Chand drew a room with one ancient bed with concave springs. Bed clean and no bugs. Many of the people, male and female, wear wooden sandals with two cross pieces across and under the instep, one and a half inches high. They stumble forward sans difficulty on these streets. Costumes are usually Japanese for women and European for men, but all sorts of mixtures of the two. Many women have faintly rouged cheeks. In the Toda-Ji temple at Nara is a Buddha 53 feet high, with face 16 feet, with 288 pounds of gold, cast in bronze in 749 AD. The present building is of wood and is the third to house the image, begun in 1699. Building said to be the world's largest wood structure, 188 × 166 × 160 feet high. 60 pillars, 15½ feet in circumference (this from guide book, doubt the measurements of pillars). The Buddha is plain and extremely effective. There are a few "guardian" figures 50–60 feet high of wood about the grounds, carved a la Tibet. Very nice. The other temples were

---

2. Genus of shrub, including heaths and some heathers.

closed. The old stone lanterns like lamp posts with a place for burning a candle were new to us. The openings of many had a fresh piece of paper pasted over them (*Ram Ko rana Piyara, Kam Ko Kana Piyara*). Many of the pedestrians wear an influenza mask over the nose. Forests of young trees outside of parks. When timber cut whole plat is cleared and reforested. Statues are scarce. In Hong Kong, the British had put up the usual Victoria sitting under a bizarre shed and a Princess of Wales with a face like a weasel in attire of 1860—bussel, full sleeves, and all. In Shanghai, more British statues on the waterfont. In Nara saw the big bell (Tsurigane), rung by a large ram, cast 1752, 13.5 feet high, 27 feet in circumference, 48 tons. The Sangatsodo is celebrated for wood sculptures but was closed. Nigatsodo has the "11 faced Kwammon"[3] an emperor always wa [indecipherable] but not shown. There is an imperial museum but closed and nearby the world's oldest wood building: Horifuji Temple, not seen. The most striking structure is the lofty Kafukuji Gyunto that stands near the park entrance, a five-storied pagoda, not old, but of fine architecture.

April 7, 1934. We started off after breakfast in a taxi, two yen per hour for a sightseeing tour. The first place was the imperial palace. First we had to go to the Chamber of Commerce and get a permit, signing forms of name, address, etc. Ordinarily it says permit can be had only through embassies in Tokyo. With the Chamber of Commerce letter we got tickets at the next place. The palace dates from [18]50s or 60s. It is architecturally simple, one-storied but pleasant. The floors are lacquered black and a few of the boards squeak. That is a specialty with Japanese buildings. The roofs are ca. 18 inches thick of small strips of cedar wood. Chains hang down from the gables at intervals for facilitating repair. The walls of the rooms are paneled with rough-stained brown ordinary wood and in the best rooms some ordinary Japanese pictures are painted on the paneling. Straw mats, padded, cover the room floors. There is a nice garden to be seen with a pond and bridges. We had to take off our shoes (slippers were provided free) and our overcoats, though it was bitter cold and no heating of course. Women could keep on theirs. I nearly froze. The Emperor's style of living appears very ordinary. There was no place for nice hangings. Rugs could be strewn, but walls were hopeless. Besides, illumination on outside wall only. From there we went to Nijo Castle, for which our permit also served. I couldn't stand taking off my coat and stayed outside. McCartney met a huge flock of old "girls" from first class Hakozaki Maru, and was delighted. The castle is nice architecture and has a fine garden. The walls are of huge hewn grey stone blocks, 10–15 feet thick and 20 feet high with a moat all around. Built 1602. Next, Kinka-Kuji or Gold Pavilion. Celebrated for its garden but as far as I could see the landscaping was done by nature and not particularly well. The shrubs and trees of Japan are all most attractive, many evergreen broadleaf. The building

---

3. The 11-faced bodhisattva, Guanyin (Avolokitesvara), today most often transliterated as Kannon.

doesn't excite admiration, but we found afterwards there were a few nice *objets d'art* in it. Now a temple. Then Chionin Temple situated on a slope of the East Mountain, seat of the Jodo Sect of Buddhists. Founded 1212, but these buildings date from ca. 1650. There were swarms of visitors and services were in full swing. A fine old monk who spoke English very well acted as our guide and showed us all around. There are many singing boards; some do it very pleasantly. One stretch called the nightingale walk. There were some pictures in panels in all the rooms, most of them magnificent. The old monk let me look as long as I liked. The men thought 50 sen would be a good tip but I didn't dare. He asked if Rup Chand were Mongol and was much impressed that he was Tibetan and told the other monks he met. The service Rup Chand [said] was very nice, but the robes were new cloth and not nearly so magnificent as our poor priest's robes in Lahul. The Maruyama Park is beside the temple. Full of people, comely view. Many cherries just in pink bud. The Heian Jingu is the highest government shrine dedicated to Emperor Kwammu (771–806), the founder of Kyoto. Erected in 1895. There is a large park with attractive open buildings called the Jui Jitsu Scooh. After lunch went to the Museum. Full of Japanese *objets d'art*, but little good porcelain. Kakiemon[4] fills several large halls, three or four pieces very lovely. Not many prints and not much else striking. From the Museum we went to a silk museum where were displayed samples of silk goods for wholesale. The proprietor was a very pleasant person speaking good English. We bought a lot of woven bags, etc. for 50 sen[5] each and the proprietor told us they were made of paper. He took us up the street a couple blocks where an acquaintance had a brocade weaving factory. He showed us how it was done. On a brocade of two colors the shuttles flew automatically. On more complicated designs the men ran the machines by hand. First the design is sketched out on a fine cross-ruled paper; then a series of long cards as long as the fabric to be woven (?) are punched with holes to correspond to the pattern on the cross-section paper. This controls the number and kind of warp threads that come above the weft, thrown by hand. We went next to the curio shops all in one street. There were six or seven very nice shops with old prints, pictures, and pottery. In one there was a beautiful old Satsuma teapot, one cup size, for 120 yen. No other pottery of interest except celadons from Korea. I didn't look at prints though they were plenty. One man seemed a good connoisseur and had plenty of American customers. I.O. Higuchi, Shinmonzen St., Kyoto. Spoke good English which he said he had learned in Hong Kong. I wanted some old religious pictures, having got a keen appetite from three or four in the Museum. He said come back in two hours. When I finished the tour of the other shops (one had a lot of old brocades) and found nil, I thought it hopeless and sent word to him that I shouldn't come back later. He called me saying that he had had a good kekemono[6] and it certainly was. Said to be over 300 years. The owner had

---

4. Enamel decorated porcelain ware, produced from mid-seventeenth century on.
5. 100 sen equals 1 yen.
6. *Kakemono*: scroll painting or calligraphy.

brought another which my friend said wasn't worth showing. Asked 120 yen. I offered 100 and he took it. Then he showed me a beautiful wood Buddha, originally lacquered and gilded but now plain black with soot. I asked it if were [indecipherable]. He showed me a picture of one said to be 500 years old. The drawing is cruder in these, he said, though the lines are certainly most graceful. The one he had he thought 200 years and asked 55 yen, which I gave. He said there was no point in faking such things at such prices, which is true enough. They duplicate much old pottery, but that much easier. He packed the Buddha for me and sent it to the hotel for one yen. Such pleasant and reliable seeming people I never dealt with. No one pesters you to buy but all are endlessly patient of your inspection. My friend said this was no time to buy *objets*. In November and December people sold them to be able to celebrate the New Year. He would be on the lookout next year and would let me know. We had sukayaki at the hotel. Rup Chand thought it great. Poor McCartney swallowed it and the others ate à l'anglais. They took us out of the main dining room and a little girl cooked the food. Then we went to the theatre to see the Miyako and Kamogawa (Odori dance). These are cherry dances held every year in Kyoto in April. In Osaka there are the Asibi and Konohana Odori. Both there are given by the geishas. The Osaka dance is modern. The Kyoto is Classical. When we got in we all sat on stools with a long board in front of us and put our ticket stubs in front of us on the boards. We were served first with a sweet like an egg on a little brown-glazed earthenware plate. All the audience wrapped both up in a paper napkin and put them in their pockets. Then the girls brought in a cup of dandy green bitter tea called mat cha. Thereafter we went to the waiting room and were shortly invited to the theater. There were three classes: First, 3.50 yen; Second, 2.50, Third, 1.0. Each was above the other, but each class, level. First only had seats! We were Second and had a nice crowd mostly young or middle aged. The theater second performance was full. As the curtain rose flanking the Third Class seats were the chorus and orchestra. To our right were a line of drums in the robes of geisha pink and blue figured and on the other in black and white a sort of mandolin chorus. Both sang but the blacks were the soloists. One or two recited (sang) solo, the narrative to or sans accompaniment. Often a sort of cat call or squeak or handclap or applause burst out from someone in the audience, a bit mal apropos according to our notions. The show consisted of a series of nine scenes of Japan: Fuji cherry blossoms, autumn colors, Kamogawa Rite, and the scenery was very deftly manipulated and very effective. The dances were mostly by the Geisha chorus, all dressed alike, sometimes little girls, and moving in perfect time. The dances were posture mostly, but the shifting scenery prevented monotony and I enjoyed it fully. There was one scene where two or three new dancers came out and got good applause. All were female, all powered white with the lower lip rouged and a pink over the region of both eyes. A brown streak below each side burn and [drawing "WW"] such a patch of brown on the back of the neck. When the tea was served only one bowl was brought by each person. On placing she gave a nice bow and shuffled off barefooted. We checked our shoes outside the theatre. After the theatre we walked toward home. The people had good fun watching us look into the shop windows. In one shop, we could have bought an ancient juniper for seven yen. They had dwarf rhododendrons, apples, and whatnot in blossom. None of the audience had the obvious perfume that our boys on the boat affected. They were all around a nice crowd. There were six or seven foreigners at the theatre too. Kyoto was for over 1000 years the capital, till 1868. Emperor is still crowned there. Population 765,000. Founded as capital in 794. As a town it is a jumble of small and large buildings. I never could guess where anything might be looked for and never knew the main street when I saw it. There was one line of cart tracks on which the street cars were decorated with paper cherry blossoms and lights, very gay.

April 8, 1934. We decided to get the 8:20 express to Tokyo, though it cost four yen each extra. It occurred to me to make inquiry for my cherry cane that I left on the train when I came from Nara, and sure enough there it was. They said they would send it to Tokyo but hustled it aboard the train instead, the dears. The train in which we sat had a lot of nice compartments. Our second class was a sleeper and we sat with our feet in the aisles. You had to keep them pulled in because of the considerable traffic down the aisles. The seats were full of mostly very nice looking people. Many took off their shoes. The road is through country almost uniformly scenic and much the same: cultivated flats with hills in the background near Kyoto (there were snow-covered peaks even. To the north of there are the Japanese Alps, with 10,000–12,000 foot peaks). Lake Biwa is visible shortly outside Kyoto and at two or three places the train comes to the seashore. At Mumazu we passed by Fuji (Saw Yama = mountain). There was a cloud bank over the peak as we approached, but it floated off to unveil the peak in its beauty. At the foot are the terraced fields of grain. In this region, much tea, nicely rounded bushes not over two foot high, plums trained like grapevines, peaches in exquisite pink and cherry trees in half bloom. The day started cloudy and dull but finished bright. There were bursts of sunshine during the day. Everyone in Japan uses an abacus. They even have pocket editions. On several of the large buildings in Kyoto and Kobe were large cloth fish floating in the breeze like flags. The farmers have haystacks much like the Serajis. We bought lunch from the station vendors—a box of warm rice and another of some dozen kinds of (cold) Japanese dishes, pickles, et al. for 30 yen. The Canadians bought some too but didn't eat it and found afterwards a box of ham sandwiches sprinkled with chopped cabbage to keep them fresh. Everything neatly and cleanly done up. In my box, a design cut out of bamboo sheath for ornament. Tea in very pretty brown glazed teapots and a small cup given for 20 sen. McCartney got out to buy ginger and got a box of Japanese mustardy salad that he couldn't eat and gave to me. I thought I was buying ginger ale and got a bottle of sake instead. Periodically, a porter swept up the aisles swishing past you all the rubbish of above. In Kyoto everyday there were swarms of pilgrims, mostly elderly women, with a flag bearer and badges traveling in schools and moving like a shoal of minnows. Rup Chand says everything looks artificial in Japan, the fields, the forests, even Fuji, no weeds nor birds nor anything wild or misplaced. Men and women are often seen in the fields, usually hoeing the wheat and other crops or digging out the trenches between the rows. Sometimes they are plowing with one ox or even a horse. There are no cattle, sheep or goats to be seen from the train. In Kyoto we saw an occasional bull, or even cow hitched to a cart. They say the cattle don't graze here but are fed in barns and most of them are in the north. We left our surplus luggage in the station Hotel and took a taxi over to the Nikko station. A train went at five and we got a round trip for three yen each. It soon got dark and we were in Nikko at nine. We learned from some passengers that the Konish Hotel was the best native inn and after lengthy bickering between my friends we took a taxi and drove over. One was for walking a bit to see if a taxi couldn't be saved but feared no taxi would be forthcoming in such a small town and off we went. There is a streetcar and a main street with several curio shops. The hotel proved to be nicely located and the staff had been informed by telephone we were coming. A most hearty welcome we got. The "boy" all smiles took off our shoes, gave us slippers and took us upstairs. The place throughout was beautifully matted over padding. All the doors and windows slide easily and noiselessly. A bouquet of cherry blossoms in our room. A pretty kakemono on the wall and everything quiet, clean, simple, and pretty. We sat each on three silk-covered cushions and they brought in two braziers filled with white ash on which a charcoal fire was built. Two "girls" came and prepared tea and chatted. All three so far could talk enough English. The girls were most interested in our matrimonial state and offered to remedy the lack of Mrs.; for 20 yen for the four. It was made out we were to pay five yen each for lodging and two meals and everyone was satisfied. They got a nice Japanese meal shortly, while we were entertained by one of the three, all so delightfully naïf that no offense could possibly be taken. After eating I wanted to go to a shop I had seen nearby and was accompanied by the rest all too scared to be left alone by the amorous ladies. They got away only after being felt over and praised as "nice" and "strong." The curio dealer "Hibira"[7] had a few interesting things on display but many were shut up in boxes in a house behind. These I got to see when I passed up the common display. There were some nice Buddhas but none better or older than the Kyoto one; some very nice porcelain and a piece of old cloisonné. I picked out a couple pieces of interesting things to be inspected in daylight and went back home. The bath was ready. A husky smiling boy took us all into the bath room, having left our clothes upstairs and put on a cotton kimono and padded silk one supplied by

---

7. Later written as Kibira.

the hotel. First we soaped ourselves and the boy threw buckets of hot water on us and scrubbed our backs with a brush and then we floated into a huge pool of hot water perfect in temperature. Such a delicious sensation I haven't had for years. Sitting on the bottom it came to our chins. The four together. We got to bed by midnight. Thick mats were spaced on the floor and abundant quilts were nicely enshrouded in white linen. The lower quilt had sleeves in it so perhaps the Japanese have a different way of sleeping.

April 9, 1934. The girls served a splendid breakfast of toast and eggs, the toast really tasty. They thought the matter over during the night and concluded we couldn't have understood them, so they tried again to arouse our masculinity. We went off on foot to see the sights. The lacquered bridge, closed to all but Mikado,[8] spans a little stream coming down from the picturesque hills. The country looks much like the low hills in Kulu, with abrupt forest-clad hills. There were still traces of snow banks and the plums were just beginning to open. The climate cooler than below. The landscape tremendously agreeable. There are lakes and waterfalls a couple miles away they said. The place was a seat of the shoguns and the temples were erected as shrines to their remains. They are splendid (not most beautiful) buildings we have seen. Heavily, over-burdeningly carved and gilded. The actual shrine of Iyemitsui[9] was very beautiful. Also that of the 30 shoguns, high up in the cryptomeria[10] forest. Two of the temples had beautifully carved wood pillars, but not much else of artistic merit. They advertise a sleeping cat and the three monkeys, but they are hardly interesting. Beside the monkey temple was a horse in a pen muzzled. You put a copper in a box and took out a dish of carrot slices for him. The temple grounds are beautifully wooded with Cryptomerias, chiefly but with other trees and shrubs. The location is a slope, by far the prettiest natural spot we have seen. Some of the trees were four feet in diameter. Birds were singing here, a wagtail mostly, though others were more or less common round about. Hordes of pilgrims in bands as in Kyoto. Said to be an average of 15,000 a day. When we got to the washroom this morning we found a good sort Englishman and his wife were there. They were enjoying the place as much as we. The man said he would stay at home if he had to have his Daily Mail and English companions. We had a most sympathetic talk and it turned out he was Wedgwood Benn.[11] There are nice broad stone block steps leading up to the high shrines. In all the temple places water is provided in fountains for the public. They drink from wood cheese box caps tied to a handle. I went back to finish the curio shops. None had anything good, thought two had some halfway nice things. One "Nakaya Fine Art Curios" I didn't see.

---

8. The emperor of Japan.
9. The Shogun Tokugawa Iemitsu (AD 1604–1651); the Tauyuin Mausoleum, apparently referred to here, holds the deified image of the ruler.
10. Japanese cedar (*Cryptomeria japonica*).
11. Wedgwood Benn, first Viscount Stansgate, was Secretary of State for India between 1929 and 1931.

I went finally to Kibira and picked out four pieces: a Chinese rice or tea bowl, a piece of ancient cloisonné, a plain bronze vase, and a Chinese porcelain-studded bronze incense burner. He was willing to let me take the things to Tokyo and send him money from there, but agreed to bring them down on the 11th. He continued showing things from his collection. He probably loans money on things because one beautiful Satsuma he said didn't belong to him but the owner would undoubtedly sell. He had an endless supply of boxes which he brought down and opened one by one. My bowl from Nara is Kio, first made 180 years ago. Satsuma 280 years ago. He had a piece that looked like cake—Asian ware 1000 years old. There were large white daisies with yellow eye and green leaves with the background of gold. The ware was bricklike underneath. Said to be very rare. Pretty but not beautiful. He had several nice Chinese bronzes. My companions spent their time in the curio shops and tremendously interesting they are. I bought two boxes that can't be opened except by pressing different parts ten times successively. Then there were trick boxes in which cigarettes disappeared. These interested the "girls" and the "boy" back at the hotel. There are paper prayers tied to trees here and at Nara. The scrub woman came into the toilet room while I was in the stool compartment, and seeing my slippers turned toward the insider reversed them to be ready for me. Boys coming home from school on the streetcar doffed their hats to one another on parting. We left at 4:30. The country for an hour continues hilly with terraced cultivation, much forest and picturesque landscape. Then one comes into the Tokyo plain. McCartney told me he had put Howard through college and gave him the round the world year. He gets more for his money than 99% of fathers do, who raise kids. A woman sitting with me got carsick. We went to the Station Hotel and got rooms at 5.50 yen, and 4.50 sans bath. After dinner, a good Table d'hote a 'Europ, we went walking in the Gunza bazaar, a place McCartney said sold singing frogs and singing insects. I don't know what is wanted of them but we didn't find any anyway. There was about everything else. One place had powdered snakes and snakes' eggs for sale (these are a standard commodity); one had puppies and kids, goldfish and mice abundant, and all sorts of nice novelties like mechanical toys. One large store had several windows of nice looking mannequins dressed in reproduction ancient attire, hunting and war. Ordinarily the mannequins are Caucasian faces. Money notes from one yen on (saw no yen notes). Silver 50 sen, nickel 10 and four sen with hole, copper 1 sen.

April 10, 1934. Saw more snakes (whole store full) pickled and live, many poisonous for sale. Also terns (?) eggs done up in nice boxes, strawberries in a layer, muskmelons, watermelons, pineapples, apples, oranges, and sanktras. Native oranges bought at Nikko very sour. This morning I went over to look at the imperial palace, near the hotel. There is a broad moat outside a wall of hewn stone blocks. Wild ducks in the water. Wall slows to the water [drawing]. A pheasant browsing inside the wall. Some peasants in the usual flocks came up to the moat edge in front of the palace, uncovered, and bowed. Everyone bows to you politely on any occasion, when you come, when you go, when you pay; nothing careless or inconsiderate surly in the island. Went then to National City Bank, New York for money. Paid S. 1–2+3/16 for 1 yen or 20 British pounds = 338.30. A nice young American at the bank who praised the Japanese for their innate breeding and consideration. Has been here four years. Dr. McCartney and R. Howard, 835 Sutherland Ave, North Vancouver, left for Kamakura at noon. Sorry. We went up to Ueno Park to see the Imperial Museum. There are a few pieces, not exceptional but good of jade, porcelain, bronze on the main floor and above drawings of a certain monastery's treasures and copies of frescoes and paintings owned by local nobles. I hope there may be some other sort of museums of old Japanese Art in the city. The park was full of people holidaying. The cherry trees were half out and the sun half shining and we spent a lovely two hours there, seeing the zoo. The animals are too crowded. There is a good assortment, even a penguin, bison, ostrich, cassowary, polar bear, elefant, hippo, seals, coon. For a copper you could buy some sliced beets for the wild boars and pour a couple minnows down a spout to the pelicans who sat with open mouths under the spout. There were some Japanese cranes that pleased me tremendously, beautiful big things with black on the throat and primaries. They ate minnows too, first catching them by the head and killing them. A seven story department store pleased Rup Chand very much. It was teeming with people. On the roof we bought some morning glory and gourd seeds [drawing] — the gourd that Japanese like for their saki carriers. There are a few birds and animals up on the roof too. A couple others in a tank got the most attention. We dumped in minnows but they caught most all them at once in their jaws and pressed their bodies against the glass to retain them. Any that escaped were easily caught. Phoned Marquis Kurada and invited to call at two tomorrow. There were a couple taxidermist shops near the park that had bird skins and mounted birds. Looked for *objets d'art* but found none. Prints can be had. In the shops some splendid imitations of old porcelain and bronze; the latter as expensive as the originals. I saw my first long tail Japanese fowls. They were parti-colored black and white, as large as a P. Rock, and had tails, a cluster of feathers 10 feet long. They had two P. Rocks in a pen and a donkey, two horses, buffaloes, bantams. Small trees (dwarfs) are used extensively as ornaments everywhere. Tokio is like the other Japanese cities we have seen, unimpressive, with nothing architecturally interesting in modern buildings. Shops have better wares, many restaurants serve only European food. Often one sees long fingernails on men. Women usually in native dress, men usually in European. I engaged a taxi once at Kyoto and waited a few minutes till my friends came. Then it turned out we didn't need a taxi, our Japanese silk merchant would walk us where we were going. I paid 50 sen on my bargain with the taxi driver and dismissed him. The merchant said payment wasn't necessary and the driver gave back the money and took it again only when I insisted. The bouquets the shops display are exquisite—a few branches, a piece of grass, and a flower or two, with a sharp color flavor. There is a German who has recently studied the arrangement and has written an English book: A Koehl.

April 11, 1934. Day cloudy. All weather in Japan so far has been half cloudy. Went in the morning and looked for old things. Hotel sent us to some big department store, nicer things than we had seen in the other, but they had nil. At one place which we stumbled on by accident we found a nice old Satsuma incense burner which I bought for 70 yen (price 85). They had a Japanese book with illustrations of Japanese and Chinese art, apparently someone's collection. I got the name and approximate cost. While there a mendicant with a bell, like our Tibetan lama beggars came to the door. Bought some dried persimmons done up in cellophane—not bad. A navel orange cost .50 yen. Went to American Embassy and got consular invoice for my Japanese purchases. Cost $2.50 and had to write cost, dealer's name and address, when made, and where and when dealer got them. Said the Japanese recently seized a package of books from an American on an American ship. Books in hold and passenger not even getting off in Japan. Put an Englishman in jail three days ago and didn't notify relatives. National City Bank took photos in Kobe for exhibition in USA and arrested photographer. Picketed bank and tried to force employees to quit. Spy crazes engineered by militarists to get appropriations. Went to see Dr. Kunda. Found man my age, gentle and sympathetic. Showed me his collection of live ducks and geese and other birds. Has Ross goose,[12] monal. skins, small series of Japanese birds and representatives of most US birds. Found he had given me two of his six Scolopax mira.[13] Considers it a species. On the same island is a jay that lays almost white eggs in holes in trees. Bird destructive to potatoes. Stayed an hour and didn't have time for tea with him. My Nikko curio (Hibira) came down at five with my four things at 150 yen. Said a friend might have kakemono so we drove over beside the Imperial Hotel. M. Nakasaw, 14 It. Chome Yurakucho. Man had a very nice old Satsuma jar [drawing] about 14" high, nicely decorated with spray of blue shrubbery for 120 yen. Also a nice Satsuma rice bowl, 22 yen. Said blue jar was 200 years and bowl 80 years. There were imitations of the ware such as my teapot but the decoration is poor in finish. Couldn't spend more money. Wanted dose castor oil. Drugstore in big department store building never heard of mixing it with lemon soda, but did it myself. Went over to the Ginza Street for supper. Accidentally got into geisha house or whorehouse. In the ice cream stalls sat young Japanese petting the girls and drinking. Four came and sat down beside us. Nice mannered but not beauties and no English. We left after 10 minutes and gave them our small change, much to their amusement. Went to a chop suey house. Waiters totally different from any seen. Found they were from Kobe. When proprietor heard we talked English came. Spent 20 years

---

12. *Chen rossii*, Ross's goose, a North American species.
13. *Scolopax mira*, Ryuku woodcock.

in California. Gave us two towels for souvenirs and two boxes matches, and I bought only ¥1.20 worth of food and had to tell them 30 sen short on bill. Man said he has to have nice looking waiters or people don't want to eat. Soil apparently deep around the area visited. Cuts in banks are always earthen. The Japs have learned to put up statues. One sees bronze men standing here and there, if not exciting, at least not as bad as the English statues in the east.

April 12, 1934. We left Tokyo at 8:15 for Yokohama. At the hotel before departure we met our German friend (Baron Von Esebeck) who wanted us to ride around Tokyo, which we declined. He can take extra passengers free. He liked the country and wanted to stay longer on account of cheapness but the next boat is completely full. Dislikes the Japs. Finds the ceremonies insincere. Arrived at Yokahama, we had so much baggage, small parcels of curios, that the passengers had to help us put them out of the window. I nearly forgot two pieces which they hastily stuffed after and bowed to me till the train was away. We took a car to Kamakura and saw the bronze Buddha, the 2nd largest in Japan. It is a beautiful figure seated in the mists of a park of cherries now in bloom. Ht. 12.89 meters, weight 100,000 kg, 681 years old. Gold eyes. The largest Buddha is at Nara, the third at Kyoto of wood, can go inside the image. The figure is constructed of slabs of metal about a yard square. There is a beautiful figure of the goddess of mercy in another temple. The God of War has a new temple. The old ones burned down. The park is a regular plantation of trees like our campus, not at all interesting. View always is foiled by dinky tin-roofed houses in and near all large towns. The taxi driver (R. Sugitani, no. 68 Kusumashi, Kanagawa Ku, Yokohama) would take me from Kobe to Nara on a 11 day tour for 299 yen and a yen a day for himself and driver for expenses. Buick car. Speaks English and is going to night school to get ready to pass exam for licensed guide. Arrived at Yokahama; inspected four curio shops. One had a beautiful old wood figure, three feet, said to be 800 years old price 320 yen; and a nice pair of srungma[14] figures in wood for 250 yen. Said to be 400 years old. The first is very fine and should be bought. Dealer K. Takahashi. The largest dealer is George Tsuboi. He had a nice Buddha that he wanted 15,000 yen for. The man who brought me there was a bit taken aback and ashamed. This is the first city where prices were "for Americans only." I should like to have the wood Buddha and the two pieces of crockery in Tokyo. The dealer said Japanese prefer wood figures to metal, softer outline. Porcelain is very scarce. In all the shops I have either bought all the good pieces or noted them; this excepting always pieces that were of good make but queer shape. These were also few. Wood Buddhas of the age of mine are rather [Expenses Japan for 2: 125 yen and 30 yen for postcards, cloths, theatre, souvenirs, Art 738 yen, 100 yen self] common, but such attractive ones are rare. Tankas of any sort are rare. First class ones only the one bought at Kioto.

---

14. Buddhist protector deities.

There were piles of wood to be made into charcoal on the way to Kioto. Met two autos of geisha girls, were out on a holiday to see the cherry trees. They had their musical instruments and were singing. In one field were two rows of rushes, sized for making mats in the houses. Farmers' houses used to store grain, cattle, and animals as at Kulu. Saw 20–30 dairy cows and oxen drawing carts on the roads. Cows and horse here usually small. Say sheep scarce. Japs don't eat mutton. My steward on Ginyo Mam (Aoki) is aboard here, also Yosokum M., 2nd class purser. My 2nd officer Hokozaki Maru had been here and brought me an image of Dáruma carved in camphor wood. I went over to Kakozaki to see him and they gave him leave till the Chichibu sailed. My bath steward on the HM brought me a potted tulip. A nice dwarf apple in the window was 4 yen at Tokyo. *Art of Japanese Flower Arrangement* by Alfred Koehn, 8 yen. 20 pp. Nara Hotel, Nara Japan.

April 13, 1934. The fish on the buildings are for the Boys' Festival. Purser says they must make out the long manifest for each passenger for US Immigration such as Rup Chand signed answering whether you are polygamist, anarchist, deportee, going to America to overthrow government, etc., and if there are errors, $2 for each misspelled word. Purser came to cabin to see my Japanese things. Thought the Buddha lovely. Said valued by Japanese for expression of features; all sculptors aspire to make a beautiful Buddha. The shawls uninteresting, but the Indian pictures attractive. Liked the old Rama the best. Said the green potter in the Han bronze very nice. Said the [indecipherable] to have an old painting mounted on brocade. All day a flock of ca. 30 white-faced, white-tailed with black tern band, rest of color brown-grey, albatrosses followed the boat. They made the journey with the greatest ease. They can soar long in all directions as skillful as vultures. Cloudy all day with a little drizzle. Temperature: air and sea 62°. Officer said received 1000 telegrams for passengers yesterday and were too busy to put up a radio bulletin today. Stewards have to pay $60 every trip for things passengers steal. Read some English short stories, old and new in a volume: *Great English Short Stories* by Dewis Melville and Reginald Hargraves, NYC, 1930. Mostly feeble.

April 14, 1934. Julius Caesar: "But men may construct things after their fashion, clean from the purpose of the things themselves." Brutus: "There is a tide in the affairs of men which taken at the flood lead on to fortune. Omitted, all the voyage of their life is bound in shallows and in miseries." This play pleased me more than any of S's others so far read. The elegance of the language is overwhelming. The Japanese purser took the India things to show the ship's officers. They liked the Rama shepherd at the green wedding procession best. He brought also a crude Nepal modern image of some demon and his consort. He was a bit shocked at it. He had seen Pompeii and his brother had written he had seen such an image in a monastery in Manchuria. Mik mema, lingtok yang met (no eye, no eye [indecipherable] either).

April 15, 1934. This morning the albatrosses had dwindled to seven and by evening there were none. Two or three petrel or shearwaters of two species were visible off the course. Distance from Yokohama to Honolulu 3394 miles. Ran 450 miles till noon from noon. Water 61°, air 65° today noon. Cloudy as usual but fits of sun. Sea calm. Read *Timon of Athens*, went to Japanese movie, probably old. Exceedingly slow, long scenes showing eating, drinking, babies close up, views and no plot.

April 16, 1934. Fast wind recorded 231 mi ph. Albatrosses of yesterday still visible this morning. My little San Francisco-born Japanese friend of Neno Park in third class. Has been two years in Japan, almost forgot English. Read *Romeo and Juliet*. "The grey-eyed morn smiles on the frowning night; chequering the eastern clouds with steaks of light; and flecked darkness like a drunkard reels. From forth day's path and Titar's fiery wheels." . . . "For nought so vile that on the earth doth live. But to the earth some special good doth give." Pursers says Japs trying to supply wood and cotton demand of empire from Manchuria. Says Japanese males no use as house servants. Consider cooking and housework as women's work. Large scale cooks, however, are men. Japanese movies as rule slow. Rup Chand's 32nd birthday today. Sunny half the day. Ship's routine as on *Hakozaki Maru*: breakfast 8, lunch 12, dinner 6:30. Here a waking gong at 7:30, not on *Hakozaki Maru*. Bill of fare a bit more elaborate here. Cabin stewards wait table. Sleep here like a stone. Probably due to the gentle vibration and light oscillation. Trap shooting at four pm; swimming pool 2–3. No orchestra.

April 17, 1934. The albatrosses (seven this morning) still coming. Cross the 180 parallel today and repeat the day. Read *Henry IIII*. Evening fireworks; skyrockets that burst with sparks. Pretty. Then an ancient Clara Bow movie, worse even than the Japanese one of the other night. Incredible that folks like such rubbish, but then I never could stand Clara Bow at any time. Says our cargo to San Francisco is cement and silk.

April 17, 1934.[15] 22 albatrosses this morning. Read *King Richard II*. Bolingbrooke: "O no! The apprehension of the good gives but the greater feeling to the worse." Cf. Dante. F. St. Mars: "Watchers of the Mud," from "*Off the Beaten Track*" 1917. Very clever. Find something else of his. A delegation of Japanese textile factory managers and owners came to see the shawls. Most excited. Admired the Nurbarabbad and the long Lucknow one with the creeping vine. Interested in pattern and color. The purser told me about a Japanese fly trap that works on the principle of two revolving cylinders and a screenbox. The bait is put on the cylinder and the flies get brushed off into the box. Japanese grow many strawberries. Bought a money order for 8.20 yen and sending tomorrow for Koehn's book on flower arrangement. Japanese immigrants come from

---

15. Koelz began two separate entries with this date.

Siberia, Mongolia, China and the East Indies. Stone age implements in Japan. Social organization: Imperial Princes, nobles: prince, marquis, count, viscount, baron. Samurai with title *shi* (feudal retainers—warrior caste), commons, butchers and chumar. The last two now merged with commoners. Marriages arranged by families. Nobles require imperial sanction. Near Nara is the Shosun with store of old things brought by the old rulers from the South. Said to be things of Persian design among them. A theater evening. Three Japanese played five mouth organ selections including a Brahms Hungarian Dance, a sword dance which pleased the Japanese audience very much. An orchestra (plays usually in Class I) played rather poorly. Lahili Lohar says "your belly may burst but don't fail to eat your share." The French Jews aboard attacked the extra sandwiches in the bar as though they hadn't eaten for days. The Frenchman into whose cabin I stumbled the other day came today in ours. Rup Chand was lying on the davenport and he stared at Rup Chand in amazement till I laughed from the other corner of the room. Sumitomo family has a large collection of old Kichizaiamong bronzes. No room 13 on boat. Japanese don't like the number four (also means "death" – "I-dai" = ouch! Air 66°.

April 18, 1934. Air 68°. Water 66°. Day bright with drifting cloud [indecipherable]. First fair day since Shanghai. Captain Arakida sent for me to show the Indian pictures. The chief engineer was there too. Captain very intelligent and fond of flowers and pictures. Both completely agreeable. Eight albatrosses this morning. Wrote Takahashi of Yokohama offering 200 yen for the old wood Buddha and to Makasama offering 120 yen for the two pieces of Satsuma. Read first part of *King Henry IV*. Prince Henry: "if all the year were playing holidays, to sport would be as tedious as to work; but when they seldom come, they wish'd for come, and nothing pleaseth but rare accidents." Tibetans never give an empty dish as a present, always a grain or dab of flour on it: The Japs managers asked to come again to see the Indian pictures. This time they brought me as a present a Japanese silk painting with a little conceit of paper (no shi). They read the inscription on the match bowl as "Kutam yaki sen." Kutam is in province of Johi Kawa. Said to be famous pottery works there. Old work rare. The little San Francisco boy came as interpreter. Flocks of very capable flying fish this forenoon. Water becoming blue now. Until now not pretty. Kurimoto says whales migrate from the coast of Borneo through China Sea to the north and outside Hondo. Small fish travel with them and the Japanese fishermen follow too. Movie at night. The most maudlin of all She by [indecipherable]. Pasteboard rocks and besmirched whites for Africans. Plot the sheerest blather and the actors as washed out a lot as you could think of.

April 19, 1934. Tulips faded today. Mr. Yamuri came and told me had a collection of old Japanese costumes at home and when I come again to Japan to bring my India paintings. Boat docked[16] at 12:30 and I went ashore. Got in car with three Japanese and drove around the city four hours for $12. Day clear and temperature delightful; cool even on the hilltops. Everything neat about the city. Homes usually small or unpretentious but well kept and with a lawn of flowers. Many kinds of shrubs and trees in bloom, exquisite in color. Water reserve of untouched vegetation and very attractive. Drove up to a hill overlooking coastal plain on the other side of the island where the wind was a good gale. View of paradise. The mountains rise perhaps 2000 feet, strongly eroded and covered with grass and in places trees. The island is visually lovely, the nicest place I have seen. The aquarium has a good assortment of tropical fish, of the gayest colors and patterns conceivable. Just now as I was reminding Rup Chand about the seahorses he had finished writing the word "home." Our companions hastened through and we didn't have time to look at the fish. The German Baron thought it uninteresting. Went to a pineapple farm and learned that pineapple slips are planted through strips of paper and bear in 18 mo. One small plat with plants two feet high were in blossom. Canned pineapple from Hawaii ca. 300 million tons a year. Not all soil suitable, in fact very little. Spray leaves with iron solution. Saw surf riders. Two Hawaiian youths dived for coins around the boat, as graceful as seals. Beautiful physiques. See many Hawaiians, Chinese, Japanese, and mixtures. Any of them may be auto drivers and all are well dressed. Garlands of fragrant flowers sold to travelers. Curio shops all rubbish: coconut men, hula shirts, etc. most of passengers went to hula dance in evening and said it was nice. In some barber shops Chinese girls do the hair cutting.

April 20, 1934. Went after money £ = $5.11. Then went out to the Bishop Museum. A large collection of ethnological material on exhibit from the Hawaiian islands and from the other Pacific islands. Most beautiful were the feather exhibits, especially the feather robes of red and yellow feathers. Design of large red triangle on a yellow background. No wood carvings of beauty seen. Went back of the scenes and saw the skin collection. They have a really nice series of skins of Hawaiian birds. The curator of collections Mr. Bryan is a delightful person. Said couldn't exchange unless got some Pacific Island things but might make a loan. Nothing can be killed on the island, even by the museum. Many species increasing, including stilt, Hawaiian goose, Laysan teal, thou this is extent in Laysans! Bryan went with the Sanford Whitney Expedition[17] and knew Beck. Said Beck was a wonder at his work. Another man, Mr. Burrows, was once in Journalism Department at AA. Nice lot and I should have stayed all day but the boat sailed at 12. Bought some mangos and fresh figs,

---

16. Honolulu, Hawai'i.
17. Whitney South Seas Expedition, from 1921 to 1935, collected bird specimens from Polynesia for the American Museum of Natural History.

papaya. Figs as good as they ever get! Mangos fair. Rock is here on way back to Tibet. Chrysanthemums, asters, roses in bloom. Took picture of Rup Chand and Hawaiian newsboy Jocob Ki Ku. Stewards on NYK get 38 yen a month and furnish their own clothes and laundry. Milton: "Truth is rarely born but, like a bastard to the shame of him who begets it." Douglas Jerrold had a story in NY Times: "The Preacher Parrot," very clever. Read some more. Story of a Feather, Chronicles of Clovernook, A Man Made of Money. Wind, sea water rose a bit in late afternoon but not uncomfortable.

April 21, 1934. Day partly clear. Sea agitated but not bad. Stewards says French, Scotch, and Jews give little tips. One said "I don't like French. I got two." Poor boy, they are French Jews to boot. Japanese give good tips and so do Americans. I told my steward I was French. "No you're not. I looked at your sheet when you got aboard." Read *Henry IV, Part 2*. K. Henry: "thy wish was father, Harry, to that thought" and uneasy lies the head —. Americans make a fuss when they get on and thereafter are easy to look after. Movie about Japan, Japan's production of silk 1931 was 76% of world's total, 77% exported. 95% to US; 75,000 bales produced. Gold fish may bring $100 each. Silk worm eggs are inspected by government before sold to farmers. Silk given elaborate government tests before allowed to go abroad.

April 22, 1934. Hawaii Island, 6,400 sq mi, chain 400 mi. Honolulu 120,000. Exceptional summer 85°, min 60°. No fans. Northeast winds. Much area 2000 feet. Two peaks Mauna Lea Kilauea 14,000 feet, with snow fall. Rain erratic, Maru on site 300", 15 mi away 25", 1930 census, 368,000, 22,000 Hawaiians, 30,000 Hawaiian. Japs, 140,000. Principal industry sugar (million tons yearly), and pineapples. Ranch of 32000 Hereford cattle. Du Puy: *Hawaii and its Race Profile*, 1932. Dead swell today. Partly fine. No albatrosses since Honolulu but occasional petrels. Kurimoto says Japanese lost faith in justice in American courts with the Tortescue trial. The courts were just. The governor pardoned. Read *Henry V* and first part of *Henry VI*. Sugar cane originated in India. Pineapple America.

April 23, 1934. A couple of albatrosses for a while today. Day overcast. Sea calm; 64° air, 62° sea. Read *Henry 6th Part II* and *III*: Suffolk: "small things make base men proud." Japanese rice is always sticky, completely abominable to the Indian who likes his dry. Saw *Esquire*, a new magazine for men. American 50 cents, 1934. Not bad. The fancy dress ball interested Rup Chand much and he stayed a good hr, watching the dance, not in admiration however.

April 24, 1934. Okura private art museum in Tokyo has 19,000 pieces according to New York Bulletin. Kurimoto came to see the Tibet paintings (Butsuga = tanka in Japanese). Baron Kuki Tokyo is said to have a beautiful collection of porcelain. Marquis Kosukawa general art collection. Mr. Hara of Yokahama has picture collection. (Nihonga = secular pictures). Read *Richard III*. Recalled the play seen with Walter in Detroit with Robert Mansfield. Rotten performance. Started *Titus Adronicus* which is the last of Shakespeare's plays to be read. NYK farewell dinner was a very nice affair. Gave each passenger a pretty little vase in a wood box, a log of the journey in which it said the sea was "rather rough" for several days, probably to make passengers feel more heroic. Our two tablemasters thought the captain should have come down to eat with us or have invited us to go to first class for tea. Why, I don't know. Kurimoto came in evening and looked at Japanese porcelain again. Read inscriptions and brought books. Kurimoto gave information about Jap porcelain and art students.

Porcelain kinds
　　　　　　　　　　Shimadzu (Lord)
Satsuma　　　　　　(Marquis Shumazu) at Kagoohima
Kutari/Kutani　　　　(Marquis Maeda) Kanazawa
Banko　　　　　　　(Coutn Kuki) on Ise Bar
Prof. Ino Dan, Tokyo Imp. University, Arts (40)
Prof. Hamada, Kioto Imp University, Korean Art and Archaeology (60)
Foreign Grad, Prof Soetsu Yanangi, Tokyo Imperial University, Oriental Arts (pottery), 40
James Scherer, *Romance of Japan*
EW Clement: *Short History of Japan*

April 25, 1934. Off the coast and flocks of California gulls. Day beautiful. At quarantine some pressmen came aboard and photographed Rup Chand and me. Took leave of my Japanese aboard with regret. Hirozawa is going to send me the Daibutsu at Kamakura. Honda wished us good health. Gave each 15 dollars. Customs looked at my Japanese antiques and consular invoice with much misgiving. Said 80% of such antiques declared were fakes, and cast out all but the Kakemono and the Han bronze. I objected and they called the expert, a very "fine" looking person. He said all were genuine, and very good at that. My Persian shawls and paintings were put on the declaration and value reduced to $80.00 the lot and passes sans consul invoice or other formality. The immigration gave Rup Chand permission for a year and said to write him if longer stay was needed. All government officials tremendously polite and yet strict. Tried to get Mike but no such luck and went to look for my berth and ticket. The Great Northern on whom the ticket order was drawn don't want to give up the ticket to the Santa Fe. No trace of a berth being paid for. That finished, took a ferry across to Berkeley and found Mike. Looks grand. Never so happy for months. The same dear. Has a very pleasant apartment. Still didn't come back after Christmas and is on way to Florida.

April 26, 1934. Rup Chand flabbergasted at magnificence of the United States. San Francisco is really a fine thing to see. Nice buildings, clean streets, well planned, on series of small hills (terraced), with suburbs; Oakland, Berkeley etc. fringing the beautiful harbor. Flowers brilliant and abundant. At night the dancing lights all around the bay, like fireflies. No fog. Clear pleasant day. Say unusual that fog doesn't descend at night. University splendid campus and magnificent new buildings. Students nice looking but Mike still yearns for Michigan. Climate here must be lovely. Days sunny and pleasant and nights cool. No [indecipherable] in evidence. Vegetation varied, shrubs and trees, rolling hills and small mts and the lovely bay. No smoke. Golden Gate a narrows opening to the ocean that will be spanned by a bridge. Working a bridge across the bay toward Richmond (?) and are tunneling through an island to save a small distance of circuiting. Amuses Rup Chand much that time is so precious. Went to San Francisco after lunch and to Chinatown or at least a street full of Oriental wares. Many curio shops but none had anything good. One Japanese had some new tankas (Tibetan and Bhutanese) said to be bought in China, $65 each. Nothing else even slightly interesting. Said to give name and he would let me know if he got anything good. While I was writing it on his card, the nice French couple from Empress of Britain whom we met in Nikko at the sword dance had accosted Rup Chand outside and asked if he hadn't been at Nikko. Delighted to see them. They are going via Panama to New York. Saw a Japanese of my Honolulu party, a young Englishman and ma coming on ferry to Berkeley and another couple with us on C.M. World small! A Hindu had a common phulkari for $85. A Japanese had the rubbish turquoise imitation bracelets, worth 8 annas in India which he bought wholesale for $1.75. Showed me the list with pictures! A flower shop was filled with rose lavender rhododendron trees in tubs. Flower shops here like dreamland. Fruit shops likewise well stocked and all cheap. Went to a Russian tea room and had a seven course dinner for $1.00. A woman at piano, three men with guitars, and another woman sang and played Russian music by requests if there were any. Every one in costume and all Russians except a French American maître d'hotel. Music first class. People were nice looking and food beyond criticism. Seven relishes all with a subtle flavoring. Russian cabbage soup with a pleasant suggestion of cabbage flavor, Russian meats, deliciously flavored, and beautifully cooked. Tremendously pleased, except decorations were predominately of lacquer red a la [indecipherable]. Rup Chand said he couldn't eat. Red is considered bad for appetite by Tibetans. An invalid is never shown raw meat even lest the color turn his stomach. Certainly, I couldn't eat in spite of the savor and I didn't know why till I had finished. Arthur U. Pope in *An introduction to Persian Art* says under "Textiles" that these are hunting "tapestries" of 16th century. In 18th century Kerman saw the development of an offshoot of the Kashmiri shawl industry. Nothing else.

April 27, 1934. Rup Chand spent morning over Mike's Rubaiyat, copying Persian verse from the illustrations. Went to meet Mike at noon and met Prof. Vaughan who is specialist in Italian dialects which are countless. Then went to Mike's advanced class at 1 and met a Mrs. Meyer, a nice little woman of culture. Saw one attractive student, Bava. At San Francisco got my tickets finally, but no word of berth from MC at Detroit, so I have to buy a berth anew. Just as well or couldn't stop off and besides berths cheaper now. Went to a Japanese shop and saw beautiful corals. Says Japanese sell raw coral from Japan and Formosa to Italy. In California Gem Company saw Nevada opals, blackish with fierce flashes of fire. Wonderful but none but large ones. A man there had a collection of opals. Went then to eat at Bernstein's Fish Place. All the waiters were middle-aged Jews and the clientele all Jews. Came two or three times to ask if the fish was all right. Anyway Rup Chand saw a collection of Hebrews. Bought the very minimum of food. Left Oakland at 11:30 AM. Mike stayed to the bitter end and thrust a loan of $15 on me, saying one can't have too much money; one never knows what will happen. Probably true. Common Lahuli Buddha $50. Rup Chand liked the advertising mannekins in the windows that wink and gesticulate. He says he can't see why I want to go away from the US. A boy in San Francisco came rushing up and said your picture is in the paper and sure enough a nice 10 inch photo with "nono says yes."

April 28, 1934. In the morning we were in some well-cultivated area which soon gave way to patches of cultivation. I never realized before how very comfortable our trains are. At Bakersfield we stopped an hour for lunch. I got a peck of huge oranges for 35 cents from a farmer who had a truckful. He had "seedless" grapefruit of excellent flavor. One had only two seeds. All the California specimens had very few seeds. The conductor E.P. Underwood of Bakersfield took a fancy to us strangers and took care of us till Bantow. He brought literature on bimetallism, showed us the new deposit of Kemsite (99% pure borax) and explained the working of the automatic brakes. The Kemsite lies in a deposit, nine miles by two and a half miles by 70–150 ft. Controlled by British. The engine pumps air into a tank. The release of this air pressure drops the brakes. The Santa Fe obeys the federal laws rigidly. The other day an express car came into the station with a safety nail damaged. It had to be unloaded on the spot. A.E. Dobbs of San Bernandino, Supervisor of Air Brake, was also very friendly. He showed us how the air cooled the dining car and dining rooms—a grand thing; how they typewrote instead of telegraphed and a typewriter at the other end writes out the message. They are working on the use of cosmic energy for train propulsion and the energy transmission by radio. The California Limited picked us up at Barstow about 4. At Needles we saw some very fat young Mojave squaws selling bead strings. Spoke excellent English. Dinner on train in cooled car very pleasant. Outside a bit stuffy. The desert very beautiful with the blue ranges on the flanks.

April 29, 1934. [indecipherable] on our train are four passengers from Chichibu! A French couple that came in second class but with whom I spoke nary a word turn out to be very nice: Marius Bovel, 10 Rue de Lille, Haiphong, Tonkin. Rup Chand got off at Isleta and saw the pueblo, which was just like British Indian houses or Spiti. I found Mr. McDevitt in the Harvey House and saw some of his things. A nice Hopi blanket, one or two pots, and an old bracelet. At Gallup in the platform was Tom Ramsay who is living there and attempting to run a flower shop. Went to see the Mathews. They have moved down on West Central near Old Town and have a small shop where they have a few pieces of junk for sale and keep going with refinishing and upholstering. Mrs. Mathews got a judgment of $14,000 against Mrs. Balduni, but lost the money somehow. She looks very well. The very plain daughter has a husband who runs a beauty parlor. A Mrs. Johnson, widow, was with them. Had supper with Mr. McDevit and showed him the Japanese things. There is an Apache grain basket, $85.00, a Zuni bracelet $85.00 that are very interesting in the Harvey House and some old European jewelry. The latter is most attractive and not expensive.

April 30, 1934. Went off to see Judge Botts first thing. Is in partnership with his son Bob just out of Harvard. Bob is a very nice sort. Miss Clark is still in the office. Had a long visit with the judge. Mr. Leverett came and said he had several people looking at the house, but event at $4000 couldn't get a bite (Frenchman recommended A Genais: *Aesculope en Chine* a physician's experiences in Cochin China). Situation more hopeful now though. The renter moved out the other day. Saw Dr. Peters and Mrs. Buck. Both look well but apparently few patients. Found Mrs. Ramsay. Her husband John Ahlgrim is a good sort and likes Polly. Bud and Junior have grown into nice looking youths. The two small children Judy and the boy are also nice. Polly is the same dear as always and is a great favorite. She stays outside summer and winter, night and day, but never leaves her own yard. She was under the house when we came. The whole house is chewed to pieces. When she wants to get into the house she chews a hole into the screen door. They put on a glass door but she ate off the frame and the glass fell out. If she gets hungry she helps herself to what she can find. A loaf of bread will have tunnel through it. We drove up and saw Mrs. Becker. I had seen Ralph in the Bank where I found $124.00 on credit in a savings account. The Coopers have the Acme Beauty shop in the Rosenwald Building and seem very prosperous. Even the Mexican and Indian boys have permanent waves. Saw the house. The yard looks nice and the house is as nice as I left it. Floors, woodwork, and walls as good as new with the gas range and drapes still there. Someone took the lightbulbs. Judge Botts said an Indian friend was complaining of the prosperity of the Indian youths under the federal relief works. They earn much and come to the cities to spend it. "Pretty soon we'll be just as bad as the white folks." Found Mr. Schweizer in his office. He showed Rup Chand some of his choicest things: the ancient string of turquoise of beautiful color blue wampum, a padlock that has two keys and a trick for exposing the keyhole, a Mexican serape with silver threads and exquisite color and design, a ship on the ocean beautifully done in beads, a carding board with teasel pods for teeth, an amulet made of clam shell (horned oyster with the horns rubbed off) inset with turquoise, some (3) incredibly fine small baskets of beautiful design. He sold many thousands of dollars of silver, pottery, blankets, baskets to the new museum in Kansas City. He still has one Santa Ana jar I want and a number of buffalo hide shields. The old European jewelry is about off the market, he says. Went up to Judge Bott's house and saw the family and the beagle puppies that almost chewed me up. Mrs. Botts likes iris but has only two kinds. Promised to send her some. The Ramsays met me at the station and saw me off. The Mathews also came; Jo the oldest son also came. I hadn't seen him. Jim is the clever furniture repairer. The *Albuquerque Journal* called and asked a few questions. Got aboard the Chief and left at 10:05 AM. Not a sad recollection of the town and a most pleasant renewal of acquaintance. The people don't look a bit poor and there are many new houses. The Veteran's Hospital and Indian Hospital and the new St. Joseph's two million dollars are all new to me. They have made a long pond down by the zoo where people can swim and hold boat races even. The zoo is growing. They raised five African lion cubs that were born there and they are splendid examples. The mountain lions as Pete Schulz fit with are dead. The bison herd is growing and a baby dromedary has just arrived.

May 1934

May 1, 1934. Arrived in Trinidad about 7 AM. Walked out to St. Raphael's Sanitarium and told the nurse we came to see Sister Rose Alexius. Sister came down expecting to find some relief workmen. We promptly went on a picnic with the senior nurses and Sister Martin, whom I knew in Albuquerque (Sr. Domenica's sister). We drove up to a place beside the stone wall about 40 miles up the creek from Trinidad and ate a splendid lunch beside a little stream. There were Kingfishers and dippers on it and the poplars were just leafing out. The valley is open and there are broad fields of pasture and cropland, flanked by low hills covered with pine and brush. The willows on the stream backs were leafing and the apples and wild flowers were in full bloom. Everywhere the sap had filled the leafless branches of the other trees and pastel shades of grey white (cotton woods), purple (mountain mahogany, oak) yellow (willow) and blue splashed the landscape. The day was typical of April, smiles and tears, but no sprinkle that disturbed our outing. From the Chamber of Commerce Park we drove on up to the Monument Lake Park (8500 feet). There is a small artificial lake with a stack about 25 feet high rising out of the water. There are trout (rainbow) in it. The city has some land there in the National Forest and raise buffalo and elk. A youth De Smitt seems very clever with rearing. Feeds on Eagle Brand [condensed milk] diluted two-thirds. One bison walks around outside. There are nine in the pen. Elk, wild turkey, mountain sheep, deer occur nearby. He has a coyote that is the most affectionate and playful thing one could imagine. He has a grey fox pregnant with a male coyote. He gave me two wild turkey eggs. Sister Rose a half dozen trout. Sister Rose thinks trout are the greatest delicacy one can have, like mother's "fruit salad." Sister Evanista went as Sister Rose's companion. There were some large anemones of lovely porcelain purple just coming out in the woods. There are a large number of CCC[1] "bums" at work. The Government has established housing centers where these vagrants are assembled.

---

1. Civilian Conservation Corps.

They are given medical attention and if fit sent to public works. At the park they are certainly well fed, roast beef, corn, tomatoes, custard, for supper when we were there. Their supervisor is Mr. Adams. His wife gave us a bouquet of the anemones. The drive back was finished at 4:30. I went to see the "museum." In one room in an old house they have a loan assemble of all manner of things: a paisley shawl, a chair made of steer horns, old china, old guns, Kit Carson's old leather Indian dress (Kit Carson lived nearby), old newspapers, mountain sheep horns, etc. A girl of 16 showed us through and said she had a brother in Guam and in Shanghai, that there was no charge but was glad for a quarter for ice cream. Sister Rose put us up in one of her three pretty guest rooms in which she allows no sick lest the rooms smell of medicaments and served a nice supper. We reminisced till 10:30. The poor dear still longs for her old home in Colorado Springs, Glockner San where she lived 19 years and which she built. She has never left a hospital in debt. At Albuquerque she fired Peter and Jan Atta for inefficiency and dishonesty, refused contributions to the advertising fund and made herself general "obnoxious" to the wayward. She began there a general renovation, painting, repairing. Sister Dominica said "Sister Rose can't abide anything tacky." Her business ability is phenomenal. A lone woman without an advising board not only runs a huge hospital but plans and builds a couple at costs of millions. No one is ever turned away. A negress came today. The doctor will do the operation she needs sans fee and Sister Rose will give free hospital. Recalled visit with Mother. Mother got her a lunch and gave her wine! and called her Rosie, and Sister Rose liked it hugely. Left the sweet old dear with tears and got on the train at 11:40 PM. She paid the taxi. Hired Stell.

May 2, 1934. Berth in the observation car, first time in my life. Train has 14 cars and three express vans. A number of Japanese from the NYK *Chichubu* are aboard here too! The lilacs and Judas tree are in bloom in Kansas and the fields are green with spring. ~~Lilac and Judas tree in bloom.~~ April sprinkles. Day overcast. Small Cuban pineapples in the shops. Arrived 4:45 PM in Kansas City and went to Hotel President. Looked at Library, which seems fair and then went home. The town is dirty enough, has a modern downtown skyline and the people look prosperous enough. Not many darkies; even porters in the station white.

May 3, 1934. Went to William Rockhill Nelson Gallery of Art, opened December 1933. Building of marble cost $2,750,000. Spent four million on objects. Rooms are built to appropriately house collection and exhibits aren't jumbled. The Harvey Indian exhibit had a considerable amount of ethnographic material, but a few pieces of beauty: a gato, belt buckles, Cochiti and Santa Ana jar, Kewa, and Shoshone Paiute basket, Navajo and Zuni squaw dress, a Germantown and Bayetta blanket. In the Persian room was a lovely Geordez rug, perfect preservation with blue field and red dominant border design. The nicest I have ever seen. There was a fine loan collection of 13th century Persian pottery by H. Kevorkian and a few velvets. The Japanese line not much. Chinese had some fine specimens of old Chinese pottery: Sung, Tang, Han, a Chinese porcelain figure of priest, a large Chinese wood image 1300, Korean and Chinese celadons. The five colonial rooms are beautiful. On the wall is a picture of flowers painted on velvet, American. Very effective. An exhibit of modern work most pleasing. Selected good medium class modern artists. Ernest Fiene is striking in his expression of the beautiful in the common place of American landscape. A signboard, dreary country house, country church, becomes interesting under his hand. A Walter Koeniger: "Old Mill" was nice: "Winter Landscape with Melting Snow." Saw Director Paul Gardner. Nice looking man with spending taste and sense. Says nothing in Japan; doesn't like Japanese porcelains. Rup Chand liked the Colonial rooms best of all, then the Italian-Spanish Renaissance room. The European primitives and some moderns most interesting in way of pictures. He bought a reproduction of Flemish unknown artist's paintings "Vision of St. Hubert" after an etching by Durer. Liked Indian things about as indicated above, including Nasca pottery. Inscription on outside wall front: "as all natures thousand changes, but one changeless god proclaiming. So in art's wide kingdom ranges one sole meaning still the same. This is truth, eternal reason which from beauty takes its dress. And serene through time and season stands for age in loveliness." Shall write Lechman Das about old Persian manuscript. When we got to the station (The Union Station is immense. They said it was the second rail center of United States. Anyhow 16 trains were going out from 8 to 12), they had 13 federal prisoners that they were shipping off to Nevada or somewhere, not Leavenworth. Federal prisoners with sentences one year one day are sent to federal penitentiaries. Otherwise government pays board in state or county institutions. A detective accompanying crowd stayed to visit with us. He gets $200 a month and says he is apt to be shot any minute. These prisoners weren't handcuffed. An old lady came up to Rup Chand and asked if he weren't Nono—whose picture she had seen in the San Francisco papers. A porter started Spanish with me and though I haven't been able to recall a Spanish word for years, I talked very much.

May 4, 1934. Arrived in Chicago at 9 but the clocks had been turned ahead and we went straight to the Field Museum. The building is so enormous, 700 feet long, that we tramped around till I was dizzy. I had no notion they had such an immensity of display. There isn't much of beauty among the Indian pottery. The Chinese things are abundant and nice. Tibetan abundant but mediocre, except for some good images in bronze. The mammals are uniformly beautifully done, likewise habitat groups. Rup Chand was specially pleased with extinct animal section, the 60 foot dinosaur. Plants fine wax reproductions. Walked heaven knows how many miles, looking at the exhibits and took Rup Chand's picture in the Stanley Field Hall beside the elephants. He bought some postcards of the exhibits: the steed horse, giant dinosaur. Then we went to the Art Institute. There are some nice Persian and Chinese things, many of the latter. Japanese things few, none in Fields. An old baronial hall from France was the most attractive of several period rooms. Went to Marshall Fields. They have some Chinese brocades of fair age and many antiques in furniture line, no rugs. Renewed my charge account and brought a nice Borsalino[2] for $10. Then went out to Yamanakas.[3] Has one good Butsuga[4] of Kamakura red but dull color. Many nice Chinese things, but very expensive. A Sung China bottle, one foot high at $1000. Japanese Butsuga, $350. No Japanese porcelain. Says have a branch at Osaka and Kioto. Went to McDonalds for supper. Yacht magnificent. Mrs. M. a very attractive and interesting person. Plays beautifully and composes in an interesting fashion. Met Baker Brownell, Professor of Contemporary Thought at Northwestern University, and wife. Nice sort. Gende had a tube of ethyl chloride that is great for local and general anesthesia. Had a little Goerz pencil-like microscope, a magnetic thermometer. Some newspaper reporters came and photographed the three of us for Hearst papers. Stayed til 11 and Gene sent us to station in his car. Wanted to give Rup Chand the microscope and invited him back to see the Fair. Gene went to Cocos Island with a radio apparatus that records presence of metal but the two men he took along couldn't work the instrument. Men thought anti-Semitism bound to spread, even to U.S.

May 5, 1934. Arrived in Ann Arbor at 8 and found my baggage from San Francisco, also the Baron's trunk. Went over to old Electric Station and got a bus to Chelsea. Phoned Glenn. He had been apprised by telegram from Chicago, surely the first telegram he ever got. Bought three trees from Harris, two cherries and a Bath pear, all dry as his things usually are. The season has been backward. The oats just out of the ground. Very dry. Poor Shotz whined and howled terrible till I let him loose and then almost choked with excitement. Mother looks frail, probably due to rheumatism. She has the house wired for electric lights. Teddie[5] committed suicide with cyanide at home April 16. No reason. Little iris just coming into bloom. Tulips look sick and the sod triumphant. Mr. Archerbonn is burlier than ever and tank 200 pounds. No one seems to have died in the old settler class: worked at the hoeing. Rup Chand liked the wine and schnitzbrod; Laura Sholz caught a bullfrog by diving after him. I had sent Mother a clipping of our picture taken in San Francisco.

May 6, 1934. Hoed all day. Fearfully dry and sod easily killed. Loyal Broesamlet and wife and father and mother, the Roszels, Ellsworth Hoffe, and Mabel Guthrie, Paul Moroney and wife and mother stopped in. Old Miss M. is a perfect dear and beautiful. Paul likes peonies and is working to give headstones to Civil War veterans. Some in our cemetery with unmarked graves. R. says my old furniture has been moved into the Museum. Very warm.

May 7, 1934. Day cool with north wind. Almost finished the hoeing in the north plot. The quack grass patch in the iris had flourished *a marvelle*. The cherries and pears came out today. Not much has died except some peaches and there

---

2. Italian fedora hat.
3. Yamanaka and Company store; store specializing in Asian antiquities, liquidated during World War II by Alien Property Custodian of the U.S. Government.
4. Buddhist painting.
5. Walter Koelz's brother.

isn't a peach blossom. Mother burned up two of the big pines in the corner and killed a few shrubs. Forsythias froze and there are no flowers. Lula says they were brilliant last year. Poor Mrs. Groshams is still bedridden and shrunken to a skeleton. The girls are plump as partridges. They have a "snakeplant" in a tin can that Elsa gave them and make bouquets out of a tin can and some tissue paper. The tin can furnishes the stalk of the bouquet. John Clark brought his usual buttermilk and rhubarb for cider. Planted two trees that mother bought: Starking apple. Said to be very grand. The pictures are certainly magnificent.

May 8, 1934. Cleaned out the old iris bed—solid sod and mulched the strip under the cherry trees where sod always comes. Took corn stalks out of the barn for the purpose. Mrs. R. asked if I wasn't afraid of falling out of the barn and said I wasn't any younger than before I went to India. Dennis Guinan plowed up a strip of the big rectangle and we planted Kleckleys Sweets and Harris Early Watermelon and Emerald Gem and Extra Early Osage Muskmelon. The cherries in full bloom today. The Roman azalea from SPJ is in full bloom with 27 rich lavender rose blooms. Mother says she knew we should put salt on the quince bushes. They bore beautifully last year. A wren has been working all day, trying to lodge sticks in a corner of the well house but they all fall on the floor. I put a piece of strawberry box to widen out the shelf and the work is proceeding better. All the tin cans I put up last year have been unoccupied and an oatmeal box under the well house eaves doesn't attract the present tenant.

May 9, 1934. Raining this morning. Drought has been severe. Rup Chand went at the sod on the Gleaner line. I decided to go to Ann Arbor and Hank took me to Chelsea after lunch. Had a nice visit with Ed Vogel, Rube Keeler, and the Burgs. Found all the things in perfect condition at Ann Arbor. The things had arrived safely and were well cared for. Had lunch with Harley and went to a movie with him. He says our plant collections are splendid, that Merrill won't get a blade of it. Parke Davis has donated their herbarium. Saw Aunt Jane and Uncle Lovie who look better than ever. Stayed with Tom and Rhea who are dearer than ever.

May 10, 1934. Saw Mrs. Faus who is well and cheerful. God knows why and how. Ran into Herman on the street and Henry and Linke afterward. Lunch with Stanleys. Aunt Jane is working wildly on pencil sketches. Henry is blooming and apparently generally appreciated. Linke didn't know me. The Perisols' yard is lovely as always with apple blossoms. Saw the Riechers and the new baby and Mrs. Day. Elizabeth looks fine! Herman had some sort of drink of cream, gin and something else. Saw Professor Winter who says he was so busy that he couldn't see the art objects I brought, didn't know what was to be done with me, but thought I'd better stay here a year. The four boxes from Karachi are in Detroit. The sky has been weird all day, heavy with dust. Sent off telegraph accepting $3150 for Albuquerque house.

May 11, 1934. Picked out suit. John Brummin came and went to lunch with Henry and me. John has some violent social ideas and though he talked long and forcibly I couldn't make out what they were. He thought I was so unreasonable and bigoted that he didn't want to let me pay for the lunch. Sorted out the Bhadwar birds. Henry is going to sing "Abendstern" tomorrow night for the governor. Saw the President.[6] He had been approached by the Roerichs to call me home and replied he was writing the British Government through the State Department as to the attitude to be taken toward me. Thereafter all was still. President disturbed by the presence of my personal effects among the museum's things. Certainly my position is queer one: I got a salary of $5500 and $1500 for expenses and now my collections that were obtained with this money are the University's. I told the President I was willing to work sans salary for the University provided I was allowed to do what I wanted to. I asked him what my status was and if I could go back and he said he'd see. Saw Maud and Mrs. R. Maud is as dear as always. Got a book by Koop: *Early Chinese Bronzes*, but couldn't find my bronze from Japan. A Hindu from Travancore waited on me in the Chinese restaurant. Speaks Malayalam and is graduating architecture in June. Says doesn't like our clothes or our life but people have been very kind. Doesn't get lonesome. Nice looking and agreeable lad. The Museum has 40,000 birds, 10,000 of my collecting.

May 12, 1934. Worked all day with Dr. March separating collection of *objets d'art* and giving catalog information. A Mrs. Weibel from the Textile Department, Detroit had seen some of the shawls and said there were none except one fragment older than 1800; some of the rarest ones were assembled from disjointed fragments of different ages due to differences in "virility of design." (Sometimes these different designs were woven together.) God help us! How is anyone to get around such blither. Had supper with the Austins. Dear Linkan was very glad and barked and rubbed and whined. Mrs. Austin looks fine. Hear Tubby is very masculine and has had to take some lady friends to the hospital. The use of gold as ornament is often as inappropriate as beauty parlor treatment to a coarse woman.

May 13, 1934. Last night cool but rainy all day. Drove out by Dixboro with P[indecipherable] and after lunch went to Waterloo. Took mother some of the loveliest freshest spies I ever ate. Must have been in cold storage. Mother says it froze ice solid the 11th. The tamaracks are nipped and the early tulips. Went over by Silver Lake where Reardon has found an old sheep farm. The country is beautifully hills and wild, with the original log house. Mayflowers and wild plums. Rup Chand has been digging sod diligently. Went to movie: Clark Gable, *Men in White*, about the hospitals. Abundant hugging. OK. Sat beside McIntyre of the Whitney. Said Pavlova here on commission, 65% hers; brother business manager theaters in New York few and dull times. Actor school closed five years.

May 14, 1934. Spent forenoon cleaning house with Rhea. Lunch with Henry. Henry would run off at any concrete proposal. Saw Raymond Coines who has

---

6. Alexander Ruthven.

grown into a nice looking lad. Fell in love with a Van Arnengigen girl and pa and ma opposed. Then girl fell in love once more with a lad as poor as Raymond. Again opposition and girl committed suicide. Went up and unpacked tankas with Peter and March. They thought the tankas were nice. Saw Harley a few minutes. Green has written a thesis on Wisconsin fish disturbances and has had to refer to my whitefish papers. He says there are "serious doubts" about the "validity" of the forms but they are accepted for the present. George is reading it for its glacial history. George has done some perfect enlargements of some of my old films. A student of Harley's was writing a description of a new Echinacerus from South Texas. Had dinner with Van Tynes. Bully supper. Helen is most sympathetic. Has beautiful wedding presents: silver plate and old pieces, china. Never saw wedding presents with so high percentage of fine things.

May 15, 1934. Went out with Van Tyne to the Lincoln School beyond [indecipherable] to see a Turkey Vulture nest. Dr. Steere went along. His grandson Russel showed us where it was. Old Doctor is 92 and as fresh and bright as a rose. He talked about his Amazon trip and about his old Peruvian pottery without a trace of senility. The nest was in a hollow stump about three feet high in a beech-elm woods. The birds sat on the ground beside the eggs and allowed us to push her about. Eggs beautifully marked. Warblers evident. Woods full of phlox and three species trillium including painted. Tundra berry is common. New to me. Say pawpaws too. Russel is a nice lad of 16, whom we left in Ann Arbor where he is going to sing in the chorus. Had dinner with the Lewises, [indecipherable] and Mr. Brick. Went to see Robert Henderson's "The Brontes" afterwards. The show was well done but the subject wasn't dramatic. The character of the girls was well-portrayed in a series of situations. The scene was really effective with the minimum of exposition—the interview with the publishers. Brick is Superintendent of Public Inst. in Virginia. Splendid sort with broad views. Says new method of English instruction in U.S. supplants Shakespeare with anything that the children will read. Came away totally charmed with the Lewises. Have always suspected them of being good.

May 16, 1934. Last night frost. Went to Detroit with Mr. Shemm to clear the shipment of birds, etc. Mr. Richardson, Customs Broker, says he has to know customs law and all customs cases come to a special U.S. court that convenes in Detroit three times a year. Mr. Fitzgerald is the inspector and Mr. Spencer an appraiser. Had opened all the boxes and had found everything in them and listed all. Then said we'd have to post an exhibition bond of $10 or pay duty of 110% on the silver and beads, 30% on the [indecipherable], 45% oriental rugs, birds free. Went to see Kopp who will frame the little paintings. He got typhoid somewhere, he thinks from a Chinese restaurant and didn't report to police. He tried for a Guggenheim fellowship and failed. Wanted to study old gilding and framing. Dinner with George. Harley came and spent the evening with us. Went riding with Maud for an hour or so. Day beautiful but cool. Oakbuds bursting and woods full of spring flowers. New [indecipherable] were delivered.

May 17, 1934. Worked on birds in morning and then went out to Waterloo with Henry. Warmer. My cherry trees from Starks had come but Mother refused them and we had to go to Munith after them. A bit dried out but suffered more by bruising in the packet. Rup Chand says when Lahulis plant a tree they put in a couple seeds of barley. Henry stayed all night.

May 18, 1934. Henry left for a ten o'clock with one half hour to get there. Talked all night. Went to Jackson with Hank and Mother. Met Eileen and found Blanche terribly overwrought. Looks 70 and seems demented. Ted looks fine but must surely go crazy in the end. Eileen says parents drove the son to suicide, that he attempted it in January. Bob looks pale and unhappy. Rup Chand went to Realys for an inspection of their farm. Roszells came at night.

May 19, 1934. Potatoes and melons just coming out. Hoed peas and onions. Lovejoys came and finished up the housecleaning. Burned up Pa's old account books except one beginning 1873. Mother asked Lovejoys to come back Monday because she felt she wouldn't live much longer. Lilacs in full bloom and more bushes were filled. Paul Niehaus came and asked me to speak at the Chelsea Alumni Banquet. Rup Chand went over and saw Nettie Howlett's house and got a picture of it on a postcard. The folks like him and freely show him all their things. He is much impressed with the ease with which we can earn a living. Our sod is easily worked and productive compared with Lahul. All heavy work done by machinery and no hardship for humans. In Lahul mainly humans get calluses and sores from carrying burdens and get muscle bound in the legs. The people here all say they hope to see you again. Lahulis never say it but to "great people," for a lama, etc.

May 20, 1934. Ellsworth Hoppe came ca. 10:30 and took me to Crooked Lake. Elsie (Mrs. Green of Grand Rapids) was there with her sister Mrs. Foster of Detroit and her old mother. Elsie is interested in the occult. Went to Chelsea and got Mabel and then to Ann Arbor where we gathered at the Roszells to wait for the Browns of Detroit. And taking Browns then went to Faculty Cottages near Lakeland and had a picnic. 11 people ate 12 pounds steak plus ice cream and cake and coffee, nuts, cherries, and candy. Ehlers were there and later Prof. Reighard and Mrs. Curtis came out. Professor Reighard has been taking some sort of serum in New York and feels better and looks better than ever. The serum is for bacteria that have been poisoning him in intestines. Harley still has traces of screwworm that he broke off in his forehead in Guatemala. The protein poison that resulted almost killed him. Castor bean size of pin head under skin. Kills you dead. Stopped to see Webbers at Whitmore Lake—Mabel's uncle. Mrs. W. has an electric belt that cures everything. Her son ain't got no sense and the daughter in law is a bad egg. She is a sister to Morris Eisenbeiser. The Hoppes have a creamery in Chelsea and the creamery inspectors test the cream to see if they have paid enough, butter for salt and water, pasteurization, etc. Rup Chand had gone with Hank to castrate lambs and cut tails at Mr. Intees and then to a ball game.

May 21, 1934.   Mrs. Lovejoy and daughter and grandson Les came. Mrs. Lovejoy is going to stay till tomorrow. Tom got us and we came to Ann Arbor. Went to see "And So to Bed," a play of Charles II and Pepys. More situations, but well done. Rup Chand thought actors very clever. Heat terrific all day but shower at night. Drought has been very severe. The tulips are burned up today. Rup Chand says our men had to throw away the pelican grease they saved. It stank to heavens and got onto their clothes.

May 22, 1934.   Stayed home and cleaned house till noon. Then went to Museum and unpacked the last four boxes that came from Karachi. The contents were in perfect condition. One board of one side of one box had been replaced somewhere. No evidence of damage by Demestres. Rup Chand and Henry were going to a movie but none good enough so came home and looked at shawls.

May 23, 1934.   Rup Chand went out with Tom to see bank, meat market, dairy farm, orchard and even drove car. Finished up the collection dispersing and gave March all the things he wanted out of my anthropological collection in return for transportation expense on my things and an exchange of a badly patched shawl. Maud came up and met Rup Chand and then I went to dinner with Miss Dewey and the Eatons. Mrs. Eaton nice. Intended to go home after dinner with Henry's car, but decided to get report off to President first. Gave alumni relations man Morrissey interview and saw Schonfeld who is going to London. Met Lyman Bothwell, the cousin of Bolivian Ruiz whom I hadn't seen since he was in High School. Grown into a nice youth. Also a brother of Carlos Guardia of Panama. Bought a box of watercolors 12 colors and box for $4.50 to mend paintings. Henry came and looked at paintings. Good rain at night.

May 24, 1934.   Very cold all day. Sent report on work in Orient to President, Gaige and Winter. Went home then with Henry and Rup Chand but Henry had to go right back to take his ma to Detroit. Covered up grapes and melons and worked on identification of Kangra birds.

May 25, 1934.   Finished the box of birds brought along but can't find the Spiti sturmia? Went to Collins' woods to get some spring flowers to plant on north side of house. Martins place is getting to be a forest. Along the woods is a mass of oaks 4–6 feet and the poplars in the lowlands are 30+ feet. Cool all day but nothing froze last night. Potatoes up big enough to hoe.

May 26, 1934.   Painted all day on the Likir tankas. Dr. Riecher came out and Pete was so glad he grabbed the mop and chewed up the handle. His grandmother was furious and says he has to be sold. Rup Chand wanted to take him to India but it is objected that the expense is too great and the Indian food wouldn't agree with Scholtz. Cool.

May 27, 1934. Cool. Nucles came in the morning. Mildred has been in hospital with heart trouble—Berenice wants to teach Waterloo School. Miss Dewey and Eatons came in the afternoon and we had a nice visit over wine and cookies. Evening Nina and Charlie G. Nina is as refreshing as ever. Cool. Continued painting.

May 28, 1934. Home all day. Retouching tankas.

May 29, 1934. Worked all morning finishing tankas. Drawing most carefully and skillfully done—shading and brush lines amazingly clever. Hank took me to Chelsea and went to Ann Arbor in time to hear Mrs. Weibal of Detroit Art Institute give lecture on tapestries. She showed some fine Swiss and Franconian examples with symbol design prior to 1400. At circa 1400 began the very elaborate weaving of great skill but less artistic merit. Met Mrs. B. Davis for first time. Weibel is Swiss and diplomatic. Linda Eberbach and Laura are keeping house at [indecipherable] while the family went to Grayling. Saw John Bean who had just blown up a dog and nearly his self in a compressed oxygen experiment. John is nephew to the Gieskes. Called on Bruces and found Sundwalls there.

May 30, 1934. Got the things Ruthven had and went to West Gallery with March to hang shawls etc. The room has a terrible black green baseboarding and is full of seats. Hung shawls and paintings on walls, latter mounted on beaver board pieces. Filled one case with wood carvings (block prints), one with Ku, three with silver jewelry, gaos, etc. Also hung two sozni. Nothing to prevent crowds from handling pictures and shawls. Whole effect very fine. March gave newspapers a splendid writeup. Finished at three. Workmen interested in textiles and curious of travel opportunity. Finished afternoon in bird division identifying large birds. Evening at home.

May 31, 1934. Mrs. Stanley and a Jewess friend Heimann came to see the display, also Henry. Mrs. Stanley liked the textiles best but saw the beauties of drawing and composition in the pictures. Pictures apparently take time for arousing appreciation: women were asking what the picture represents. What was in his hands, etc. Got back Isle Royale report from Gaige. Hubbs had rewritten parts and inserted corrections and addenda. Henry had brought Kopp's framed miniature from Detroit (Sidney S. Koppe, 638 W. Palmer), nicely done in gold and silver leaf. $25. Mrs. Kern called and wanted me to appraise an Indian shawl she has. Henry and I went to Waterloo at 4. Made shortcake. Rup Chand saw three big dogfish at bridge and thinking must be bass went down. Glenn gave us a one and half pound bass. Iris beginning to bloom. Rup Chand says Lahuli Tibetans grind their discarded human teeth (pious people do) mix with clay, mold cakes and put half cakes at river edge, half at mountain peak, Former to be worn off by wave action, latter by wind. Alva Beeman caught a barn owl in his barn today.

Appendix: Descriptions of Thangkas in the George Lauff, the Harley H. Bartlett, and the Mr. and Mrs. Alexander Ruthven Collections

by T. Joseph Leach and Rebecca Bloom

The thangkas described here were collected by Walter Koelz, and donated to the University of Michigan Museum of Anthropology by George Lauff, Harley H. Bartlett, and Mr. and Mrs. Alexander Ruthven.

George Lauff Collection

2007-22-1 Avalokiteshvara

The central figure in this unusual silver-ground thangka is a form of Avalokiteshvara. Avalokiteshvara is the bodhisattva of compassion and a particularly important figure for Tibetan Buddhists. Although the small image of Amitabha Buddha resting on the head of the central figure unmistakably identifies him as Avalokiteshvara, the overall iconography is unusual and does not adhere to any of the standard representations of the bodhisattva's many forms. His body is golden colored with six arms, two legs, and one face. He sits on a moon disk resting in the center of a lotus, framed by a blue aureole and a green nimbus. His left leg is tucked in and his right leg is partially extended in *lalitasana*, or the posture of royal ease. His central pair of hands makes a variation of the *dharmachakra mudra* or teaching gesture in front of his chest. His upper left hand holds a trident with unusual offset points and his lower left hand holds a water pot. His upper right hand holds a *mala* (string of beads) and his lower right hand makes the *varada* mudra or giving gesture.

This particular arrangement for a six-armed figure resembles a form of Avalokiteshvara called Amoghapasha, the Unfailing Lasso. As in this image, Amoghapasha is seated in lalitasana, holds a vase and makes the varada mudra. However, Amoghapasha usually also holds a lasso, from which he derives his name. The absence of a lasso here makes it difficult to identify this figure as Amoghapasha. Another possible identification of the central figure is Sugatisandarshana Lokeshvara. This six-armed form of Avalokiteshvara is usually standing, makes the varada and *abhaya* mudras, and holds a rosary, lotus, water-pot, and trident.[1] The central figure of the present image is seated rather than standing and does not hold a lotus, but is otherwise quite similar.

The area around Avalokiteshvara further complicates a more specific identification of the central figure. The surrounding landscape is in outline against a silver ground and depicts Potalaka, the pure land of Avalokiteshvara. He floats within a walled courtyard filled with plants that opens at a central gate, and an elaborate multi-tiered palace rises behind. Representations of Avalokiteshvara in Potalaka usually depict his four-armed form, Avalokiteshvara Chaturbhuja.[2] In fact, the courtyard and palace in this painting are quite similar to those in another image of Chaturbhuja's pure land.[3] However, the differing number of arms and the different objects each form of Avalokiteshvara holds makes it unlikely that this image is meant to be Avalokiteshvara Chaturbhuja.

There are two figures inside the courtyard just underneath Avalokiteshvara. On the viewer's right is a seated figure probably identifiable as Hayagriva.[4] He is in a crouching position with his right hand raised in front of his face, palm in and index finger pointing up in the *tarjani* mudra or "threatening pointer" gesture.[5] He has a distinctive mustache and beard, heavy round earrings, and a piece of cloth wrapped around his head. Opposite Hayagriva is a robed monk with bare feet who holds peacock feathers. Both the monk and Hayagriva gaze upward at Avalokiteshvara. A table with offerings stands between them inside the courtyard, while just outside the gate a lotus emerges from a body of water. A small figure, perhaps the painting's donor, is seated on the lotus, holding an offering in his hands. Each of these figures within Avalokiteshvara's Potalaka pure land is rendered in outline against a silver background.

Avalokiteshvara is surrounded by four other figures, all Gelug teachers. In the upper right is Tsong kha pa (1357–1419), the founder of the Gelug tradition of Tibetan Buddhism. He faces outward, like the central figure, and wears monastic robes and the characteristic pointed yellow hat of the Gelug tradition. He sits on a lotus with his hands in the dharmachakra mudra or two-handed teaching gesture. He also holds the stems of two flowers that support a sword and a book on either side of him. Opposite Tsong kha pa is a Gelug monk depicted turned toward the central figure. He holds an alms bowl in his left hand and makes the *bhumisparsha* or earth-touching gesture with his right. This is possibly one of the Dalai Lamas. In the bottom corners are two additional Gelug monks, also possibly representations of the Dalai Lamas or alternatively depicting Tsong kha pa's two main disciples, Gyaltsap Dharma Rinchen (1364–1432) and Kedrup Geleg Pal Zangpo (1385–1438).

The central figure Avalokiteshvara and each of these four surrounding Gelug teachers are fully painted colored figures set against a silver background. Both the use of a silver ground and the combination of fully painted figures with outlined figures are quite unusual

---

1. Donaldson, Thomas E., *Iconography of the Buddhist Sculpture of Orissa* (New Delhi: Indira Gandhi National Centre for the Arts, 2001), 207.
2. "Buddhist Deity–Avalokiteshvara (Potalaka Pureland)," Himalayan Art Resources, accessed February 7, 2013, http://www.himalayanart.org/search/set.cfm?setID=1289.
3. "Avalokiteshvara-Chaturbhuja," Himalayan Art Resources, accessed February 7, 2013, http://www.himalayanart.org/image.cfm/80.html.
4. In his earlier forms, Hayagriva often accompanies four- and six-armed forms of Avalokiteshvara as a diminutive, bearded and mustached figure. Robert N. Linrothe, "Ruthless Compassion: Wrathful Deities," in *Early Indo-Tibetan Esoteric Buddhist Art* (Boston: Shambhala, 1999), 97, 103.
5. Robert Beer, *The Handbook of Tibetan Buddhist Symbols* (Chicago: Serindia, 2003), 229.

Appendix: Descriptions of Thangkas in Collections 283

2007-22-1 Avalokiteshvara

features of this painting. While there is a tradition of painting in outline on gold, red, or black ground, there are few known examples of outlined figures on a silver background. Additionally, gold-, red-, and black-ground paintings typically only use outlined figures with some full-color highlights, rather than combining outlined figures with fully painted figures. Coupled with the ambiguous identity of the central figure, this silver-ground painting is quite unusual.

Early 19th century

Image: 22" x 14.5";
total dimensions with mount: 39" x 19.5"

Acquired by George Lauff at Christie's Auction, October 1990 (October 3, 1990, Christie's Catalog, p. 88)

(*above left*) Detail of monk holding peacock feathers

(*left*) Detail of Hayagriva

2007-22-2 Akshobhya Buddha

Akshobhya (the "Immovable One") is one of the five principal buddhas of Tibetan Buddhism, each associated with a color and direction. Akshobhya traditionally occupies the eastern direction and is linked with the color blue. Since he is usually depicted in the company of the four other principal buddhas—Vairochana, Ratnasambhava, Amitabha, and Amoghasiddhi—it is possible that this thangka was originally part of a set of five paintings that would have been hung together.

Akshobhya is recognized by his blue body and the upright *vajra*—the thunderbolt or diamond scepter symbolizing the power of the Vajrayana teachings—held in his left hand. He is seated upon a moon disk and lotus and encircled by a bicolor halo and *mandorla* (aureole) and surrounded by blooming peonies. His right hand extends downward in the earth-touching gesture (bhumisparsha mudra). Below him sits Amitabha (the Buddha of Infinite Light) on the left and the Medicine Buddha (Bhaisajyaguru) on the right. Luxurious gold pigment is used to depict the rays of light emanating from each of the three buddhas, as well as decorative details in their robes and haloes.

The silk brocade mounting of this painting is still intact.

Late 19th century

Image: 34.0" x 20";
total dimensions with mount:
41" x 24"

Acquired by George Lauff at Christie's Auction, October 1990 (October 3, 1990, Christie's Catalog, p. 87)

2007-22-2
Akshobhya Buddha, with preserved covering cloth

2007-22-2 Akshobhya Buddha

2007-22-3 Arhat Subhuti

This painting is one of a set depicting the lives of the Panchen ("Great Pundit") Lama, an important incarnate lama of the Tibetan tradition. This painting depicts Subhuti, an *arhat* (enlightened disciple) of the historical Buddha Shakyamuni, shown stroking the head of a *garuda* (mythical bird). It is based on a famous set of thirteen woodblocks representing the third Panchen Lama (1738–1780) and all of his previous incarnations. The blocks were carved at Narthang monastery in Tibet in the eighteenth century, and were widely reproduced in Tibet as well as in the royal atelier of China's Qianlong emperor. It seems that the set was augmented to accommodate subsequent incarnations of Panchen Lama as well. This version of the Subhuti portrait exemplifies the artistic variations and interpretations made possible by the compositional ambiguities of woodblock reproduction. In other paintings based on the same woodblock, the stream that emerges from the landscape in the middle of the painting next to Subhuti twists and turns from the horizon line at the top of the painting, curving between the figures and garudas in the area.

The layout and background of the painting were influenced by courtly painting styles of the Chinese Qing dynasty (1644–1912), and are quite different from traditional Tibetan thangka design of Nepalese-Newari tradition. The central figure is seated off-center in a highly stylized landscape influenced by traditional Chinese blue-green landscape painting. The late-seventeenth-century establishment of this hybridized New Menri style is credited to Chojing Gyatso, and is characterized by exuberant use of vivid color, overlapping forms, subtle modeling and lush landscape elements to express a multitude of emotional states and enlightened qualities.

(See also UMMA 47159 and 47160; Mr. and Mrs. Alexander Ruthven Collection.)

Image: 27" x 16.75"; total, in partial mount: 33" x 21.5"

19th century

Acquired by Lauff from Carolyn Copeland (who acquired from Koelz)

Detail of 2007-22-3 Arhat Subhuti

2007-22-3 Arhat Subhuti

Appendix: Descriptions of Thangkas in Collections

Mr. and Mrs. Alexander Ruthven Collection

47157 Shakyamuni Buddha with Jataka (Previous Lives) Scenes

This painting features a large central figure of Shakyamuni Buddha, the historical founder of Buddhism, surrounded by *jataka* scenes from his previous lives. The scenes are based on the [bodhisattva] Avadanakalpalata, an eleventh-century text by Ksemendra. The text recounts 108 stories or *avadanas* about the former lives of the Buddha. The present painting was probably part of a larger set that depicted all 108 stories. Sets of paintings that feature a large central image of Shakyamuni Buddha surrounded by scenes from the avadanas are a relatively common genre of Tibetan art. In addition to depicting the same content, this painting is quite similar in composition to other known sets of avadana paintings.[6] Of particular interest is the striking similarity between this image and one acquired by the Italian Tibetologist Giuseppe Tucci. Tucci traveled to many of the same places as Koelz at roughly the same time. Tucci acquired a set of avadana paintings during his travels, one of which resembles this image.[7]

The central Shakyamuni Buddha figure in this painting is seated crossed-legged on a lotus throne. He wears monastic robes and makes the dharmachakra mudra or two-handed teaching gesture. The numerous narrative scenes around him are set within a natural landscape and are much smaller in scale. A Tibetan inscription identifies each of the scenes, some of which are still legible despite the present state of the painting.

This painting has been patched together with several fragments of other paintings. The added fragments appear to be from the same set of avadana paintings. This is particularly evident in the lower left corner and in the vertical strip along the right side (see detail photographs). The painting and fragments have all been glued onto a cloth backing. Despite the visible disjuncture between the composite pieces, the primary image is largely intact.

19th century

Image: 34" x 21"

---

6. The Tibet House in India has a set of 31 avadana paintings, one of which is compositionally similar to the present image ("Shakyamuni Buddha–Jataka (Previous Lives)," Himalayan Art Resources, accessed February 6, 2013, http://www.himalayanart.org/image.cfm/72030.html).

7. Koelz referred to Tucci several times in his journals; see, for example, entries for September 9, 1933, and October 5, 1933. See Giuseppe Tucci, *Tibetan Painted Scrolls* (Roma: Libreria dello Stato, 1949).

47157 Shakyamuni Buddha with jataka (previous lives) scenes

Appendix: Descriptions of Thangkas in Collections

(*above and below*) Detail showing joins of painting segments

47158 Shakyamuni Buddha in Bhumisparsha Mudra

This Gelug lineage "field of accumulation" or "assembly field" genre painting features Shakyamuni Buddha at the center surrounded by Tibetan and Indian teachers, with a register of protector deities along the bottom.[8] This genre of painting typically depicts a lineage of teachers spatially oriented around a central figure, usually Tsong kha pa or Shakyamuni in the Gelug lineage. Here, the center is occupied by Shakyamuni Buddha in a standard iconographic form. He is seated cross-legged in vajra posture on a lotus supported by a lion throne. He wears the robes of a monk and has stretched earlobes indicating his renunciation of a princely lifestyle. He makes the bhumisparsha mudra or earth-touching gesture with his right hand; his left hand rests in his lap in meditation or *dhyana* mudra. The two figures to Shakyamuni's immediate left and right are the bodhisattvas Maitreya and Manjushri, both distinguished by the attributes resting on the flowers they hold.

The rest of the image is populated by several clusters of figures hovering on clouds in a landscape surrounding the central Buddha and a register of protector deities along the bottom of the painting. The group on the upper left depicts the "lineage of extensive deeds" with Asanga, a founder of the Yogacara school, at the center. Asanga is depicted frontally while the Indian figures surrounding him are in three-quarter profile facing the central Shakyamuni. Opposite this group in the upper right corner is the "lineage of profound view" with Nagarjuna, the founder of the Madhyamaka school, in the center. Nagarjuna, identifiable by the hood of serpents visible behind his head, is depicted frontally parallel to Asanga, while the figures surrounding him are in three-quarter profile, facing the central figure of Shakyamuni.

Between these two groups of Indian teachers is a central cluster of three vertically arranged groups. In the uppermost cluster of five figures is Vajradhara. Vajradhara, or the vajra (thunderbolt scepter) holder, is blue and holds two vajras in his hands with his arms crossed at the chest. Vajradhara is considered to be the primordial Buddha by the three "new schools" of Tibetan Buddhism, the Gelug, Sakya, and Kagyu.[9] As such, he is the progenitor of the *vajrayana* teachings and the embodiment of the *dharmakaya* or truth body.[10] Vajradhara usually appears in assembly field compositions associated with one of the three new schools. He is surrounded by two female figures and two bare-chested Indian siddhas, possibly Tilopa and Naropa. Below Vajradhara the first set of three figures depicts Atisha (982–1054) with his primary student Dromton (1004–1064) on the right and perhaps Atisha and Dromton's disciple Potowa Rinchen Sal (1027–1105).[11] Another set of three figures below them depicts Tsong kha pa (1357–1419), the founder of the Gelug tradition of Tibetan Buddhism, flanked by his two primary disciples, Gyaltsap Dharma Rinchen (1364–1432) and Kedrup Geleg Pal Zangpo (1385–1438).

---

8. "Field of accumulation" and "assembly field" are both translations of the Tibetan term *tshogs zhing*. This genre is also often referred to as a "refuge tree," perhaps due to the tree-like organization of later assembly field paintings or because field and tree are homophones in Tibetan (*zhing* and *shing*).
9. Conversely, the Nyingma or "old school" considers Samantabhadra to be the primordial Buddha.
10. In Tibetan Buddhism, *vajrayana* or *rdo rje theg pa* is one of the three routes or "vehicles" to enlightenment. The *dharmakaya* or truth body is one of the three bodies of the Buddha (*trikaya*). The other two are the *sambogakaya* and *nirmanakaya*.
11. The figure on Atisha's left could also be one of Dromton's other two main disciples, Puchungwa Shonnu Gyaltsen (1031–1106) and Chennga Tsultrim Bar (1038–1103); however, Potowa Richen Sal seems to appear more frequently than the other two disciples alongside Dromton and Atisha.

Tibetan teachers are arranged in four groups on either side of the central Shakyamuni. On the left side of the painting is a group of five figures with a group of eight figures below. In the upper group, the five figures are shown in three-quarter profile facing downward. All wear yellow hats, identifying them as Gelug teachers, with the distinctive folded hat of one of the figures identifying the Panchen Lama Lobzang Chokyi Gyaltsen (1570–1662). In the group of eight figures below, seven are in three-quarter profile facing down while the uppermost right figure in this group is frontally oriented. The two groups on the right side of the painting parallel these figures, and are also organized in two groups of five and eight. A frontally oriented monk makes the dharmachakra mudra or teaching gesture and is hatless. The last group of three Gelug teachers appears below the central Shakyamuni and above the row of protector deities. The teacher at the center is frontally oriented with two Gelug figures facing him on either side. These three are possibly the teachers in the lineage who were alive at the time of the painting's creation.[12]

The bottom register of the painting depicts five protector deities. Starting on the left is yellow Vaishravana, the guardian of the northern direction. He holds a banner and a mongoose and rides a white snow lion. Moving right is the blue-black six-armed form of Shadbuja Vajra Mahakala.[13] He holds a string of skulls (mala), drum, and chopper on his right and a trident, skull cup and possibly a lasso on the left, while trampling a white elephant. Next to Mahakala is Vajrabhairava with his consort Vajra Vetali.[14] He has nine faces, thirty-four hands, and sixteen legs, and holds a skull cup and chopper in his principal hands. Next to this pair is blue Yama Dharmaraja, who, alongside Shadbuja Mahakala and Vaishravana, is one of the three primary protectors of the Gelug tradition. Yama Dharmaraja has the head of a buffalo, holds a lasso and bone stick, and rides on the back of a buffalo trampling a human body. The last figure on the right is Magzor Gyalmo, the queen who repels armies. She holds a skull cup in her left hand and a vajra staff in her right hand. She sits atop a flayed human skin while riding a mule in a sea of blood. All five of these protector deities are associated with the Gelug tradition and appear in other paintings Koelz collected.

The placement, orientation, and scale of each of these figures are significant with respect to the overall teaching lineage depicted. Typical of most Tibetan paintings, the central figure of Shakyamuni Buddha is significantly larger than the surrounding figures. Vajradhara occupies the uppermost central position as the primordial Buddha. Vajradhara, Atisha, Tsong kha pa, and the teacher beneath Shakyamuni all appear on a central axis with the main figure of Shakyamuni. These four figures are also depicted frontally along with Asanga, Nagarjuna, and two additional Tibetan teachers. In the upper half of the painting, Atisha and Tsong kha pa's position between Asanga and Nagarjuna is important. Atisha in the center integrated the teachings of Asanga on the left and Nagarjuna on the right within the Kadam lineage, which was continued by his student Dromton.[15] Tsong kha pa continued Atisha's teachings by incorporating aspects of the Kadam lineage into the Gelug tradition. Each of these frontally depicted figures in the upper half of the painting is a core teacher in the overall lineage depicted in this assembly field. Additionally,

---

12. "Refuge Field (Buddhist)–Gelug Lineage," Himalayan Art Resources, accessed February 7, 2013, http://www.himalayanart.org/image.cfm/53405.html.
13. This is a wrathful form of the white six-armed Mahakala in the bottom right of painting UMMA 47309.
14. This form of Vajrabhairava is iconographically the same as the central figure in painting UMMA 47308.
15. Bstan-'dzin-rgya-mtsho, and Thupten Jinpa, "Origin of Lamrim Instructions," in *Path to Bliss: A Practical Guide to Stages of Meditation* (Ithaca, N.Y.: Snow Lion Publications, 1991), 20–34.

the protector deities typically occupy the lowest register, as is the case in this painting.[16] Thus, the hierarchical organization of the lineage is represented in the painting in several different ways: from the top to the bottom, along a central axis, frontally oriented rather than oriented toward the center, and in terms of scale.

Although present in other Tibetan traditions, assembly field compositions are much more commonly Gelug productions.[17] While this painting features many common characteristics of Gelug "field of accumulation" or "assembly field" compositions, the clustered arrangement of teachers is an unusual layout compared to other paintings in this genre.[18] Gelug assembly field paintings typically feature the lineage of teachers grouped together in a tree shape that rises from a large lotus-stem base, with the figures branching out, in terms of both placement and importance, from the center to the periphery of the painting. Another common arrangement groups the "lineage of extensive deeds" and "lineage of profound view" on the left and right respectively, as is the case in this painting, and organizes the rest of the teachers fanned out below the central figure (usually Tsong kha pa or Shakyamuni). The diffused organization of the teachers in this painting and the presence of many smaller groups as opposed to a typical arrangement of fewer and more clearly delineated groups makes this painting a rather unique representation of the genre.

Date unknown

Image: 29"x 19.5"; total dimensions with mount: 45.5" x 27"

---

16. David Paul Jackson, and Janice A. Jackson, *Tibetan Thangka Painting: Methods & Materials* (Ithaca, N.Y.: Snow Lion Publications, 1988), 26–27.
17. "Subject: Refuge Field Main Page," Himalayan Art Resources, accessed February 8, 2013, http://www.himalayanart.org/search/set.cfm?setID=157.
18. The unusual composition of the present painting is evident when compared with the set of assembly field images at: "Subject: Refuge Field Main Page," Himalayan Art Resources, accessed January 31, 2013, http://www.himalayanart.org/search/set.cfm?setID=157.

Appendix: Descriptions of Thangkas in Collections

47158 Shakyamuni Buddha in Bhumisparsha mudra

47159 Go Lotsawa

This painting is one of a set depicting the lives of the Panchen ("Great Pundit") Lama, an important incarnate lama of the Tibetan tradition. This painting depicts Go Lotsawa (1392–1481), the great translator and historian, overseeing the translation of Sanskrit scriptures into Tibetan. It is based on a famous set of thirteen woodblocks representing the third Panchen Lama (1738–1780) and all of his previous incarnations. The blocks were carved at Narthang monastery in Tibet in the eighteenth century, and were widely reproduced in Tibet as well as in the royal atelier of China's Qianlong emperor. It seems that the set was augmented to accommodate subsequent incarnations of Panchen Lama as well. Each portrait contains allusions to both the form and biography of the central figure, encompassing the outer and inner lives of the lineage members. The worldly translation activities of Go Lotsawa are therefore depicted in the portraits of his tutelary deity in the upper right, his guru in the upper left, and his protector deity in the lower left. His disciples inhabit the mountainous landscape around him.

The layout and background of the painting were influenced by courtly painting styles of the Chinese Qing dynasty (1644–1912), and are quite different from traditional Tibetan thangka design of Nepalese-Newari tradition. The central figure is seated off-center in a highly stylized landscape influenced by traditional Chinese blue-green landscape painting. The late-seventeenth-century establishment of this hybridized New Menri style is credited to Chojing Gyatso, and is characterized by exuberant use of vivid color, overlapping forms, subtle modeling and lush landscape elements to express a multitude of emotional states and enlightened qualities.

(See also UMMA 47160, Mr. and Mrs. Alexander Ruthven Collection, and 2007-22-3, George Lauff Collection.)

19th century

Image: 27" x 17"; total, in partial mount: 32" x 21.5"

Detail of
47159 Go Lotsawa

Appendix: Descriptions of Thangkas in Collections 297

47159 Go Lotsawa

47160 Bhavaviveka

This painting is one of a set depicting the lives of the Panchen ("Great Pundit") Lama, an important incarnate lama of the Tibetan tradition. This painting depicts Bhavaviveka, an important sixth-century Indian philosopher of the Madhyamaka or "Middle Way" school, regarded as a previous incarnation of the Panchen Lama. It is based on a famous set of thirteen woodblocks representing the third Panchen Lama (1738–1780) and all of his previous incarnations. The blocks were carved at Narthang monastery in Tibet in the eighteenth century, and were widely reproduced.

The layout and background of the painting were influenced by courtly painting styles of the Chinese Qing dynasty (1644–1912), and are quite different from traditional Tibetan thangka design of Nepalese-Newari tradition. The central figure is seated off-center in a highly stylized landscape influenced by traditional Chinese blue-green landscape painting. The late-seventeenth-century establishment of this hybridized New Menri style is credited to Chojing Gyatso, and is characterized by exuberant use of vivid color, overlapping forms, subtle modeling and lush landscape elements to express a multitude of emotional states and enlightened qualities.

As evident in the lower right corner, someone (perhaps Walter Koelz) performed some very poor repairs to replace a missing section of the original painting.

(See also UMMA 47159, Mr. and Mrs. Alexander Ruthven Collection, and 2007-22-3, George Lauff Collection.)

19th century

Image: 27" x 16"; total dimensions with mount: 44.5" x 25.5"

Detail of Bhavaviveka

Appendix: Descriptions of Thangkas in Collections

"Repairs" in lower right corner of Bhavaviveka (47160) painting

Harley H. Bartlett Collection

47308 Vajrabhairava (Yamantaka) with Consort Vajra Vetali

The central figure of this thangka is Vajrabhairava in a nine-headed form. Vajrabhairava is the wrathful form of the bodhisattva Manjushri and is generally identified by his buffalo head. Vajrabhairava can have one or more faces and in nine-headed forms, the faces can be arranged in two different ways. In the first arrangement there are three vertically stacked sets of three faces, with the buffalo head as the center face of the first register. In the second method, which is used in this image, two sets of three faces flank the central buffalo head with two additional faces stacked on top. This latter arrangement was popularized by Tsong kha pa (1357–1419) and is generally associated with the Gelug tradition.[19] In either arrangement, the orange face of Manjushri typically appears as the uppermost central face. In addition to the nine faces, this form of Vajrabhairava has sixteen legs and thirty-four hands, with the main hands holding a skull cup and chopper.

Vajrabhairava is accompanied by his consort Vajra Vetali. Vetali is wrathful with one face and two hands, and holds a skull cup in her left hand. She stands in the center embracing Vajrabhairava. Vajrabhairava and Vetali stand on a raised lotus and trample several figures underfoot, including human forms, several animals, and Hindu deities.

---

19. "Buddhist Deity: Vajrabhairava Main Page," Himalayan Art Resources, accessed January 31, 2013, http://www.himalayanart.org/search/set.cfm?setID=166.

Appendix: Descriptions of Thangkas in Collections

Three robed teachers float on clouds above the main figure. Although only partially visible, the central teacher is likely Tsong kha pa. He wears monastic robes, is seated on a lotus base, and makes the dharmachakra mudra or teaching gesture. The lotus stems rising on either side of the teacher likely support a sword and book, implements associated with both Tsong kha pa and Manjushri. Since Tsong kha pa popularized this form of Vajrabhairava, he often appears in the upper register of Vajrabhairava paintings, making such an identification of the central teacher likely. The two monks on the right and left of the central teacher also wear the yellow hat typical of Gelug teachers. The teacher on the viewer's left holds a vase in his left hand and makes the abhaya mudra or "no fear" gesture with his right hand. The teacher on the viewer's right holds a book in his left hand and also makes the abhaya mudra with his right hand. These two figures are possibly Tsong kha pa's two primary disciples, Gyaltsap Dharma Rinchen (1364–1432) and Kedrup Geleg Pal Zangpo (1385–1438).

There are four figures in the lower register of the image, from left to right: Magzor Gyalmo, Yama Dharmaraja, Shadbuja Mahakala, and Vaishravana. Magzor Gyalmo is a protector deity with a special connection to the Gelug tradition. She is blue and rides a mule in a sea of blood. To the right of Magzor Gyalmo is Yama Dharmaraja riding a black buffalo with his consort Chamunda. Yama Dharmaraja is a protector of the Vajrabhairava Tantra and one of the three primary protectors of the Gelug tradition.[20] To his right is a six-armed form of the protector Mahakala called Shadbuja Mahakala.[21] This blue-black form of Shadbuja Mahakala is also frequently associated with Vajrabhairava and the Gelug tradition. On the far right is the protector Vaishravana, who is yellow and rides a white snow lion. He has two hands and bears a banner on the right and a mongoose on the left. Vaishravana also often appears in Vajrabhairava images.

Date unknown

Image: 20.5" x 15"; total dimension, in partial mount: 23" x 18"

Detail of lower figures: Yama Dharmaraja (*center left*) and Shadbuja Mahakala (*center right*)

---

20. "Yama Dharmaraja (Buddhist Protector)," Himalayan Art Resources, accessed January 31, 2013, http://www.himalayanart.org/image.cfm/406.html.
21. "Mahakala: Shadbuja Main Page," Himalayan Art Resources, accessed January 31, 2013, http://www.himalayanart.org/search/set.cfm?setID=85.

47308 Vajrabhairava (Yamantaka) with consort Vajra Vetali

Appendix: Descriptions of Thangkas in Collections 303

47309 Tsong kha pa

The central figure of this thangka is Tsong kha pa (1357–1419), the founder of the Gelug tradition of Tibetan Buddhism. He is seated on a lotus flower that rises from a body of water, in front of which is a bowl of offerings. Tsong kha pa wears monastic robes and the typical yellow pandita-style hat. His hands are in the dharmachakra mudra or teaching gesture and hold the stems of two lotus flowers that rise on either side of his shoulders. The lotus on his right supports a flaming sword while the lotus on his left supports a book, both typical attributes of Tsong kha pa.

Tsong kha pa is surrounded by nine other figures. Directly above Tsong kha pa is Shakyamuni Buddha. Shakyamuni wears monastic robes and is seated on a lotus atop three jewels. His right hand makes the bhumisparsha mudra or earth-touching gesture and he holds a bowl in his left hand. To Shakyamuni's left and right are two Gelug teachers, both wearing monastic robes, a yellow hat, and making the abhaya mudra or "no fear" gesture. The teacher on the viewer's right holds a book in his left hand while the teacher on the left holds a bowl in his left hand.

Four female figures flank Tsong kha pa. On the upper right is the goddess Prajnaparamita, "perfection of wisdom." She is yellow with four arms, seated on a lotus, and holds a vase, flower, and book. Below her is Green Tara, with one face and two arms, seated on a lotus. She holds the stem of a lotus between the thumb and forefinger of her left hand while her right hand makes the varada mudra or gesture of giving. Opposite these two figures on Tsong kha pa's right side are two goddesses. The upper figure is white with two arms and holds a parasol, possibly a form of Sitatapatra. Below her is White Tara, who is seated with her legs crossed in vajra posture. She holds the stem of a white utpala flower between her left thumb and forefinger and makes the varada mudra or gesture of giving with her right hand. A distinct feature of White Tara is that she has five eyes in addition to the usual two for a total of seven eyes: two on the palms of her hands, two on the soles of her feet, and one on her forehead.

The bottom register is occupied by Yama Dharmaraja and Shadbhuja Sita Mahakala. Yama Dharmaraja is one of the three primary protectors of the Gelug tradition and is particularly associated with Tsong kha pa.[22] He is blue, has the head of a buffalo, and holds a lasso in his left hand and bone stick in his right hand. He rides a blue buffalo trampling a human body. Next to Yama Dharmaraja is a form of Mahakala called Shadbhuja Sita Mahakala. He is white with one face, six arms, and two legs, trampling two white elephants underfoot. On the left he holds a chopper, drum, and wish-fulfilling jewel. On the right he holds a trident, a hook, and a skull cup filled with jewels. Evidenced by the jewels held in his hands, this form of Mahakala is a wealth deity. In this particular form, he is also associated with the Gelug tradition.[23]

Date unknown

Image: 22.5" x 14.5"; total dimensions with mount: 42" x 23"

---

22. "Yama Dharmaraja (Buddhist Protector)," Himalayan Art Resources, accessed January 31, 2013, http://www.himalayanart.org/image.cfm/406.html.
23. "Mahakala: Shadbhuja, White (6 Hands)," Himalayan Art Resources, accessed January 31, 2013, http://www.himalayanart.org/search/set.cfm?setID=420.

Appendix: Descriptions of Thangkas in Collections 305

Detail of
Shakyamuni Buddha

Detail of
Yama Dharmaraja

47310 Tsong kha pa

The central figure of this thangka is Tsong kha pa (1357–1419), the founder of the Gelug tradition of Tibetan Buddhism. He is depicted according to his standard iconography. He wears monastic robes and the characteristic pointed yellow hat of the Gelug tradition. He sits on a lotus with his hands in the dharmachakra mudra or two-handed teaching gesture. He also holds the stems of two flowers that support a sword and a book. These implements are associated with both Tshong kha pa and Manjushri, the bodhisattva of wisdom.

Above Tsong kha pa are three buddhas. In the center is Shakyamuni Buddha, who wears monastic robes, holds an alms bowl, and makes the bhumisparsha mudra or earth-touching gesture. The two buddhas on either side of Shakyamuni make various forms of the dharmachakra mudra or teaching gesture. All three are seated on lotuses floating in the sky.

Below the three buddhas on either side of Tsong kha pa are two Gelug teachers. On the left is one of the Panchen Lamas, identifiable by his hat. The Panchen Lama is an important reincarnation lineage within the Gelug tradition. He holds a vase in his left hand and makes the *vitarka* mudra or one-handed teaching gesture with his right. On the other side is a Gelug teacher who is probably one of the Dalai Lamas. He wears the pointed yellow hat and holds a book and makes the varada mudra or gift-giving gesture. Two more unidentified Gelug teachers appear below Tsong kha pa on either side. These are possibly Tsong kha pa's two primary disciples, Gyaltsap Dharma Rinchen (1364–1432) and Kedrup Geleg Pal Zangpo (1385–1438).

The figure underneath Tsong kha pa in the bottom of the image is Yama Dharmaraja. Yama Dharmaraja is one of the three primary protectors of the Gelug tradition and is particularly associated with Tsong kha pa.[24] He is blue, has the head of a buffalo, and holds a lasso in his left and bone stick in his right hand. He rides a buffalo trampling a human body. He appears here with his consort Chamunda, who holds a skull cup in her left hand and trident in her right. Yama Dharmaraja appears in several other paintings in the Koelz collection, including 47309 and 47308.

Date unknown

Image: 18" x 11"; total dimensions with mount: 32" x 17"

---

24. "Yama Dharmaraja (Buddhist Protector)," Himalayan Art Resources, accessed January 31, 2013, http://www.himalayanart.org/image.cfm/406.html.

47310 Tsong kha pa

(*left*) Detail of Tsong kha pa (47310)

(*below*) Detail of Yama Dharmaraja and unidentified Gelug teacher (47310)